Risk Modelling in General Insurance

Knowledge of risk models and the assessment of risk is a fundamental part of the training of actuaries and all who are involved in financial, pensions and insurance mathematics. This book provides students and others with a firm foundation in a wide range of statistical and probabilistic methods for the modelling of risk, including short term risk modelling, model based pricing, risk sharing, ruin theory and credibility.

It covers much of the international syllabuses for professional actuarial examinations in risk models, but goes into further depth, with numerous worked examples and exercises (answers to many are included in an appendix). A key feature is the inclusion of three detailed case studies that bring together a number of concepts and applications from different parts of the book and illustrate how they are used in practice. Computation plays an integral part: the authors use the statistical package *R* to demonstrate how simple code and functions can be used profitably in an actuarial context.

The authors' engaging and pragmatic approach, balancing rigour and intuition, and developed over many years of teaching the subject, makes this book ideal for self-study or for students taking courses in risk modelling.

ROGER J. GRAY was a Senior Lecturer in the School of Mathematical and Computer Sciences at Heriot-Watt University, Edinburgh, until his death in 2011.

SUSAN M. PITTS is a Senior Lecturer in the Statistical Laboratory at the University of Cambridge.

INTERNATIONAL SERIES ON ACTUARIAL SCIENCE

The *International Series on Actuarial Science*, published by Cambridge University Press in conjunction with the Institute and Faculty of Actuaries, contains textbooks for students taking courses in or related to actuarial science, as well as more advanced works designed for continuing professional development or for describing and synthesising research. The series is a vehicle for publishing books that reflect changes and developments in the curriculum, that encourage the introduction of courses on actuarial science in universities, and that show how actuarial science can be used in all areas where there is long-term financial risk.

A complete list of books in the series can be found at www.cambridge.org/statistics. Recent titles include the following:

RISK MODELLING IN GENERAL INSURANCE

INSURANCE

From Principles to Practice

ROGER J. GRAY

Heriot-Watt University, Edinburgh

SUSAN M. PITTS

University of Cambridge

CAMBRIDGE
UNIVERSITY PRESS

CAMBRIDGE
UNIVERSITY PRESS

University Printing House, Cambridge CB2 8BS, United Kingdom

One Liberty Plaza, 20th Floor, New York, NY 10006, USA

477 Williamstown Road, Port Melbourne, VIC 3207, Australia

314-321, 3rd Floor, Plot 3, Splendor Forum, Jasola District Centre, New Delhi - 110025, India

79 Anson Road, #06-04/06, Singapore 079906

Cambridge University Press is part of the University of Cambridge.

It furthers the University's mission by disseminating knowledge in the pursuit of education, learning and research at the highest international levels of excellence.

www.cambridge.org
Information on this title: www.cambridge.org/9780521863940

© Roger J. Gray and Susan M. Pitts 2012

First published 2012

A catalogue record for this publication is available from the British Library

Library of Congress Cataloging in Publication data
Gray, Roger J.
Risk modelling in general insurance : from principles to practice /
Roger J. Gray, Susan M. Pitts.
p. cm.
ISBN 978-0-521-86394-0 (hardback)
1. Risk (Insurance) – Mathematical models. I. Pitts, Susan M. II. Title.
HG8054.5.G735 2012
368´.01–dc23
2012010344

ISBN 978-0-521-86394-0 Hardback

Additional resources for this publication at www.cambridge.org/9780521863940

To the memory of
Roger J. Gray
1946–2011

Contents

vii

Preface

My co-author Roger died in March 2011. His tragic death was a terrible shock, and he is, and will be, greatly missed by me and, I am sure, by all who knew him.

The original plan for writing this book was that Roger and I would each write our own chapters separately. We then planned to go through the whole book together, chapter by chapter, and make various changes as necessary when we had each read what the other had written. Unfortunately, and very sadly, Roger died before this process was completed. At the time of his death, the draft versions of Chapters 2 to 7 and Appendix A were written, and we had a very preliminary sketch of Chapter 1. However, only two chapters had been discussed in detail by both of us together. Fred Gray (Roger's brother), David Tranah (Cambridge University Press) and I were unanimous that Roger would have wanted the book to be completed, and so I began to put Roger's and my draft chapters together, to complete Chapter 1, and to edit the whole book in order to unify our two approaches, to fill obvious gaps, and to avoid too much repetition. My aim was that the result would be in line with what Roger would have wanted, and I very much hope that the finished book stands as a fitting tribute to his memory.

There are many people to thank for their help during the production of this book. First and foremost, thanks are due to everyone at Cambridge University Press. Special thanks go to David Tranah, who has been most helpful, with great patience and kindness at every stage. Thanks also go to Irene Pizzie for her careful and efficient copy-editing.

During our discussions Roger told me that he had a long list of people to thank in connection with the book, but unfortunately the conversation moved on without any names being mentioned. I know that David Wilkie, Iain Currie and Edward Kinley would have been on Roger's list, and I would like to take this opportunity to thank them. I would also like to thank everyone else who was helpful to Roger in the writing of his parts of the book, but whose names are unknown to me.

For my own part, I have been fortunate in having had excellent teachers, co-workers and students over the years, and my understanding of the subject matter of the book, and of effective ways to teach it, would not have been possible without them. I would like to thank them all. In addition, my thanks go to all those who were so supportive of my efforts to complete the book after Roger's death. Among these, I am especially grateful to David Tranah (whose wise advice and generous practical help were invaluable), Alan and Brenda Cole, Brigitte Snell and Rita McLoughlin. Finally, but most importantly of all, I thank my husband, Andrew, for his unfailingly good-humoured support and encouragement throughout the writing of this book.

Susan M. Pitts

1

Introduction

1.1 The aim of this book

Knowledge of risk models and the assessment of risk will be of great importance to actuaries as they apply their skills and expertise today and in the future. The title of this book "Risk Modelling in General Insurance: From Principles to Practice" reflects our intention to present a wide range of statistical and probabilistic topics relevant to actuarial methodology in general insurance. Our aim is to achieve this in a focused and coherent manner, which will appeal to actuarial students and others interested in the topics we cover.

We believe that the material is suitable for advanced undergraduates and students taking master's degree courses in actuarial science, and also those taking mathematics and statistics courses with some insurance mathematics content. In addition, students with a strong quantitative/mathematical background taking economics and business courses should also find much of interest in the book. Prerequisites for readers to benefit fully from the book include first undergraduate-level courses in calculus, probability and statistics. We do not assume measure theory.

Our aim is that readers who master the content will extend their knowledge effectively and will build a firm foundation in the statistical and actuarial concepts and their applications covered. We hope that the approach and content will engage readers and encourage them to develop and extend their critical and comparative skills. In particular, our aim has been to provide opportunities for readers to improve their higher-order skills of analysis and synthesis of ideas across topics.

A key feature of our approach is the inclusion of a large number of worked examples and extensive sets of exercises, which we think readers will find stimulating. In addition, we include three case studies, each of which brings

1

together a number of concepts and applications from different parts of the book.

While the book covers much of the international syllabuses for professional actuarial examinations in risk models, it goes further and deeper in places.

The book includes appropriate references to the open source (free and easily downloadable) statistical software package *R* throughout, giving readers opportunities to learn how simple code and functions can be used profitably in an actuarial context.

1.2 Notation and prerequisites

The tools of probability theory are crucial for the study of the risk models in this book, and, in §1.2.1, we give an overview of the required basic concepts of probability. This overview also serves to introduce the notation that we will use throughout the book. In §1.2.2 and §1.2.3, we indicate the assumed prerequisites in statistics and simulation, and finally in §1.2.4 we give information about the statistical software package *R*.

1.2.1 Probability

We start with definitions and notation for basic quantities related to a random variable X. Our first such quantity is the distribution function (or *cumulative distribution function*) F_X of X, given by

$$F_X(x) = \Pr(X \le x), \quad x \in \mathbb{R}.$$

The function F_X is non-decreasing and right-continuous. It satisfies $0 \le F_X(x) \le 1$ for all x in $\mathbb{R}$, $\lim_{x \to \infty} F_X(x) = 1$ and $\lim_{x \to -\infty} F_X(x) = 0$. Most of the random variables in this book are non-negative, i.e. they take values in $[0, \infty)$. If V is a non-negative random variable, then we assume without comment that $F_V(v) = 0$ for $v < 0$. For a non-negative random variable V, the *tail* of F_V is $\Pr(V > v) = 1 - F_V(v)$ for $v \ge 0$.

A *continuous* random variable Y has a *probability density function* f_Y, which is a non-negative function f_Y, with $\int_{-\infty}^{\infty} f_Y(y)dy = 1$, such that the distribution function of Y is

$$F_Y(y) = \int_{-\infty}^{y} f_Y(t)dt, \quad y \in \mathbb{R}.$$

This means that F_Y is a continuous function. The probability that Y is in a set A is $\Pr(Y \in A) = \int_A f_Y(y)dy$. (For those readers who are familiar with measure theory, note that we will tacitly assume the word "measurable" where necessary. Those readers who are not familiar with measure theory may ignore this remark, but may like to note that a rigorous treatment of probability theory requires more careful definitions and statements than appear in introductory courses and in this overview.)

Let N be a *discrete* random variable that takes values in $\mathbb{N} = \{0, 1, 2, \ldots\}$. Then $\Pr(N = x)$, $x \in \mathbb{R}$, is the *probability mass function* of N. We see that $\Pr(N = x) = 0$ for $x \notin \mathbb{N}$, so that, for a discrete random variable concentrated on $\mathbb{N}$, the probability mass function is specified by $\Pr(N = k)$ for $k \in \mathbb{N}$. We then have $\sum_{k=0}^{\infty} \Pr(N = k) = 1$. The distribution function of N is

$$F_N(x) = \sum_{\{k : k \leq x\}} \Pr(N = k), \quad x \in \mathbb{R},$$

and the graph of F_N is a non-decreasing step function, with an upward jump of size $\Pr(N = k)$ at k for all $k \in \mathbb{N}$. The probability that N is in a set A is

$$\Pr(N \in A) = \sum_{\{k : k \in A\}} \Pr(N = k).$$

We use the notation $\mathbb{E}[X]$ for the *expected value* (or *expectation*, or *mean*) of a random variable X. The expectation of the continuous random variable Y is

$$\mathbb{E}[Y] = \int_{-\infty}^{\infty} y f_Y(y) dy,$$

while for the discrete random variable N taking values in $\mathbb{N}$, the expectation is

$$\mathbb{E}[N] = \sum_{k=0}^{\infty} k \Pr(N = k).$$

We note that there are various possibilities for the expectation: it may be finite, it may take the value $+\infty$ or $-\infty$, or it may not be defined. The expectation of a non-negative random variable is either a finite non-negative value or $+\infty$.

For a real-valued function h on $\mathbb{R}$ and a continuous random variable Y, the expectation of $h(Y)$ is

$$\mathbb{E}[h(Y)] = \int_{-\infty}^{\infty} h(y) f_Y(y) dy,$$

whenever the integral is defined, and for a discrete random variable N taking values in $\mathbb{N}$, the expectation of $h(N)$ is

$$\mathbb{E}[h(N)] = \sum_{k=0}^{\infty} h(k) \Pr(N = k).$$

For $r \geq 0$, the *rth moment* of X is $\mathbb{E}[X^r]$, when it is defined. The rth moment of a continuous random variable Y is

$$\int_{-\infty}^{\infty} y^r f_Y(y) \, dy,$$

and the rth moment of the discrete random variable N taking values in $\mathbb{N}$ is

$$\sum_{k=0}^{\infty} k^r \Pr(N = k).$$

Recall that if $\mathbb{E}[|X|^r]$ is finite for some $r > 0$, then $\mathbb{E}[|X|^s]$ is finite for all $0 \leq s \leq r$. Throughout the book, when we write down a particular moment such as $\mathbb{E}[N^3]$, then, unless otherwise stated, we assume that this moment is finite.

The *rth central moment* of a random variable X is $\mathbb{E}[(X-\mathbb{E}[X])^r]$. The second central moment of X is called the *variance* of X, and is denoted by Var$[X]$. The variance of X is given by

$$\mathrm{Var}[X] = \mathbb{E}[(X - \mathbb{E}[X])^2] = \mathbb{E}[X^2] - (\mathbb{E}[X])^2.$$

The *standard deviation* of X is SD$[X] = \sqrt{\mathrm{Var}[X]}$. We define the *skewness* of X to be the third central moment, $\mathbb{E}[(X - \mathbb{E}[X])^3]$, and the *coefficient of skewness* to be given by

$$\mathbb{E}[(X - \mathbb{E}[X])^3]/((\mathrm{SD}[X])^3). \tag{1.1}$$

We define the *coefficient of kurtosis* of X to be

$$\mathbb{E}[(X - \mathbb{E}[X])^4]/((\mathrm{SD}[X])^4), \tag{1.2}$$

but note that various definitions are given in the literature; see the discussion in §2.2.5.

The *covariance* of random variables X and W is given by

$$\mathrm{Cov}[X, W] = \mathbb{E}[(X - \mathbb{E}[X])(W - \mathbb{E}[W])] = \mathbb{E}[XW] - \mathbb{E}[X]\mathbb{E}[W].$$

The *correlation* between random variables X and W (with Var$[X] > 0$ and Var$[W] > 0$) is given by

$$\mathrm{Corr}[X, W] = \frac{\mathrm{Cov}[X, W]}{\sqrt{\mathrm{Var}[X]\,\mathrm{Var}[W]}}.$$

For random variables $X_1, \ldots, X_n$ we have

$$\mathrm{Var}[X_1 + \cdots + X_n] = \sum_{i=1}^{n} \mathrm{Var}[X_i] + 2\sum_{i<j} \mathrm{Cov}[X_i, X_j].$$

Random variables $X_1, \ldots, X_n$ are independent if, for all $x_1, \ldots, x_n$ in $\mathbb{R}$,

$$\Pr(X_1 \leq x_1, \ldots, X_n \leq x_n) = \Pr(X_1 \leq x_1) \ldots \Pr(X_n \leq x_n).$$

For independent random variables $X_1, \ldots, X_n$ and functions $h_1, \ldots, h_n$, we have

$$\mathbb{E}[h_1(X_1) \ldots h_n(X_n)] = \mathbb{E}[h_1(X_1)] \ldots \mathbb{E}[h_n(X_n)].$$

This means that, for independent random variables $X_1, \ldots, X_n$, we have

$$\text{Var}[X_1 + \cdots + X_n] = \text{Var}[X_1] + \cdots + \text{Var}[X_n],$$

because, for $i \neq j$, the independence of X_i and X_j implies that $\text{Cov}[X_i, X_j] = 0$. Random variables $X_1, X_2, \ldots$ are independent if every finite subset of the X_i is independent. We say $X_1, X_2, \ldots$ are independent and identically distributed (iid) if they are independent and all have the same distribution.

Conditioning is one of the main tools used throughout this book, and it is often the key to a neat approach to derivation of properties and features of the risk models considered in later chapters. The conditional expectation of X given W is denoted $\mathbb{E}[X \mid W]$. The very useful *conditional expectation formula* states that

$$\mathbb{E}[\mathbb{E}[X \mid W]] = \mathbb{E}[X]. \tag{1.3}$$

The *conditional variance* of X given W is defined to be

$$\text{Var}[X \mid W] = \mathbb{E}\left[(X - \mathbb{E}[X \mid W])^2 \mid W\right]$$
$$= \mathbb{E}[X^2 \mid W] - (\mathbb{E}[X \mid W])^2.$$

The *conditional variance formula* is

$$\text{Var}[X] = \mathbb{E}[\text{Var}[X \mid W]] + \text{Var}[\mathbb{E}[X \mid W]]. \tag{1.4}$$

This may be seen by considering the terms on the right-hand side of (1.4). We have

$$\mathbb{E}[\text{Var}[X \mid W]] = \mathbb{E}\left[\mathbb{E}[X^2 \mid W] - (\mathbb{E}[X \mid W])^2\right]$$
$$= \mathbb{E}[X^2] - \mathbb{E}\left[(\mathbb{E}[X \mid W])^2\right],$$

where we have used the conditional expectation formula, and

$$\text{Var}[\mathbb{E}[X \mid W]] = \mathbb{E}\left[(\mathbb{E}[X \mid W])^2\right] - (\mathbb{E}[\mathbb{E}[X \mid W]])^2$$
$$= \mathbb{E}\left[(\mathbb{E}[X \mid W])^2\right] - (\mathbb{E}[X])^2,$$

on using the conditional expectation formula again. Adding these terms it is easy to see that the right-hand side of (1.4) is equal to the left-hand side.

We assume that *moment generating functions, probability generating functions* and their properties are familiar to the reader. The moment generating function of a random variable X is denoted

$$M_X(r) = \mathbb{E}[e^{rX}], \tag{1.5}$$

and this may not be finite for all r in $\mathbb{R}$. For every random variable X, we have $M_X(0) = 1$, and so the moment generating function is certainly finite at $r = 0$. If $M_X(r)$ is finite for $|r| < h$ for some $h > 0$, then, for any $k = 1, 2, \ldots$, the function $M_X(r)$ is k-times differentiable at $r = 0$, with

$$M_X^{(k)}(0) = \mathbb{E}[X^k], \tag{1.6}$$

with $\mathbb{E}[|X|^k]$ finite. If random variables X and W have $M_X(r) = M_W(r)$ for all $|r| < h$ for some $h > 0$, then X and W have the same distribution.

The moment generating function of a continuous random variable Y is

$$M_Y(r) = \int_{-\infty}^{\infty} e^{ry} f_Y(y) dy.$$

The moment generating function of a discrete random variable N concentrated on $\mathbb{N}$ is

$$M_N(r) = \sum_{k=0}^{\infty} e^{rk} \Pr(N = k).$$

The *probability generating function* of N is

$$G_N(z) = \mathbb{E}[z^N] = \sum_{k=0}^{\infty} z^k \Pr(N = k), \tag{1.7}$$

for those z in $\mathbb{R}$ for which the series converges absolutely. Since the series converges for $|z| \leq 1$ (and possibly for a larger set of z-values), we see that the radius of convergence of the series is greater than or equal to 1. If $\mathbb{E}[N] < \infty$ then

$$\mathbb{E}[N] = G_N'(1),$$

and if $\mathbb{E}[N^2] < \infty$ then

$$\text{Var}[N] = G_N''(1) + G_N'(1) - (G_N'(1))^2,$$

where $G_N^{(k)}(1) = \lim_{z \uparrow 1} G_N^{(k)}(z)$ if the radius of convergence of G_N is 1. From (1.5) and (1.7) we have

$$G_N(z) = M_N(\log(z)) \text{ and } M_N(r) = G_N(e^r),$$

where here, and throughout the book, when we write down relationships between generating functions, we assume the phrase "for values of the argument for which both sides are finite".

Moment generating functions and probability generating functions are both examples of *transforms*. Transforms are useful for calculations involving sums of independent random variables. Let $X_1, \ldots, X_n$ be independent random variables, and let M_{X_i} be the moment generating function of X_i, $i = 1, \ldots, n$. Then the moment generating function of $T = X_1 + \cdots + X_n$ is the product of the moment generating functions of the X_i:

$$M_T(r) = M_{X_1}(r) \ldots M_{X_n}(r). \tag{1.8}$$

Similarly, let $N_1, \ldots, N_n$ be independent discrete random variables taking values in $\mathbb{N}$, and let G_{N_i} be the probability generating function of N_i, $i = 1, \ldots, n$. Then the probability generating function of $M = N_1 + \cdots + N_n$ is

$$G_M(z) = G_{N_1}(z) \ldots G_{N_n}(z). \tag{1.9}$$

Sums of independent random variables play an important role in the models in this book, so transform methods will be important for us.

The *cumulant generating function* $K_X(t)$ of a random variable X is given by

$$K_X(t) = \log{(M_X(t))},$$

and this is discussed further in §2.2.5.

In the above discussion, we have given separate expectation formulae for continuous random variables and for discrete random variables. We now introduce a more general notation that covers both of these cases (and other cases as well). For a general random variable X with distribution function F_X, we write

$$\mathbb{E}[X] = \int xF_X(dx). \tag{1.10}$$

This is a Lebesgue–Stieltjes integral. We can think of the integral as shorthand notation for $\int xf_X(x)dx$ if X is continuous with density f_X, and as shorthand for $\sum_{k=0}^{\infty} k \Pr(X = k)$ if X is discrete and takes values in $\{0, 1, 2, \ldots\}$. This notation means we can give just one formula that covers both continuous and discrete random variables. However, it also covers more general random variables. Later in this book we will meet and use random variables which are neither purely continuous, nor purely discrete, but which have both a discrete part and a continuous part. To make this precise, suppose that there exist real numbers $x_1, \ldots, x_m$ and $p_1, \ldots, p_m$, where $0 \le p_k \le 1$ for $k = 1, \ldots, m$, and where $\sum_{k=1}^{m} p_k \leq 1$, and suppose there also exists a non-negative function f, with $\int_{-\infty}^{\infty} f(t)dt \leq 1$, such that the distribution function of X is

$$F_X(x) = \Pr(X \leq x) = \sum_{\{k:x_k \leq x\}} p_k + \int_{-\infty}^{x} f(t)dt. \tag{1.11}$$

Of course, we must have

$$\sum_{k=1}^{m} p_k + \int_{-\infty}^{\infty} f(x)dx = 1.$$

In this case, the distribution of X consists of a discrete part, specified by the x_k and the p_k (with $\Pr(X = x_k) = p_k$), and also a continuous part, specified by f. The distribution function F_X has an upward jump of size p_k at x_k, $k = 1, \ldots, m$, and is continuous and non-decreasing (and not necessarily flat) between these jumps. We say that the distribution of X has an atom at x_k (of size p_k), for $k = 1, \ldots, m$. For this X, and for a set A, we have

$$\Pr(X \in A) = \int_A F_X(dx) = \sum_{\{k:x_k \in A\}} p_k + \int_A f(x)dx. \tag{1.12}$$

As in (1.10), the expectation of X is $\mathbb{E}[X] = \int x F_X(dx)$, and, with F_X as in (1.11), the integral is

$$\int x F_X(dx) = \sum_{k=1}^{m} k p_k + \int_{-\infty}^{\infty} x f(x)dx. \tag{1.13}$$

In general, for a function h, we have

$$\mathbb{E}[h(X)] = \int h(x) F_X(dx), \tag{1.14}$$

and, when $h(x) = e^{rx}$, we find that the moment generating function of X is

$$M_X(r) = \mathbb{E}[e^{rX}] = \int e^{rx} F_X(dx). \tag{1.15}$$

With F_X as in (1.11), the equations (1.14) and (1.15) become

$$\mathbb{E}[h(X)] = \sum_{k=1}^{m} h(k) p_k + \int_{-\infty}^{\infty} h(x) f(x)dx$$

and

$$M_X(r) = \int e^{rx} F_X(dx) = \sum_{k=1}^{m} e^{rk} p_k + \int_{-\infty}^{\infty} e^{rx} f(x)dx.$$

Note that a Lebesgue–Stieltjes integral over an interval $(a, b]$, $a \leq b$, is written

$$\int_{(a,b]} \ldots F_X(dx),$$

where ... is to be replaced by the required function to be integrated. Finally, we have, from (1.12),

$$\int_{(a,b]} F_X(dx) = \Pr(X \in (a, b]) = F_X(b) - F_X(a^-),$$

where $F_X(a^-)$ denotes $\lim_{x \to a^-} F_X(x)$, and $x \to a^-$ means that x converges to a from the left.

In this subsection, we have given a brief overview of probability. For more discussion and details, see, for example, Grimmett and Stirzaker (2001), Gut (2009) and the more advanced Gut (2005).

1.2.2 Statistics

We assume that the reader has met point estimation and properties of estimators (for example, the idea of an unbiased estimator), confidence intervals and hypothesis tests (for example, t tests, χ^2 tests, Kolmogorov–Smirnov test). We further assume a working knowledge of maximum likelihood estimators and their large sample properties. Familiarity with plots, such as histograms and quantile (or Q–Q) plots, is assumed, in addition to familiarity with the empirical distribution function. Useful references are DeGroot and Schervish (2002) and Casella and Berger (1990). The introduction to §2.4 contains an overview of some ideas and methods in statistics. At various points in the book we use more advanced statistical ideas – whenever we do this, references to appropriate texts are given.

1.2.3 Simulation

We take as prerequisite some knowledge of simulation of observations from a given distribution using a pseudo-random number generator and various techniques, such as the inverse transform (or inversion or probability integral transform) method. For more details and background, see, for example, chapter 11 in DeGroot and Schervish (2002) and chapter 6 in Morgan (2000).

1.2.4 The statistical software package *R*

The simulations, statistical analyses and numerical approximations in this book are carried out using the statistical software package *R*. We assume familiarity with how **R** works and with basic commands in **R**. Useful references are Venables and Ripley (2002) and Verzani (2005). The package **R** is available for (free) download; see `http://cran.r-project.org/`.

There is an add-on actuarial package `actuar`, and this can be installed using the `Installpackage(s)` submenu of the `Packages` menu. Choose a convenient CRAN mirror, and then select the package `actuar` for installation. It only has to be installed once, but it must be attached to the **R** workspace at the beginning of each **R** session, using the **R** command `library(actuar)`.

2

Models for claim numbers and claim sizes

In a portfolio of general insurance risks, such as a portfolio of motor insurance policies, two obvious quantities of interest are the *number of claims* arriving in a fixed time period and the *sizes* of those claims. We model these quantities as random variables with appropriate probability distributions, and this modelling process is the subject of this chapter.

There are many probability distributions available as potential models for both claim numbers and claim sizes in general insurance. Suitable models for claim numbers are "counting distributions"; that is, distributions of discrete random variables that can assume some or all of the values in $\mathbb{N} = \{0, 1, 2, \ldots\}$. The most suitable and widely used models for claim sizes are distributions of continuous random variables that assume positive values only and have "fat tails" (or "heavy tails"), that is distributions which allow for occasional occurrences of very large values.

In this chapter we consider the principal models used in practice. We review the properties of the distributions one by one, illustrate how they are fitted to data on claim numbers and sizes, and consider how we assess the success of the models in reflecting the variation and distribution of the data.

In §2.1 and §2.2 we give summaries of the relevant properties of the various distributions – our aim is that these two sections will provide a useful reference for the reader, and will also fix notation for these distributions. In §2.1 we consider three families of counting random variables used as models for claim numbers, namely the one-parameter *Poisson* family, the two-parameter *negative binomial* family (which includes the one-parameter *geometric* subfamily), and the two-parameter *binomial* family. In this section, we also include a discussion of the *Poisson process*.

In §2.2 we consider eight families of continuous random variables used as models for claim sizes. The first three, while not providing good models for claim sizes in most practical situations, are useful for reference and comparison purposes, and are included for completeness – these families are the two-parameter *normal* (*Gaussian*) family, the one-parameter *exponential* family and the two-parameter *gamma* family (which includes the exponential as a sub-family). The five families of important distributions used as models in practice and considered here are four two-parameter families, namely the *lognormal, Pareto, Weibull* (of which the exponential is a sub-family) and *loggamma* families, and the three-parameter *Burr* family (of which the Pareto is a sub-family). All these distributions, except the normal distribution, are for positive random variables. A normally distributed random variable can take negative values as well as positive values, and so, strictly speaking, it is not an appropriate model for a positive claim size. However, it is included here because the normal distribution may be used as an approximation to many distributions and also because it arises as a limiting distribution, for example in the Central Limit Theorem.

In §2.3 we consider mixture distributions, which arise in a Bayesian context when we extend claim-size distribution models to allow for heterogeneity of risks within a portfolio. We do this by recognising that there is uncertainty in the value of a parameter in a claim-size distribution, and then adopting a probability distribution, called a "prior" or "mixing" distribution, to model that uncertainty. We then derive the overall, unconditional (marginal) distribution of the claim-size (or claim-number) random variable, here called a "mixture" distribution. The approach provides further motivation for the use of particular families of claim-size distributions and is also itself a source of fat-tailed distributions.

In §2.4 we consider the fitting of models to data on claim numbers and claim sizes. We will fit all the distributions introduced in §2.1 to data on claim numbers and all the distributions introduced in §2.2 to data on claim sizes, using the method of maximum likelihood and several other approaches to the estimation of model parameters. We will assess the goodness of fit of each of the models using various informative visual displays and appropriate test statistics.

2.1 Distributions for claim numbers

The most widely used model for the process which gives rise to claims in a portfolio of business in general insurance is a Poisson process, for which (informally, and in the simplest case) claims arise "at random", one after another through time and at a constant intensity (the rate per unit time). In this case the number of claims which occur in a given time interval has a Poisson distribution with appropriate mean. A more formal description of the Poisson process and its properties is given in §2.1.1 and in §2.2.3. Other distributions used for the number of claims in a given time interval include the two-parameter family of negative binomial distributions, which allows for a heavier tail than the Poisson and may provide a better fit to claim-number data in certain cases (for example in motor insurance). In certain cases, when it is appropriate to declare that the number of claims cannot exceed some known number, it can be appropriate to adopt a binomial distribution for the number of claims. This case can arise, for example, when we are dealing with a portfolio consisting of a known number of similar policies on each of which at most a single claim can arise.

Let $\{N_t\}_{t \geq 0}$ denote the claim-number process, where N_t is the number of claims which arise up to and including time t. We sometimes write $N(t)$ instead of N_t. Unless otherwise stated, we will consider a time period of length 1 and write N for N_1.

2.1.1 Poisson distribution

The Poisson family of distributions has a single parameter, usually denoted λ (> 0), which represents the mean of the distribution: that is, the expected number of claims per unit time in a Poisson process.

Notation $N \sim \text{Poi}(\lambda)$ or $N \sim \text{Poisson}(\lambda)$.

The probability mass function is given by

$$\Pr(N = n) = e^{-\lambda} \frac{\lambda^n}{n!}, \quad n = 0, 1, 2 \ldots .$$

The probability generating function $G_N(z) = \mathbb{E}[z^N]$ is given by

$$G_N(z) = \exp\{\lambda(z - 1)\}. \tag{2.1}$$

The moment generating function $M_N(t) = \mathbb{E}[e^{tN}]$ is given by

$$M_N(t) = \exp\{\lambda(e^t - 1)\}, \tag{2.2}$$

from which we find $\mathbb{E}[N] = \lambda$ (confirming λ as the mean) and $\mathbb{E}[N^2] = \lambda + \lambda^2$, giving $\text{Var}[N] = \lambda$. Note that the mean and variance of a Poisson distribution are equal.

The third central moment of the distribution is $\mathbb{E}[(N - \mathbb{E}[N])^3] = \lambda$ (this follows easily from the results of Exercise 2.5). It follows that the coefficient of skewness (see (1.1)) for the distribution is $1/\sqrt{\lambda}$. As λ increases, the coefficient of skewness decreases and the distribution becomes more symmetrical.

Simulation We can simulate a random sample of size n from $N \sim \text{Poi}(\lambda)$ in *R* using the command `sample = rpois(n, lambda)`, where the

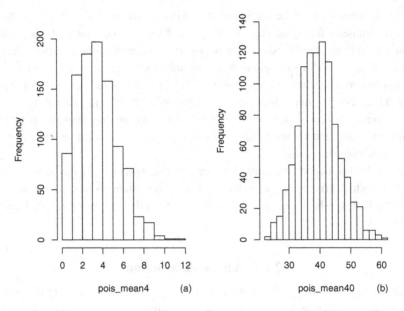

Figure 2.1. Histograms of samples simulated from Poisson distributions with parameters $\lambda = 4$ (mean = 4) (a) and $\lambda = 40$ (mean = 40) (b).

R objects n and lambda contain the values of n and λ, respectively. Here, the *R* object we create is a vector of observations of length n called "sample" (see §2.1.5).

The histograms in Figure 2.1 display 1000 claim numbers simulated from Poi(4) and Poi(40) distributions in *R*.

The commands used were of the form

```
pois_mean4=rpois(1000,4)
hist(pois_mean4)
summary(pois_mean4)
var(pois_mean4)
```

The sample means and variances, and the minimum and maximum values that were observed, were as follows:

sample from Poi(4): mean 3.986, variance 3.950, min 0, max 12;
sample from Poi(40): mean 40.117, variance 40.716, min 23, max 61.

The reader will note that the Poi(40) is much more symmetrical than the Poi(4).

The sum of independent Poisson random variables is a Poisson random variable (with mean equal to the sum of the component means). This is seen as follows. Let $N_1, \ldots, N_k$ be independent Poisson random variables, with $N_i \sim \text{Poi}(\lambda_i)$, $i = 1, \ldots, k$, and let $N = N_1 + \cdots + N_k$. By (1.9), the probability generating function of N is

$$G_N(z) = \prod_{i=1}^{k} \exp\left(\lambda_i(z-1)\right) = \exp\left(\left(\sum_{i=1}^{k} \lambda_i\right)(z-1)\right),$$

so that $N \sim \text{Poi}\left(\sum_{i=1}^{k} \lambda_i\right)$.

Example 2.1 Claims arise on two portfolios, A and B, independently of one another. The number of claims which arise on portfolio A in a week has a Poi(λ_1) distribution; for portfolio B the distribution is Poi(λ_2).

Suppose $\lambda_1 = 5$ and $\lambda_2 = 3$. Let T denote the combined number of claims on both portfolios in a week: it follows that $T \sim \text{Poi}(\lambda_1 + \lambda_2) \sim \text{Poi}(8)$. Then the probability that a total of ten or more claims occur in a week is

$$\Pr(T \geq 10) = 1 - \Pr(T \leq 9) = 0.2834,$$

using the **R** command 1 - ppois(9,8).

The Poisson process

The Poisson distribution is a key building block for the Poisson process, which we now describe. Consider a process where events occur at points in time; for example, consider the process of claim arrivals at an insurance company. Let $N(t)$ be the number of events in the time interval $(0, t]$, and define $N(0) = 0$. The collection of random variables $\{N(t) : t \geq 0\}$ is a stochastic process that models the number of events over time. For $t \geq 0$ and $s > 0$, the random variable $N(t + s) - N(t)$ is the *increment* of the process $\{N(t) : t \geq 0\}$ over the interval $(t, t + s]$, and this gives the number of events in $(t, t + s]$. A process $\{N(t) : t \geq 0\}$ that satisfies the three properties below is called a *Poisson process* with rate (or intensity) λ (> 0).

(a) **Independent increments** For $k = 2, 3, \ldots$, the numbers of events in k disjoint intervals (given by the increments of $\{N(t) : t \geq 0\}$ over these intervals) are independent.

(b) **Stationary increments** For all $h > 0$ and for all $t \geq 0$, the distribution of the increment $N(t + h) - N(t)$ depends only on h (and not on t), i.e. the distribution of the number of events in an interval depends only on the length of that interval and not on its left end point.

(c) **Poisson distribution** For all $t \geq 0$, the random variable $N(t)$ has a Poisson distribution with mean λt.

We can use the above three properties to deduce, for example, that, for $s > 0$ and $t \geq 0$, the number of events $N(t + s) - N(t)$ in $(t, t + s]$ and the number of events $N(t)$ in $(0, t]$ are independent Poisson random variables with means λs and λt, respectively.

2.1.2 Negative binomial distribution

The negative binomial family of distributions has two parameters, usually denoted k (> 0) and p ($0 < p < 1$).

Notation $N \sim \mathrm{nb}(k, p)$.

The probability mass function is given by

$$\Pr(N = n) = \frac{\Gamma(k + n)}{\Gamma(n + 1)\Gamma(k)} q^n p^k, \quad n = 0, 1, 2 \ldots,$$

where $q = 1 - p$, and where Γ is the gamma function, defined by $\Gamma(\alpha) = \int_0^\infty x^{\alpha-1} e^{-x} \, dx$ for $\alpha > 0$. We recall that $\Gamma(\alpha + 1) = \alpha \Gamma(\alpha)$, and that, for $n = 1, 2, \ldots$, we have $\Gamma(n) = (n - 1)!$.

In the case that k is an integer, the distribution models the number of failures before the kth "success" occurs in a series of independent Bernoulli trials (that is, trials which can be regarded as having only two possible outcomes, which we call "success" and "failure"), each with $\Pr(\text{success}) = p$. The probability mass function can now be expressed as follows:

$$\Pr(N = n) = \binom{k + n - 1}{n} q^n p^k = \frac{(k + n - 1)!}{n!(k - 1)!} q^n p^k, \quad n = 0, 1, 2 \ldots.$$

The probability generating function $G_N(z)$ exists for $|z| < 1/q$, and is given by

$$G_N(z) = \left(\frac{p}{1 - qz} \right)^k. \tag{2.3}$$

The moment generating function $M_N(t)$ is finite for $t < -\log q$ and is given by

$$M_N(t) = \left(\frac{p}{1 - qe^t} \right)^k, \tag{2.4}$$

from which we find $\mathbb{E}[N] = kq/p$ and $\mathrm{Var}[N] = kq/p^2$. Note that, since $p < 1$, the variance is greater than the mean – a feature not available for the Poisson family. This observation indicates that, for Poisson and negative binomial distributions with the same mean, the latter allows for greater spread – and in particular a heavier tail – than the Poisson. This feature helps to explain why the negative binomial distribution sometimes fits claim-number data better than the Poisson does – this can occur when the observed frequency distribution of the number of claims tails off to include some rather high values (which a fitted Poisson model cannot capture); see Example 2.2.

The third central moment is given by $[kq(2 - p)]/p^3$ (readers wishing to verify this result are recommended to use the results of Exercise 2.5). It follows that the coefficient of skewness (see (1.1)) for the distribution is always positive and is given by $(2 - p)/\sqrt{kq}$. For fixed p, the coefficient of skewness decreases as k increases, and the distribution becomes more symmetrical. For fixed k, the coefficient of skewness increases as p increases.

Simulation We can simulate a random sample of size n from $N \sim$ nb(k, p) in
R using the command `sample = rnbinom(n,k,p)`, where the R
objects n, k and p contain the values of n, k and p, respectively.

The histograms in Figure 2.2 display 1000 claim numbers simulated from nb(2, 1/3) and nb(20, 1/3) distributions, which have means 4 and 40, and variances 12 and 120, respectively.

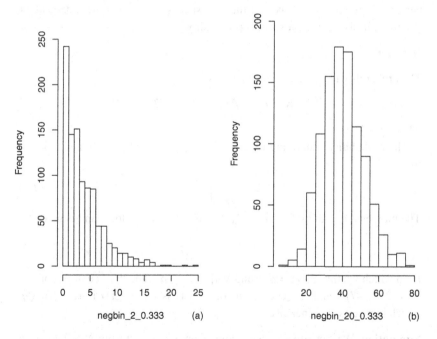

Figure 2.2. Histograms of samples simulated from negative binomial distributions with parameters $k = 2$, $p = 1/3$ (mean = 4) (a) and $k = 20$, $p = 1/3$ (mean = 40) (b).

The commands used for the simulations were

```
negbin_2_0.333=rnbinom(1000,2,1/3)
negbin_20_0.333=rnbinom(1000,20,1/3)
```

The sample means and variances, and the minimum and maximum values that were observed, were as follows:

sample from nb(2, 1/3): mean 4.187, variance 13.003, min 0, max 25;
sample from nb(20, 1/3): mean 40.5, variance 128.256, min 9, max 78.

The reader will note that the nb(20, 1/3) is much more symmetrical than the nb(2, 1/3). For a related distribution, sometimes also called a negative binomial distribution in the literature, see §3.4.3

2.1.3 Geometric distribution

The geometric family is a sub-family of the negative binomial family, namely the special case given by setting $k = 1$; the geometric family thus has a single parameter, denoted p (where $0 < p < 1$). The distribution models the number of failures that occur before the first success in a series of independent Bernoulli trials, each with success probability p.

Notation $N \sim \text{geo}(p)$.

The probability mass function is given by

$$\Pr(N = n) = q^n p, \quad n = 0, 1, 2 \ldots,$$

where $q = 1 - p$.

The probability generating function $G_N(z)$ exists for $|z| < 1/q$ and is given by

$$G_N(z) = \frac{p}{1 - qz}. \tag{2.5}$$

The moment generating function $M_N(t)$ is finite for $t < -\log q$ and is given by

$$M_N(t) = \frac{p}{1 - qe^t}, \tag{2.6}$$

from which we find $\mathbb{E}[N] = q/p$ and $\text{Var}[N] = q/p^2$. The third central moment is $[q(2 - p)]/p^3$ and the coefficient of skewness (see (1.1)) is $(2 - p)/\sqrt{q}$, which increases as p increases.

Simulation We can simulate a random sample of size n from $N \sim \text{geo}(p)$ in **R** using the command sample = rgeom(n,p), where the **R** objects n and p contain the values of n and p, respectively.

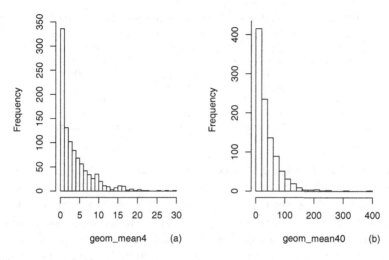

Figure 2.3. Histograms of samples simulated from geometric distributions with parameters $p = 0.2$ (mean = 4) (a) and $p = 1/41$ (mean = 40) (b).

The histograms in Figure 2.3 display 1000 claim numbers simulated from geo(0.2) and geo(1/41) distributions, which have means 4 and 40, and variances 20 and 1640, respectively.

The commands used for the simulation were

```
geom_mean4=rgeom(1000,0.2)
geom_mean40=rgeom(1000,1/41)
```

The sample means and variances, and the minimum and maximum values that were observed, were as follows:

sample from geo(0.2): mean 4.164, variance 19.753, min 0, max 30;
sample from geo(1/41): mean 38.89, variance 1634.66, min 0, max 385.

For a related distribution, also sometimes called a geometric distribution in the literature, see §3.4.3.

For integer k, the random variable $N \sim \text{nb}(k, p)$ can be represented as the sum of k independent, identically distributed (iid) random variables, each distributed geo(p). This is because the probability generating function for the nb(k, p) distribution is equal to the probability generating function for the geo(p) distribution raised to the power of k; see (2.3) and (2.5) (and similarly for the moment generating functions; see (2.4) and (2.6)).

Table 2.1. *Comparison of tail probabilities for four distributions*

	$\Pr(N \geq 10)$	$\Pr(N \geq 12)$	$\Pr(N \geq 15)$
Poi(3)	0.0011	0.000071	0.00000067
nb(3, 0.5)	0.019	0.0065	0.0012
nb(2, 0.4)	0.030	0.013	0.0033
nb(1, 0.25)	0.056	0.032	0.013

Example 2.2 In Table 2.1 we compare selected tail probabilities for $N \sim$ Poi(3), $N \sim$ nb(3, 0.5), $N \sim$ nb(2, 0.4) and $N \sim$ nb(1, 0.25) $\equiv$ geo(0.25) distributions, all of which have mean $\mathbb{E}[N] = 3$, and have increasing variances (3, 6, 7.5, and 12 respectively). The probabilities were obtained from *R*, using, for example, the command 1 - pnbinom(11,3,0.5), which evaluates $\Pr(N \geq 12)$ for nb(3, 0.5), and returns 0.006469727. In each case, the tail probability increases as we move from the Poisson distribution to successive negative binomial distributions.

2.1.4 Binomial distribution

The binomial family of distributions has two parameters, n and p, where n is a positive integer and $0 < p < 1$. The distribution models the number of successes which occur in a series of n independent Bernoulli trials, each with $\Pr(\text{success}) = p$ and $\Pr(\text{failure}) = q = 1 - p$. Unlike the distributions in the preceding situations, the values which can be assumed by the binomial random variable are restricted – to a maximum of n – so in the context of modelling numbers of claims this distribution is only appropriate in situations in which we know in advance the maximum possible number of claims. As mentioned earlier, an area of possible application is the situation in which we are dealing with a fixed number of policies (n), on each of which a maximum of one claim can arise.

Notation $N \sim \text{bi}(n, p)$.

The probability mass function is given by

$$\Pr(N = x) = \binom{n}{x} p^x (1 - p)^{n-x}, \quad x = 0, 1, \ldots, n.$$

The probability generating function $G_N(z)$ is given by

$$G_N(z) = (q + pz)^n. \tag{2.7}$$

The moment generating function M_N is given by

$$M_N(t) = (q + pe^t)^n, \tag{2.8}$$

from which we find $\mathbb{E}[N] = np$ and $\mathbb{E}[N^2] = n(n-1)p^2 + np$, giving $\text{Var}[N] = npq$. Note that the variance of N is lower than the mean.

The third central moment is $npq(q - p)$ (readers wishing to verify this result are recommended to use the results of Exercise 2.5). It follows that the coefficient of skewness (see (1.1)) for the distribution is $(q - p)/\sqrt{npq}$. As p increases from 0 to 1, the coefficient of skewness decreases from positive values through zero (at $p = 0.5$) to negative values.

Simulation We can simulate a random sample of size m from $N \sim \text{bi}(n, p)$ in **R** using the command sample = rbinom(m,n,p), where the **R** objects m, n and p contain the values of m, n and p, respectively.

The histograms in Figure 2.4 display 10 000 claim numbers simulated from bi(50, 0.1), bi(50, 0.5) and bi(50, 0.9) distributions, which have means 5, 25, and 45, and variances 4.5, 12.5, and 4.5, respectively. The command used for the first simulation was

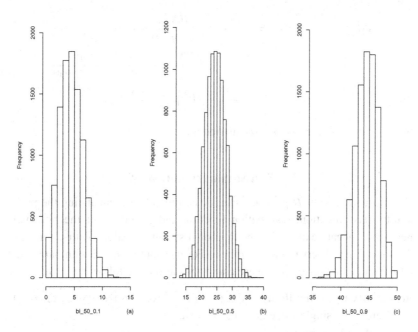

Figure 2.4. Histograms of samples simulated from binomial distributions with parameters $n = 50$ and $p = 0.1$ (mean = 5) (a), $p = 0.5$ (mean = 25) (b), and $p = 0.9$ (mean = 45) (c).

```
bi_50_0.1=rbinom(10000,50,0.1)
```

The $\text{bi}(1, p)$ distribution is called the Bernoulli(p) distribution and has moment generating function $q + pe^t$, mean p and variance pq. The random variable $N \sim \text{bi}(n, p)$ can be represented as the sum of n iid Bernoulli(p) random variables. This follows from (1.9), because the probability generating function of the $\text{bi}(n, p)$ distribution is the same as the probability generating function of the $\text{bi}(1, p)$ raised to the power n.

Example 2.3 Suppose claims arise according to a Poisson process with intensity λ (per hour), and let N_t denote the number of claims which arise in the period $(0, t]$, of length t hours. Then $N_t \sim \text{Poi}(\lambda t)$.

Suppose we know that m claims arise in $(0, t]$; that is, we know $N_t = m$.

Consider N_s, the number of claims which arise in the period $(0, s]$, where $0 < s < t$. For $n = 0, 1, \ldots, m$

$$\Pr(N_s = n \mid N_t = m) = \frac{\Pr(N_s = n, N_t = m)}{\Pr(N_t = m)}.$$

Using the independence properties of the Poisson process (see §2.1.1) we have

$$\begin{aligned}
\Pr(N_s = n \mid N_t = m) &= \frac{\Pr(N_s = n) \Pr(N_{t-s} = m - n)}{\Pr(N_t = m)} \\
&= \frac{e^{-\lambda s}(\lambda s)^n}{n!} \frac{e^{-\lambda(t-s)}(\lambda(t-s))^{m-n}}{(m-n)!} \frac{m!}{e^{-\lambda t}(\lambda t)^m} \\
&= \frac{m!}{n!(m-n)!} \left(\frac{s}{t}\right)^n \left(1 - \frac{s}{t}\right)^{m-n}.
\end{aligned}$$

So $N_s \mid (N_t = m)$ has a $\text{bi}(m, s/t)$ distribution.

2.1.5 A summary note on *R*

The basic version of *R* provides specific commands for calculating the probability mass function, distribution function and quantiles for the binomial, geometric, negative binomial and Poisson distributions, along with commands for simulating observations from them. For distribution dist with parameter p1 (or parameters p1, p2), these commands are of the following form.

probability mass function: ddist(x, p1) (or ddist(x, p1, p2)), where x is a single value or a vector of values;
distribution function: pdist(x, p1);
quantiles: qdist(p, p1), where p is a single probability or a vector of probabilities;

simulated values: rdist(n, p1), where n is the required number of observations.

For example:

dpois(3,2) returns 0.1804470 and this is Pr($N = 3$) where $N \sim$ Poi(2);

ppois(c(1,2,3),2) returns 0.4060058 0.6766764 0.8571235 (here c(1,2,3) is the vector (1, 2, 3), and *R* calculates Pr($N \le 1$), Pr($N \le 2$) and Pr($N \le 3$) where $N \sim$ Poi(2));

dnbinom(3,2,1/3) returns 0.1316872 and this is Pr($N = 3$), where $N \sim$ nb(2, 1/3);

pnbinom(v,2,1/3), with v=c(0,1,2), returns 0.1111111 0.2592593 0.4074074 (for example, 0.4074074 is Pr($N \le 2$), where $N \sim$ nb(2, 1/3));

qgeom(h,0.1), with h=c(0.25,0.5,0.75), returns the quartiles 2 6 13 (for example, the upper quartile of the geo(0.1) distribution is 13, the smallest value of x for which Pr($N \le x) \ge 0.75$, where $N \sim$ geo(0.1));

rbinom(100,20,0.4) returns a sample of 100 observations randomly generated from a bi(20, 0.4) distribution.

2.2 Distributions for claim sizes

We are concerned with modelling the financial losses which can be suffered by individuals and insurance companies as a result of insurable events such as storm or fire damage to property, theft of personal property and vehicle accidents. When an insured event occurs, the cost to the insurer is referred to as an *insurance loss*, and distributions used to model the costs are often called *loss distributions*.

An insurance company's individual loss on a policy is not only non-negative, but can also (in many cases) potentially be very high. So, to be suitable as models for claim sizes, probability distributions must allow for the occurrence in practice of very high values – distributions which do allow for this are described as having "fat tails" or "heavy tails". Such distributions are positively skewed and, in addition, have relatively high probabilities in the right-hand tails. They are discussed further in §2.2.5.

We begin with a brief review of three probability distributions which are not fat-tailed – the normal, exponential and gamma distributions. Although these distributions are thin-tailed (and the normal is not restricted to non-negative values), we sometimes use them as *reference distributions* for comparing results with those using genuinely fat-tailed distributions, so we include here brief summaries of their properties. We follow these with a consideration of a series of positive distributions generally regarded as being fat-tailed:

- lognormal distribution;
- Pareto distribution;
- Weibull distribution;
- Burr distribution;
- loggamma.

Hogg and Klugman (1984) is a classic text on models for insurance losses.

2.2.1 A further summary note on *R*

The basic *R* package provides specific commands for calculating the probability density function, the distribution function and quantiles for the exponential, gamma, lognormal, normal and Weibull distributions, along with commands to simulate observations from them. The names used by *R* command are as in dexp(x, lambda) for the exponential distribution, where lambda contains the value of λ, pnorm(x, mu, sigma) for the normal distribution, where mu contains the value of μ and sigma contains the value of σ, and similarly we have qgamma(p, alpha, lambda) for the gamma distribution, rlnorm(n, mu, sigma) for the lognormal distribution and rweibull(n, alpha, beta) for the Weibull distribution – see the following sections for further details.

The corresponding commands for the Burr, loggamma and Pareto distributions are not supported in the basic *R* package, but are available in an add-on package called actuar. Commands such as dburr, plgamma and rpareto can then be used. Most of the work in the sections below has been achieved without resorting to the facilities of actuar, but examples of calculations that require it are also included.

2.2.2 Normal (Gaussian) distribution

The normal (Gaussian) family of distributions has two parameters, usually denoted μ and σ (> 0), which (as the notation suggests) represent the mean and standard deviation, respectively, of the distribution.

Notation $X \sim N(\mu, \sigma^2)$.

The probability density function is given by

$$f(x) = \frac{1}{\sigma \sqrt{2\pi}} \exp\left\{-\frac{1}{2}\left(\frac{x-\mu}{\sigma}\right)^2\right\}, \quad -\infty < x < \infty. \qquad (2.9)$$

Note that the density function tails off exponentially in x^2 as $x \to \infty$; this is a very fast tail-off. The distribution is very thin-tailed.

The distribution function $F(x) = \Pr(X \le x)$ is not expressible in a convenient closed form, but particular values are available from *R*, and tables are widely available. For example, to find $\Pr(X \le 2)$ in the case $X \sim N(1, 0.5^2)$ in *R*, we use the command pnorm(2,1,0.5), which returns the value 0.9772499. The distribution function of a $N(0, 1)$ random variable is denoted $\Phi(x)$.

The moment generating function M_X is given by

$$M_X(t) = \exp\left(\mu t + \frac{1}{2}\sigma^2 t^2\right), \qquad (2.10)$$

from which we find $\mathbb{E}[X] = \mu$, $\text{Var}[X] = \sigma^2$, $\mathbb{E}[(X-\mu)^3] = 0$, $\mathbb{E}[(X-\mu)^4] = 3\sigma^4$.

In the context of two-tailed, symmetrical distributions, μ is a location parameter because the density function (2.9) has the form $g(x-\mu)$, where the function g does not depend on μ; we note $X - \mu \sim N(0, \sigma^2)$. In addition, σ is a scale parameter because the density function of $Y = X - \mu$ is of the form $\frac{1}{\sigma}h(y/\sigma)$, where h does not depend on σ.

A linear function of a normal random variable is also normal – in fact

$$X \sim N(\mu, \sigma^2) \Rightarrow a + bX \sim N(a + b\mu, b^2\sigma^2),$$

which follows because the moment generating function of $a + bX$ is

$$\mathbb{E}[e^{(a+bX)t}] = e^{at}M_X(bt) = \exp\left((a + b\mu)t + \frac{1}{2}\sigma^2 b^2 t^2\right).$$

In particular, note that

$$X \sim N(\mu, \sigma^2) \Rightarrow Z = \frac{X - \mu}{\sigma} \sim N(0, 1). \qquad (2.11)$$

The transformation of X to Z given in (2.11) is known as *standardisation*.

Simulation We can simulate a random sample of size n from $X \sim N(\mu, \sigma^2)$ in *R* using the command sample = rnorm(n, mu, sigma), where n, mu and sigma contain the values of n, μ and σ, respectively.

2.2.3 Exponential distribution

The exponential family has a single parameter, usually denoted λ (> 0).

Notation $X \sim \text{Exp}(\lambda)$.

The probability density function is given by

$$f(x) = \lambda e^{-\lambda x}, \quad x > 0. \qquad (2.12)$$

Note that the density function tails off exponentially as $x \to \infty$. The distribution function is given by

$$F(x) = 1 - e^{-\lambda x}, \quad x > 0. \tag{2.13}$$

The tail probability is $\Pr(X > x) = e^{-\lambda x}$, $x \geq 0$; the distribution has exponential tails and will generally underestimate the probability of large claims.

It is easy to show, by integration of $x^r f(x)$, that

$$\mathbb{E}[X^r] = \frac{\Gamma(r + 1)}{\lambda^r}, \quad r > 0. \tag{2.14}$$

The moment generating function $M_X(t)$ is finite for $t < \lambda$, and is given by

$$M_X(t) = \left(1 - \frac{t}{\lambda}\right)^{-1}, \quad t < \lambda, \tag{2.15}$$

from which we find that $\mathbb{E}[X^r] = r!/\lambda^r$, $r = 1, 2, 3, \ldots$. In particular, we have $\mathbb{E}[X] = 1/\lambda$ and $\text{Var}[X] = 1/\lambda^2$. Note that the mean and standard deviation are equal. The third and fourth central moments and the coefficients of skewness and kurtosis (see (1.1), (1.2) and §2.2.5) are considered in Example 2.5.

The distribution is often parameterised using the mean $\mu = 1/\lambda$ as the parameter; in this form, the density function (2.12) becomes

$$f(x) = \frac{1}{\mu} e^{-x/\mu}, \quad x > 0,$$

which is of the form $\frac{1}{\mu} h(x/\mu)$. Thus the parameter μ is a scale parameter.

The rth moment has the property $\mathbb{E}[X^r] \propto \mu^r$, and

$$X \sim \text{Exp}(\lambda) \Rightarrow kX \sim \text{Exp}(\lambda/k), \quad \text{for } k > 0. \tag{2.16}$$

The flexibility of the exponential family as a model for data is restricted by the fact that it has only one parameter.

Simulation A simulated sample of size n from the exponential distribution Exp(λ) can be obtained from *R* using `sample = rexp(n,lambda)`, where n and lambda contain the values of n and λ, respectively. Alternatively, we can use a command based on the inverse transform method of simulating values from a probability distribution. In this case, the method (solving $u = F(x)$ for x, where u is an observation of a uniform distribution on (0,1)) gives a result which can be implemented most easily using the command `x = - log(u)/lambda`. Hence, to generate a random sample of size n from the Exp(λ) distribution, we can use `sample = - log(runif(n))/lambda`, where `runif(n)` simulates a sample of size n from a uniform distribution on the interval $(0, 1)$.

The exponential distribution possesses an important *lack of memory* property, which will be useful in certain reinsurance calculations later on in the book. The property states that, for $x > 0$, $w > 0$,

$$\Pr(X > x + w \mid X > w) = \Pr(X > x), \tag{2.17}$$

as is easily demonstrated. Using the notation $X - w \mid X > w$ to represent the random variable $X - w$ conditional on X assuming a value greater than w, we can express the lack of memory property as

$$X \sim \text{Exp}(\lambda) \Rightarrow X - w \mid X > w \sim \text{Exp}(\lambda),$$

and also

$$X - w \mid X > w \equiv X,$$

where $\equiv$ means "has the same distribution as". The geometric distribution has an equivalent property in "discrete time" for the observed numbers of trials/claims – it is a consequence of the independence of the trials involved (see Exercise 2.3).

The Poisson process: inter-event times

The exponential distribution also plays an important role in the Poisson process. Let $\{N(t) : t \geq 0\}$ be a Poisson process with rate λ, as in §2.1.1. Let T_1 be the time from $t = 0$ to the first event, and, for $j > 1$, let T_j be the time between the $(j - 1)$st event and the jth event. The T_j are the inter-event times.

For $t \geq 0$, it follows that

$$\Pr(T_1 > t) = \Pr(N(t) = 0) = e^{-\lambda t},$$

so that T_1 has an exponential distribution with mean $1/\lambda$. For $t \geq 0$ and $s \geq 0$, we have, using properties (b) and (c) of the Poisson process (see §2.1.1),

$$
\begin{aligned}
\Pr(T_1 > s, T_2 > t) &= \int_s^\infty \Pr(T_2 > t \mid T_1 = v)\lambda e^{-\lambda v} \, dv \\
&= \int_s^\infty \Pr(\text{no events in } (v, v + t])\lambda e^{-\lambda v} \, dv \\
&= \int_s^\infty e^{-\lambda t}\lambda e^{-\lambda v} \, dv \qquad \text{by (b) and (c)} \\
&= e^{-\lambda(t+s)}.
\end{aligned}
\tag{2.18}
$$

Put $s = 0$ in the final expression to get

$$\Pr(T_2 > t) = e^{-\lambda t},$$

so that T_2 is also exponentially distributed with mean $1/\lambda$. Further, (2.18) shows that T_1 and T_2 are independent. Proceeding by induction, it is easy to show that $T_1, T_2, \ldots$ are independent exponentially distributed random variables with mean $1/\lambda$.

The lack of memory property (2.17) of the exponential distribution means that, given that a time w has already elapsed since the last event, the probability that a further time x elapses before the next event is simply the original unconditional probability that the inter-event time is at least x; the distribution "does not remember" that time w has already elapsed.

2.2.4 Gamma distribution

The gamma family has two parameters, usually denoted α (> 0) and λ (> 0).

Notation $X \sim \text{gamma}(\alpha, \lambda)$ or $X \sim \gamma(\alpha, \lambda)$.

The probability density function is given by

$$f(x) = \frac{\lambda^\alpha}{\Gamma(\alpha)} x^{\alpha-1} e^{-\lambda x}, \quad x > 0. \tag{2.19}$$

The exponential family above is the one-parameter sub-family for the case $\alpha = 1$.

The distribution function is

$$F(x) = \Pr(X \le x) = \int_0^x \frac{\lambda^\alpha}{\Gamma(\alpha)} t^{\alpha-1} e^{-\lambda t}\, dt = \frac{1}{\Gamma(\alpha)} \int_0^{\lambda x} u^{\alpha-1} e^{-u}\, du.$$

We see that $F(x)$ is expressible as an incomplete gamma function; that is, as a gamma-type integral, but with finite upper limit. Particular values are available from **R**. For example, to find $\Pr(X \le 200)$ in the case $X \sim \text{gamma}(2, 0.02)$, we use the command `pgamma(200,2,0.02)`, which returns the value 0.9084218.

For $\alpha > 1$, the gamma density tails off more slowly than the exponential distribution, and having two parameters allows the family more flexibility than a one-parameter family (for example the exponential) as a model for data.

By integration we find

$$\mathbb{E}[X^r] = \frac{1}{\lambda^r} \frac{\Gamma(\alpha + r)}{\Gamma(\alpha)}, \quad r > 0. \tag{2.20}$$

The moment generating function $M_X(t)$ is finite for $t < \lambda$, and is given by

$$M_X(t) = \left(1 - \frac{t}{\lambda}\right)^{-\alpha}, \quad t < \lambda, \tag{2.21}$$

from which we find $\mathbb{E}[X] = \alpha/\lambda$ and $\mathrm{Var}[X] = \alpha/\lambda^2$. It is useful to note that the parameters can be found easily from the mean and variance by noting $\lambda = \mathbb{E}[X]/\mathrm{Var}[X]$ and then $\alpha = \lambda\mathbb{E}[X]$.

The third and fourth central moments are $2\alpha/\lambda^3$ and $(6\alpha + 3\alpha^2)/\lambda^4$, from which it follows that the coefficients of skewness and kurtosis (see (1.1) and (1.2)) are $2/\sqrt{\alpha}$ and $3 + 6/\alpha$, respectively (see Exercise 2.7).

The parameter $1/\lambda$ is a scale parameter, and

$$X \sim \mathrm{gamma}(\alpha, \lambda) \Rightarrow kX \sim \mathrm{gamma}(\alpha, \lambda/k), \quad \text{for } k > 0. \tag{2.22}$$

The parameter α is a shape parameter – it determines the skewness and kurtosis – as α increases, the distribution becomes more and more symmetrical and approaches a normal distribution.

Simulation We simulate a random sample of size n from $X \sim \mathrm{gamma}(\alpha, \lambda)$ in *R* using the command `sample = rgamma(n, alpha, lambda)`, where n, alpha and lambda contain the values of n, α and λ, respectively.

The histograms in Figure 2.5 display 1000 claim sizes simulated from Exp(0.01) and gamma(2, 0.02) distributions, which have means 100 and

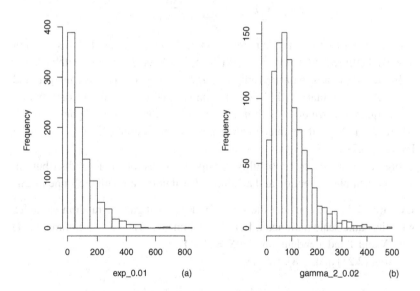

Figure 2.5. Histograms of samples simulated from an exponential distribution with parameter $\lambda = 0.01$ (mean = 100) (a) and from a gamma distribution with $\alpha = 2$, $\lambda = 0.02$ (mean = 100) (b).

variances 10 000 and 5000, respectively. The commands used for the simulation were

```
exp_0.01=rexp(1000,0.01)
gamma_2_0.02=rgamma(1000,2,0.02)
```

The sample means and variances, and the minimum and maximum values that were observed, were as follows:

sample from Exp(0.01): mean 101.5, variance 10301, min 0.03, max 826.5;
sample from gamma(2, 0.02): mean 98.17, variance 4882.3, min 2.10, max 490.7.

In the case that α is an integer, the gamma(α, λ) distribution is also known as the *Erlang* distribution, after the Danish mathematician, statistician and engineer A. K. Erlang (1878–1929), a pioneer in the study of telecommunications networks. In this case, the random variable $X \sim$ gamma(α, λ) can be represented as the sum of α iid Exp(λ) random variables by (1.8), because the gamma(α, λ) moment generating function is the αth power of the Exp(λ) moment generating function in (2.15). The distribution function can be expressed as a finite sum as follows:

$$F(x) = 1 - \sum_{j=0}^{\alpha-1} e^{-\lambda x} \frac{(\lambda x)^j}{j!}$$

(as the reader may verify by repeated integration by parts). The distribution models the time from $t = 0$ to the αth event in a Poisson process with rate λ.

In the case that α is of the form $\alpha = n/2$, where n is a positive integer, and $\lambda = 1/2$, the gamma($n/2, 1/2$) distribution is usually called the χ_n^2 distribution ("chi-squared distribution with n degrees of freedom"). If $X \sim$ gamma(α, λ), where 2α is a positive integer, then the random variable $2\lambda X \sim \chi_{2\alpha}^2$ (see Exercise 2.7).

Figure 2.6 shows the probability density functions for a normal distribution, an exponential distribution and a gamma distribution, all with the same mean.

Example 2.4   We consider the sum of independent gamma random variables with the same scale parameter. Let $X \sim$ gamma(α_1, λ) and $Y \sim$ gamma(α_2, λ) with X and Y independent, and let $S = X + Y$.

Using (2.21) and (1.8), we have

$$M_S(t) = M_X(t)M_Y(t) = \left(1 - \frac{t}{\lambda}\right)^{-\alpha_1} \left(1 - \frac{t}{\lambda}\right)^{-\alpha_2}$$
$$= \left(1 - \frac{t}{\lambda}\right)^{-(\alpha_1+\alpha_2)}.$$

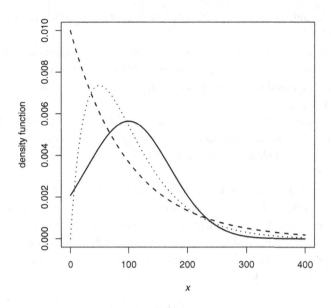

Figure 2.6. Probability density functions, for $0 < x < 400$, of normal (solid line), exponential (dashed line) and gamma (dotted line) distributions, each with mean 100; the variances are 5000, 10 000, and 5000, respectively.

We identify this as the moment generating function of a gamma($\alpha_1 + \alpha_2, \lambda$) random variable, so we can conclude that $S = X + Y \sim$ gamma($\alpha_1 + \alpha_2, \lambda$). As an illustration, suppose that claim sizes arising in one portfolio have a gamma(10, 2) distribution and that claim sizes from another portfolio have, independently, a gamma(6, 2) distribution. If we add the sizes of a randomly selected claim from each portfolio, we find the sum $S \sim$ gamma(16, 2). If we want the value of $\Pr(S > 12)$, it can be found using ***R***: the command `1 - pgamma(12,16,2)` returns the value 0.03440009.

2.2.5 Fat-tailed distributions

The concept of "fat tails" (or "heavy tails") can be characterised formally in several ways. For a non-negative distribution, one general description is to state that a fat-tailed distribution is one whose tail is not "exponentially bounded"; that is, one whose tail probabilities are *not* of the form $\Pr(X > x) \leq be^{-ax}$, $x > 0$, for any positive constants a and b. In the case of a non-negative distribution (as in the case of models for claim sizes), the concept applies to the single right-hand tail, but it can be applied to one or both tails in the case of a two-tailed distribution. We may also say that a distribution F_X, where X is a non-negative random variable, is fat-tailed if the moment generating function $M_X(t) = \mathbb{E}[e^{tX}]$ is not finite for any $t > 0$.

The literature on fat tails is extensive, and authors take different, but related, approaches to the concept. A discussion of different classes of fat-tailed distributions is beyond the scope of this book; see Embrechts *et al.* (1997) for a detailed treatment.

An example of a family of very thin-tailed distributions is the normal (Gaussian) family (see §2.2.2). An example of a family of fat-tailed distributions with two tails is the "Student's *t*" family. Figure 2.7 shows the density functions of one of each type.

Another concept related to fat tails is a characteristic of the shape of a probability density function known as *kurtosis*, which is a measure of the "peakedness" of a density relative to that of a normal distribution. The *coefficient of kurtosis* is defined for distributions with finite fourth moments and is based on the fourth central moment, namely $\mathbb{E}[(X - \mathbb{E}[X])^4]$. Recall from (1.2) that the coefficient of kurtosis is given by the scaled quantity

$$\mathbb{E}[(X - \mathbb{E}[X])^4]/\{\mathrm{Var}[X]\}^2, \tag{2.23}$$

which, for a normal distribution, equals 3. Distributions with a kurtosis coefficient greater than 3 are said to be *leptokurtic* and have a more sharply

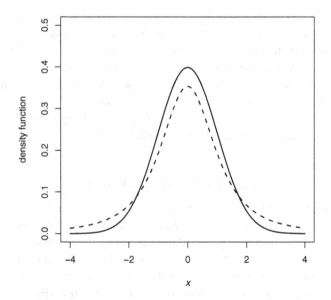

Figure 2.7. Probability density functions of a "standard" normal distribution (very thin-tailed, solid line) and a t_2 distribution (fat-tailed, dashed line).

peaked density function than the normal distribution (as for the Student's t distribution in Figure 2.7 – lower probability in the centre, higher in the tails).

While the coefficient of kurtosis can be calculated for skewed distributions (with fourth moments), its interpretation is straightforward for symmetrical, peaked distributions (comparable to the normal distribution). Its numerical value for skewed distributions, while it can be interesting, is not a particularly useful guide.

The coefficient of kurtosis is scale-free; that is, kX has the same kurtosis as X. So, for example, the kurtosis of $\gamma(\alpha, \lambda)$ does not depend on the scale parameter – the two-parameter $\gamma(\alpha, \lambda)$ has the same kurtosis as the one-parameter $\gamma(\alpha, 1)$.

In the case that X has a finite moment generating function in a neighbourhood of the origin, central moments of orders 2, 3 and 4 can be found conveniently from the *cumulant generating function* $K_X(t)$, given by

$$K_X(t) = \log M_X(t). \tag{2.24}$$

Let κ_j be the coefficient of $t^j/j!$ in the power series expansion of $K_X(t)$; we call κ_j the jth cumulant of the distribution of X, and it can be found from the power series expansion, or by differentiation, on noting that $\kappa_j = K^{(j)}(0)$, the jth derivative of $K_X(t)$ evaluated at $t = 0$.

It transpires (see Exercise 2.5) that

$$\mathbb{E}[(X - \mathbb{E}[X])^2] = \kappa_2, \quad \mathbb{E}[(X - \mathbb{E}[X])^3] = \kappa_3, \quad \mathbb{E}[(X - \mathbb{E}[X])^4] = \kappa_4 + 3\kappa_2^2,$$

giving the coefficient of skewness as $\kappa_3/(\kappa_2)^{3/2}$ and the coefficient of kurtosis as $\left(\kappa_4/\kappa_2^2\right) + 3$.

Note Some authors work with the *excess kurtosis*, which is the kurtosis as defined above in excess of 3 (and therefore such that the normal distribution is said to have kurtosis 0).

An example of a family of fat-tailed distributions is the *lognormal* family (see §2.2.6). Figure 2.8 shows the tails of normal and lognormal distributions (with the same means and variances) for comparison.

Note When displaying data in the context of fat-tailed models, plotting $\log x$ instead of x on the horizontal axis can produce a helpful alternative display.

Before examining some families of fat-tailed distributions, we present two examples which examine kurtosis relating to distributions already considered.

Example 2.5 We find the coefficient of kurtosis for $X \sim \text{Exp}(\lambda)$ using the cumulant generating function. By (2.15) and (2.24), we have

$$M_X(t) = \lambda/(\lambda - t) \quad \Rightarrow \quad K_X(t) = \log \lambda - \log(\lambda - t)$$

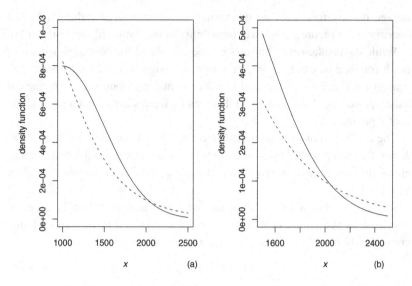

Figure 2.8. Tails of normal (solid line) and lognormal (dashed line) densities with mean 1000 and standard deviation 500: above 1000 (a) and above 1500 (magnified) (b).

and hence

$$K_X'(t) = (\lambda - t)^{-1}, \qquad K_X''(t) = (\lambda - t)^{-2},$$
$$K_X'''(t) = 2(\lambda - t)^{-3}, \qquad K_X^{(4)}(t) = 6(\lambda - t)^{-4}.$$

So

$$\kappa_2 = K_X''(0) = \lambda^{-2},$$
$$\kappa_3 = K_X'''(0) = 2\lambda^{-3},$$
$$\kappa_4 = K_X^{(4)}(0) = 6\lambda^{-4}.$$

Hence the fourth central moment is $6\lambda^{-4} + 3\lambda^{-4} = 9\lambda^{-4}$ and, by (2.23), the coefficient of kurtosis is $9\lambda^{-4}/(\lambda^{-2})^2 = 9$.

The exponential distribution may have a relatively high numerical value for its coefficient of kurtosis, but the density function is decreasing (with its mode at $x = 0$) and there is no flexibility in its shape – the exponential family does not accommodate a non-zero mode.

Note   The coefficient of skewness is $2\lambda^{-3}/(\lambda^{-2})^{3/2} = 2$.

Example 2.6   A source of fat-tailed distributions (with two tails) is found in the difference between two independent random variables of certain types. We examine here the kurtosis of the symmetrical variable $W = X - Y$, where X and Y are independent gamma$(\alpha, 1)$ random variables.

By (2.20) we have $\mathbb{E}[X] = \alpha$, $\mathbb{E}[X^2] = \alpha(\alpha + 1)$, $\mathbb{E}[X^3] = \alpha(\alpha + 1)(\alpha + 2)$ and $\mathbb{E}[X^4] = \alpha(\alpha + 1)(\alpha + 2)(\alpha + 3)$. Using these results for X, we find that $\mathbb{E}[W] = 0$ and $\text{Var}[W] = \text{Var}[X] + \text{Var}[Y] = 2\alpha$. We also have $\mathbb{E}[W^3] = 0$ (by the symmetry of W), and

$$\mathbb{E}[W^4] = \mathbb{E}[(X - Y)^4] = \mathbb{E}[X^4 - 4X^3Y + 6X^2Y^2 - 4XY^3 + Y^4]$$
$$= 2\alpha(\alpha + 1)(\alpha + 2)(\alpha + 3) - 8\alpha^2(\alpha + 1)(\alpha + 2) + 6\alpha^2(\alpha + 1)^2$$
$$= 12\alpha(\alpha + 1).$$

Hence the coefficient of kurtosis of W is given by

$$12\alpha\frac{(\alpha + 1)}{(4\alpha^2)} = 3\left(1 + \frac{1}{\alpha}\right),$$

and, since $\alpha > 0$ this is greater than the coefficient of kurtosis for the normal distribution.

Notes

(1) We can also derive the result using the cumulant generating function of $W = X - Y$, which is easily found to be $K_W(t) = -\alpha\{\log(1 - t) + \log(1 + t)\}$.
(2) The result in Example 2.6 applies to the more general case for which X and Y are independent $\gamma(\alpha, \lambda)$ random variables, since then $X - Y$ can be written as $(1/\lambda)(X_1 - Y_1)$, where X_1 and Y_1 are $\gamma(\alpha, 1)$ variables.

2.2.6 Lognormal distribution

The lognormal family has two parameters, usually denoted μ and σ (> 0).

Notation $X \sim \text{lognormal}(\mu, \sigma)$.

The probability density function is given by

$$f(x) = \frac{1}{\sigma\sqrt{2\pi}}\frac{1}{x}\exp\left\{-\frac{1}{2}\left(\frac{\log x - \mu}{\sigma}\right)^2\right\}, \quad x > 0.$$

The name "lognormal distribution" arises from the fact that

$$X \sim \text{lognormal}(\mu, \sigma) \Leftrightarrow Y = \log X \sim N(\mu, \sigma^2). \tag{2.25}$$

So, if $X \sim \text{lognormal}(\mu, \sigma)$ then $\log X$ has a normal distribution with mean μ and standard deviation σ. The lognormal parameters μ and σ are sometimes referred to as the "meanlog" and "sdlog" parameters.

The distribution function (and hence tail probabilities) are found using the log transformation to normal and the distribution function of the standard normal distribution $Z \sim N(0, 1)$:

$$F(x) = \Pr(X \le x) = \Pr(\log X \le \log x) = \Pr\left(Z \le \frac{\log x - \mu}{\sigma}\right). \qquad (2.26)$$

There is no problem regarding the existence (finiteness) of moments (but a moment generating function $M_X(t)$ is not finite for any positive value of t). The general moment $\mathbb{E}[X^r]$ is easily found from the moment generating function of $Y = \log X$, which is $M_Y(t) = \exp\left(t\mu + \frac{1}{2}t^2\sigma^2\right)$, as follows, on noting that $X = e^Y$:

$$\mathbb{E}[X^r] = \mathbb{E}[e^{rY}] = M_Y(r) = \exp\left(r\mu + \frac{1}{2}r^2\sigma^2\right). \qquad (2.27)$$

This gives

$$\mathbb{E}[X] = M_Y(1) = \exp\left(\mu + \frac{1}{2}\sigma^2\right), \qquad \mathbb{E}[X^2] = M_Y(2) = \exp\left(2\mu + 2\sigma^2\right).$$

So

$$\mathbb{E}[X] = \exp\left(\mu + \frac{1}{2}\sigma^2\right), \qquad \mathrm{Var}[X] = \exp\left(2\mu + \sigma^2\right)(e^{\sigma^2} - 1). \qquad (2.28)$$

It can be useful to note (from (2.28)) that the mean and variance are related by

$$\mathrm{Var}[X] = \{\mathbb{E}[X]\}^2(e^{\sigma^2} - 1). \qquad (2.29)$$

Particular values of the distribution function are available from **R**. For example, to find $\Pr(X \le 200)$ in the case $X \sim \text{lognormal}(4.5, 0.6325)$, we use the command `plnorm(200,4.5,0.6325)`, which returns the value 0.8965547.

The lognormal density tails off more slowly than the exponential and gamma distributions.

Simulation We can simulate a random sample of size n from $X \sim$ lognormal(μ, σ) in **R**, using `sample = rlnorm(n, mu, sigma)` or, less simply, `sample = exp(rnorm(n, mu, sigma))`, where n, mu and sigma contain the values of n, μ and σ, respectively.

Figure 2.9(a) shows the probability density functions of two lognormal distributions. Figure 2.9(b) shows a histogram of 1000 claim sizes simulated from lognormal(4.5, 0.6325), which has mean 110 and variance 5950 approximately.

The command used for the simulation was

```
lnorm_4.5_0.6325=rlnorm(1000,4.5,0.6325)
```

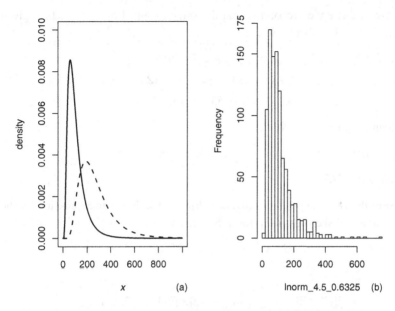

Figure 2.9. (a) Probability density functions of lognormal(4.5, 0.6325) (solid line) and lognormal(5.5, 0.5) (dashed line). (b) Histogram of sample simulated from a lognormal distribution with parameters $\mu = 4.5$ and $\sigma = 0.6325$ (mean ≈ 110).

The sample mean and variance, and the minimum and maximum values that were observed, were as follows:

sample from lognormal(4.5, 0.6325): mean 110.2, variance 6645, min 8.45, max 750.5.

Example 2.7 Let $X \sim$ lognormal(0.8, 0.3). We find the mean and standard deviation of the distribution and evaluate some tail probabilities.

By (2.28) and (2.29) we have

$$\mathbb{E}[X] = e^{0.845} = 2.3280, \quad \mathrm{SD}[X] = e^{0.845}(e^{0.09} - 1)^{1/2} = 0.7144.$$

By (2.26) we have (writing Z for a N(0, 1) random variable)

$$\Pr(X > 5) = \Pr(\log X > \log 5) = \Pr\left(Z > \frac{\log 5 - 0.8}{0.3}\right)$$
$$= 1 - \Phi(2.6981) = 0.00349.$$

Note The **R** command 1 - plnorm(5,0.8,0.3) returns 0.003486548.

We evaluate the probability that the value assumed by X exceeds 5, given that it exceeds 4, as follows:

$$\Pr(X > 5 \mid X > 4) = \Pr(X > 5 \text{ and } X > 4)/\Pr(X > 4)$$
$$= \Pr(X > 5)/\Pr(X > 4) = \Pr(Z > 2.6981)/\Pr(Z > 1.9543)$$
$$= 0.00349/0.0253 = 0.138.$$

Confirming this using **R**,

```
(1 - plnorm(5,0.8,0.3))/(1 - plnorm(4,0.8,0.3))
```

returns 0.1376341.

Example 2.8 Suppose $X \sim$ lognormal(μ, σ). We derive an expression for the coefficient of skewness as follows. By (2.28),

$$\mathbb{E}[(X - \mathbb{E}[X])^2] = e^{2\mu+\sigma^2}(e^{\sigma^2} - 1),$$

and by (2.27) we have

$$\mathbb{E}[(X - \mathbb{E}[X])^3] = \mathbb{E}[X^3] - 3\mathbb{E}[X]\mathbb{E}[X^2] + 2(\mathbb{E}[X])^3$$
$$= e^{3\mu+(9\sigma^2/2)} - 3e^{3\mu+(5\sigma^2/2)} + 2e^{3\mu+(3\sigma^2/2)}$$
$$= e^{3\mu+(3\sigma^2/2)}\left(e^{3\sigma^2} - 3e^{\sigma^2} + 2\right).$$

Hence the coefficient of skewness is given by

$$\frac{e^{3\mu+(3\sigma^2/2)}\left(e^{3\sigma^2} - 3e^{\sigma^2} + 2\right)}{e^{3\mu+(3\sigma^2/2)}(e^{\sigma^2} - 1)^{3/2}} = \frac{(e^{\sigma^2} + 2)(e^{\sigma^2} - 1)^2}{(e^{\sigma^2} - 1)^{3/2}}$$
$$= (e^{\sigma^2} + 2)(e^{\sigma^2} - 1)^{1/2}.$$

The skewness increases rapidly as σ^2 increases.

Example 2.9 Suppose that claims occur on a portfolio of general insurance policies independently of one another and are of two types; each claim is classified as being of "type A" or "type B". Type A claim sizes have an exponential distribution with mean 2, while type B claim sizes have an exponential distribution with mean 8, and 75% of claims are of type A.

We examine some properties of X, the size of a randomly chosen claim arising on the portfolio: we have

$$\Pr(X > 10) = \Pr(X > 10 \mid A)\Pr(A) + \Pr(X > 10 \mid B)\Pr(B)$$
$$= 0.75e^{-5} + 0.25e^{-10/8} = 0.0767$$

and

$$\mathbb{E}[X] = \mathbb{E}[X \mid A]\Pr(A) + \mathbb{E}[X \mid B]\Pr(B) = 0.75(2) + 0.25(8) = 3.5.$$

Noting that the second moment of $V \sim \text{Exp}(1/\mu)$ is $\mathbb{E}[V^2] = 2\mu^2$, we have

$$\mathbb{E}[X^2] = 0.75(8) + 0.25(128) = 38,$$

and so $\text{Var}[X] = 25.75$.

Suppose now we replace X by (a) a single exponential variable Y with the same mean as X, so $Y \sim \text{Exp}(1/3.5)$, and (b) a single lognormal variable W with the same mean and variance as X, and in each case repeat the tail probability calculation.

(a) $\Pr(Y > 10) = e^{-10/3.5} = 0.0574$;
(b) setting $3.5^2(\exp(\sigma^2) - 1) = 25.75$ and then $\mu + \sigma^2/2 = \log(3.5)$ yields

$$W \sim \text{lognormal}(0.68673, 1.06398).$$

Then we have $\Pr(W > 10) = \Pr(Z > 1.5187) = 0.0644$, where $Z \sim N(0, 1)$.

We note that $\Pr(X > 10)$ is greater than both $\Pr(Y > 10)$ and $\Pr(W > 10)$. By using a single "compromise" variable to model the claim sizes, be it exponential or a (fat-tailed) lognormal, we are failing to take proper account of the heterogeneity in the portfolio (the presence of two types of claims with different properties), and as a result we under-estimate the probability of large claims. This can have serious consequences for the under-pricing of insurance products and the consequent financial health of the insurance company.

Example 2.10 An insurance analyst is examining the distribution of "large" and "very large" claims, that is claims in excess of £20 000 and £35 000 respectively. Let X represent claim size (in units of £1000).

We calculate the conditional tail probability $\Pr(\text{"very large"} \mid \text{"large"})$; that is, $\Pr(X > 35 \mid X > 20)$ using three models with similar means: (a) $X \sim \text{Exp}(1/10)$; (b) $X \sim \text{gamma}(0.5, 0.05)$; (c) $X \sim \text{lognormal}(1.45, 1.30)$. First note that $\Pr(X > 35 \mid X > 20) = \Pr(X > 35)/\Pr(X > 20)$.

(a) $\Pr(X > 35 \mid X > 20) = e^{-3.5}/e^{-2} = e^{-1.5} = 0.223$. (Note that we can get the result $\Pr(X > 35 \mid X > 20) = \Pr(X > 15) = e^{-1.5}$ directly by appealing to (2.17), the lack of memory property of the exponential distribution.)
(b) $X \sim \text{gamma}(0.5, 0.05) \Rightarrow 0.1X \sim \chi_1^2$ (see Exercise 2.7)

$$\Rightarrow \Pr(X > 35 \mid X > 20) = \Pr(\chi_1^2 > 3.5)/\Pr(\chi_1^2 > 2)$$

$$= 0.0614/0.1573 = 0.390.$$

(c) $\Pr(X > 35 \mid X > 20) = \Pr(\log X > \log 35)/\Pr(\log X > \log 20)$

$$= \Pr(Z > 1.6195)/\Pr(Z > 1.189)$$

$$= 0.0527/0.1172 = 0.450,$$

where $Z \sim N(0, 1)$.

We note the increasing conditional tail probabilities as we go from the exponential to the gamma to the lognormal model.

2.2.7 Pareto distribution

The Pareto family is a wide one, with several sub-families. Here we consider the sub-family most widely used in general insurance as a model for claim sizes (sometimes called the "American Pareto"), which has two parameters, usually denoted α (> 0) and λ (> 0), and which allows for all positive values to be realised. Its density and distribution functions exhibit "power law" decay. The family of distributions is named after the Italian economist V. F. D. Pareto (1848–1923).

Notation $X \sim \text{Pa}(\alpha, \lambda)$ or $X \sim \text{Pareto}(\alpha, \lambda)$.

The probability density function is given by

$$f(x) = \frac{\alpha \lambda^\alpha}{(\lambda + x)^{\alpha+1}}, \quad x > 0. \tag{2.30}$$

The distribution function is given by

$$F(x) = 1 - \left(\frac{\lambda}{\lambda + x}\right)^\alpha, \quad x > 0. \tag{2.31}$$

Moments of the form $\mathbb{E}[X^r]$, $r = 1, 2, 3, \ldots$, exist only for $r < \alpha$. For example, in the case $0 < \alpha \le 1$ no moments exist; in the case $1 < \alpha \le 2$ only the first moment $\mathbb{E}[X]$ exists; in the case $2 < \alpha \le 3$ only the first two moments $\mathbb{E}[X]$ and $\mathbb{E}[X^2]$ exist; and so on. The mean exists only in the case $\alpha > 1$, and the variance exists only in the case $\alpha > 2$. Since the number of moments which exist is restricted, the moment generating function is $+\infty$ for all positive arguments.

The general moment $\mathbb{E}[X^r]$ is given by

$$\mathbb{E}[X^r] = \frac{\Gamma(\alpha - r)\Gamma(1 + r)}{\Gamma(\alpha)} \lambda^r, \quad 0 < r < \alpha, \tag{2.32}$$

which may be shown by induction using integration by parts. The results for the mean and variance are

$$\mathbb{E}[X] = \frac{\lambda}{\alpha - 1}, \ \alpha > 1; \quad \text{Var}[X] = \frac{\alpha \lambda^2}{(\alpha - 1)^2(\alpha - 2)}, \ \alpha > 2. \tag{2.33}$$

It can be useful to note that, in the case that the variance exists, the mean and variance are related by

$$\text{Var}[X] = \frac{\alpha}{\alpha - 2}\{\mathbb{E}[X]\}^2. \tag{2.34}$$

The Pareto distribution tails off more slowly than the exponential, gamma and lognormal distributions.

The probability density function (2.30) can be written as

$$f(x) = \frac{\alpha}{\lambda} \frac{1}{(1 + x/\lambda)^{\alpha+1}}, \quad x > 0.$$

The parameter λ is a scale parameter, and

$$X \sim \text{Pa}(\alpha, \lambda) \Rightarrow kX \sim \text{Pa}(\alpha, k\lambda), \quad \text{for } k > 0. \tag{2.35}$$

The parameter α is a shape parameter – it determines the skewness (see Exercise 2.13). Particular values of the distribution function are available from the basic **R** package – as an illustration we find $\Pr(X \leq 200)$ in the case $X \sim$ Pa(4, 300) by evaluating the distribution function directly. Noting (2.31), we use the explicit command

```
1-(300/500)^4
```

which returns the value 0.8704.

Simulation Simulated values of the distribution can be obtained from the basic **R** package using a command based on the inverse transformation method of simulating values from a probability distribution – in this case the method gives a result which can be implemented most easily as x = lambda*(u^(-1/alpha) - 1), where lambda and alpha contain the values of λ and α, respectively, and where u contains an observation from a uniform distribution on (0, 1). We can thus simulate a random sample of size n from $X \sim \text{Pa}(\alpha, \lambda)$ using the command sample = lambda*(runif(n)^(-1/alpha)-1), where n contains the value of n (and see the note below for simulation using the **R** package actuar).

The histograms in Figure 2.10 display 1000 claim sizes simulated from Pa(4, 300), which has mean 100 and variance 20 000 approximately. The command used for the simulation was

```
par_4_3000 = 300*(runif(1000)^{-1/4}-1)
```

The sample mean and variance, and the minimum and maximum values that were observed, were as follows:

sample from Pa(4, 300): mean 104.4, variance 21288, min 0.02, max 1757.

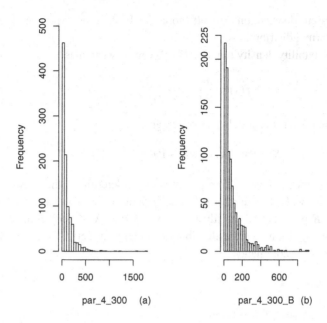

Figure 2.10. Histograms of a sample simulated from a Pareto distribution with parameters $\alpha = 4$ and $\lambda = 300$. (a) Histogram of the full data set; (b) histogram of all observations less than 1000. There were four observations above 1000, namely 1106, 1237, 1588 and 1757.

Note Using the facilities of the **R** add-on package actuar we can reproduce the results above: for $\Pr(X \leq 200)$, in the case $X \sim \mathrm{Pa}(4, 300)$, we can use the command ppareto(200,4,300), which returns 0.8704; a sample of 1000 simulated claim sizes from $\mathrm{Pa}(4, 300)$ is obtainable using the command rpareto(1000,4,300).

The Pareto distribution possesses an important *conditional tail* property, which is useful in certain reinsurance calculations. The property states that, for $X \sim \mathrm{Pa}(\alpha, \lambda)$, the variable $X - w \mid X > w$ also has a Pareto distribution, with the same first parameter and shifted second parameter. In fact,

$$X \sim \mathrm{Pa}(\alpha, \lambda) \Rightarrow X - w \mid X > w \sim \mathrm{Pa}(\alpha, \lambda + w). \qquad (2.36)$$

So, given that a claim size exceeds an amount w, the distribution of the amount by which it exceeds w has this other Pareto distribution. The property is easily demonstrated (see Exercise 2.13).

The Pareto distribution can be derived as follows. Let $X_1 \sim \mathrm{Exp}(1)$ and $X_2 \sim \mathrm{gamma}(\alpha, \lambda)$ with X_1 and X_2 independent. Using standard methods for transformation of densities (see, for example, sect. 4.7 of Grimmett and Stirzaker (2001)), $Y = X_1/X_2 \sim$ Pa(α, λ); the Pareto distribution representing the ratio of the exponential and the gamma distributions has a fatter tail than those of the two component distributions.

The Pareto distribution also arises in the context of *mixture distributions*; see §2.3.

Figure 2.11 shows the probability density functions for a lognormal distribution and a Pareto distribution with the same means and variances.

Figure 2.12 shows the probability density functions for three Pareto distributions with different parameters – in one case the mean does not exist, and in the other two cases the mean is 10.

Example 2.11 Suppose that a claim size X is modelled as $X \sim$ Pa$(5, 360)$ and that we want to calculate the mean sizes of claims greater than 80 and less than 80.

From (2.36) we have $X - 80 \mid X > 80 \sim$ Pa$(5, 440)$, so by (2.33) we have $\mathbb{E}[X - 80 \mid X > 80] = 440/4 = 110$, and thus $\mathbb{E}[X \mid X > 80] = 190$. By (2.31) and (2.33) we know that Pr$(X > 80) = (360/440)^5$ and $\mathbb{E}[X] = 360/4 = 90$.

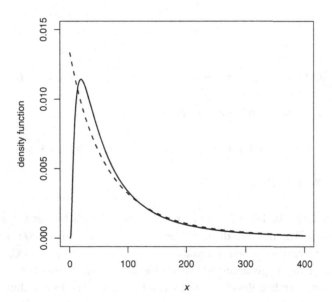

Figure 2.11. Probability density functions, for $0 < x < 400$ only, of lognormal$(4.056, 1.048)$ (solid line) and Pa$(4, 300)$ (dashed line), both of which have mean 100 and variance 20 000.

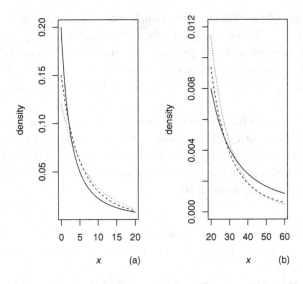

Figure 2.12. Probability density functions of Pa(1, 5) (solid line), Pa(3, 20) (dashed line) and Pa(6, 50) (dotted line). (a) Densities in the range $0 < x < 20$. (b) Densities in the range $20 < x < 60$, and it is clear that the Pa(1, 5) distribution (which does not have any finite moments) crosses over the other densities and has a fatter tail.

Then

$$\mathbb{E}[X] = \mathbb{E}[X \mid X > 80] \Pr(X > 80) + \mathbb{E}[X \mid X \leq 80] \Pr(X \leq 80).$$

Substituting values from above, we have

$$90 = 190(360/440)^5 + \mathbb{E}[X \mid X \leq 80]\{1 - (360/440)^5\},$$

and so $\mathbb{E}[X \mid X \leq 80] = 32.11$.

Example 2.12 In Table 2.2 we compare some tail probabilities for selected distributions. The first five distributions all have mean $\mathbb{E}[X] = 1000$. The first four have increasing standard deviations, 1000, 1414, 2000 and 3000, respectively (most values given are rounded). The first Pareto distribution does not have a finite standard deviation; the second and third Pareto distributions do not have any finite moments. The table illustrates the wide range of behaviours we can capture with these distributions.

Table 2.2. *Some tail probabilities*

	Pr($X > 5000$)	Pr($X > 10\,000$)	Pr($X > 15\,000$)
Exp(0.001)	0.00674	4.54×10^{-5}	3.06×10^{-7}
gamma(0.5, 0.0005)	0.0253	0.00157	0.000108
lognormal(6.10304, 1.26864)	0.0285	0.00716	0.00281
lognormal(5.75646, 1.51743)	0.0344	0.0114	0.00549
Pa(1.5, 500)	0.0274	0.0104	0.00579
Pa(0.8, 500)	0.147	0.0875	0.0641
Pa(0.5, 1000)	0.408	0.302	0.250

2.2.8 Weibull distribution

The Weibull family has two parameters, which we will denote c (> 0) and γ (> 0). The family of distributions is named after the Swedish engineer and mathematician, E. H. W. Weibull (1887–1979).

Notation $X \sim \text{Wei}(c, \gamma)$ or $X \sim \text{Weibull}(c, \gamma)$.

The probability density function is given by

$$f(x) = c\gamma x^{\gamma-1} e^{-cx^\gamma}, \quad x > 0.$$

The distribution function is given by

$$F(x) = 1 - e^{-cx^\gamma}, \quad x > 0. \tag{2.37}$$

The tail probability is given by Pr($X > x) = e^{-cx^\gamma}$, which shows that the distribution can be regarded as a generalisation of the exponential distribution, since putting $\gamma = 1$ gives the tail probability for an exponential distribution; in fact, $\text{Wei}(c, 1) \equiv \text{Exp}(c)$.

The Weibull random variable is also obtainable as a power function of an exponential variable – if $X \sim \text{Exp}(\lambda)$ then $Y = X^{1/\gamma} \sim \text{Wei}(\lambda, \gamma)$, or, equivalently, if $X \sim \text{Wei}(c, \gamma)$ then $X^\gamma \sim \text{Exp}(c)$ (see Example 2.13).

Simulation for $X \sim \text{Wei}$ We can simulate a random sample of size n from $X \sim \text{Wei}(c, \gamma)$ in *R* using the command

```
sample = rexp(n,c)^(1/gamma),
```

where n, c and gamma contain the values of n, c and γ, respectively. With a different Weibull parameterisation, we can use a simpler command rweibull (see below).

It is clear from the form of the tail probability that in the case $\gamma \geq 1$ the tails will be as thin as, or thinner than, exponential tails – so to qualify as a fat-tailed distribution we restrict the parameter value to the case $\gamma < 1$.

There is no problem regarding the existence of moments (but in general there is not a convenient, usable moment generating function). The general moment $\mathbb{E}[X^r]$ can be found easily using the transformation to exponential:

$$\mathbb{E}[X^r] = \mathbb{E}[X^{\gamma(r/\gamma)}] = \mathbb{E}[Y^{r/\gamma}],$$

where $Y \sim \text{Exp}(c)$. Hence, by (2.14), we have

$$\mathbb{E}[X^r] = \Gamma\left(1 + \frac{r}{\gamma}\right) c^{-r/\gamma}, \quad r > 0. \tag{2.38}$$

This general expression is rather awkward in certain cases, but is easy to use in other cases – for example in the case $\gamma = 1/2$ we have $\mathbb{E}[X] = 2/c^2$, $\mathbb{E}[X^2] = 24/c^4$, giving $\text{Var}[X] = 20/c^4$.

There is an alternative parameterisation which we will denote as

$$X \sim \text{Wei2}(\alpha, \beta).$$

The relationship between the parameters in the second version and those in the first, $X \sim \text{Wei}(c, \gamma)$, is $\alpha = \gamma$, $\beta = c^{-1/\gamma}$.

For $\text{Wei2}(\alpha, \beta)$, the probability density function is given by

$$f(x) = \frac{\alpha}{\beta} \left(\frac{x}{\beta}\right)^{\alpha-1} e^{-(x/\beta)^\alpha}, \quad x > 0; \tag{2.39}$$

the distribution function is given by

$$F(x) = 1 - e^{-(x/\beta)^\alpha}, \quad x > 0,$$

and the moment $\mathbb{E}[X^r]$ is given by

$$\mathbb{E}[X^r] = \Gamma\left(1 + \frac{r}{\alpha}\right) \beta^r, \quad r = 1, 2, 3, \ldots.$$

For example, in the case $\alpha = 1/2$ we have $\mathbb{E}[X] = 2\beta$, $\mathbb{E}[X^2] = 24\beta^2$, giving $\text{Var}[X] = 20\beta^2$.

The second parameterisation (2.39) reveals that the parameter β is a scale parameter – in fact, $X \sim \text{Wei2}(\alpha, \beta) \Rightarrow kX \sim \text{Wei2}(\alpha, k\beta)$, for $k > 0$, and is the parameterisation used in **R**. The parameter α is a shape parameter.

Particular values of the distribution function are available – for example, to find $\Pr(X \leq 200)$ in the case $X \sim \text{Wei2}(0.5, 50)$ we use the command `pweibull(200,0.5,50)`, which returns the value 0.8646647.

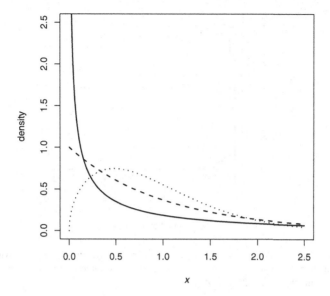

Figure 2.13. Probability density functions of Wei(1, 0.5) ≡ Wei2(0.5, 1) (solid line), Wei(1, 1) ≡ Wei2(1, 1) (dashed line) and Wei(1, 1.5) ≡ Wei2(1.5, 1) (dotted line).

Figure 2.13 shows the probability density functions for three Weibull distributions.

Simulation for $X \sim$ Wei2 We can simulate a random sample of size n from $X \sim \text{Wei2}(\alpha, \beta)$ in *R* using the command

```
sample = rweibull(n, alpha, beta)
```

where n, alpha and beta contain the values of n, α and β, respectively.

The histogram in Figure 2.14 displays 1000 claim sizes simulated from Wei2(0.5, 50), which has mean 100 and variance 50 000. The command used for the simulation was

```
weibull2_0.5_50=rweibull(1000,0.5,50)
```

The sample mean and variance, and the minimum and maximum values that were observed, were as follows:

sample from Wei2(0.5, 50): mean 94.80, variance 43283, min 0.000, max 3484.

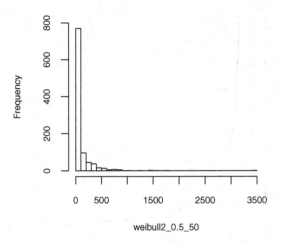

weibull2_0.5_50

Figure 2.14. Histogram of sample simulated from a Weibull distribution with parameters (version 2) $\alpha = 0.5, \beta = 50$. There were three observations above 1500, namely 1517, 1725 and 3484.

Example 2.13 Suppose $X \sim \text{Wei}(c, \gamma)$. We verify that $X^\gamma \sim \text{Exp}(c)$. By (2.37) we have $\Pr(X \leq x) = 1 - e^{-cx^\gamma}$. Letting $Y = X^\gamma$ yields

$$F_Y(y) = \Pr(Y \leq y) = \Pr(X^\gamma \leq y) = \Pr(X \leq y^{1/\gamma})$$
$$= 1 - e^{-c(y^{1/\gamma})^\gamma} = 1 - e^{-cy},$$

which is the distribution function of the $\text{Exp}(c)$ distribution.

2.2.9 Burr distribution

The Burr family has three parameters, usually denoted α (> 0), λ (> 0) and τ (> 0). The family of distributions is named after the American academic statistician and quality control specialist I. W. Burr (1908–1989).

Notation $X \sim \text{Burr}(\alpha, \lambda, \tau)$.

The probability density function is given by

$$f(x) = \frac{\alpha \tau \lambda^\alpha x^{\tau-1}}{(\lambda + x^\tau)^{\alpha+1}}, \quad x > 0. \tag{2.40}$$

The distribution function is given by

$$F(x) = 1 - \left(\frac{\lambda}{\lambda + x^\tau}\right)^\alpha, \quad x > 0. \tag{2.41}$$

The Pareto family of §2.2.7 is the two-parameter sub-family for the case $\tau = 1$. The Burr random variable is obtainable as a power function of a Pareto random variable – if $X \sim \mathrm{Pa}(\alpha, \lambda)$ then $Y = X^{1/\tau} \sim \mathrm{Burr}(\alpha, \lambda, \tau)$, or, equivalently, if $X \sim \mathrm{Burr}(\alpha, \lambda, \tau)$ then $X^\tau \sim \mathrm{Pa}(\alpha, \lambda)$. For this reason, the Burr distribution is sometimes called a "transformed Pareto" distribution.

The general moment $\mathbb{E}[X^r]$ exists only for $r < \alpha\tau$, and can be found easily using the transformation to Pareto:

$$\mathbb{E}[X^r] = \mathbb{E}[X^{\tau(r/\tau)}] = \mathbb{E}[Y^{r/\tau}],$$

where $Y \sim \mathrm{Pa}(\alpha, \lambda)$. Hence, using results for the Pareto distribution, we have

$$\mathbb{E}[X^r] = \Gamma\left(\alpha - \frac{r}{\tau}\right)\Gamma\left(1 + \frac{r}{\tau}\right)\frac{\lambda^{r/\tau}}{\Gamma(\alpha)}, \quad 0 < r < \alpha\tau. \tag{2.42}$$

The mean and variance can be obtained from (2.42) in particular cases for which the parameter values satisfy $\alpha\tau > 2$. The reader can verify that, in the case $\alpha = 6$, $\lambda = 20$, $\tau = 0.5$, we have $\mathbb{E}[X] = 40$, $\mathbb{E}[X^2] = 32\,000$ and $\mathrm{Var}[X] = 30\,400$.

Figure 2.15 shows the probability density functions for three $\mathrm{Burr}(\alpha, \lambda, \tau)$ distributions.

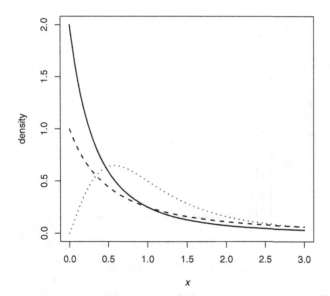

Figure 2.15. Probability density functions of Burr$(2, 1, 1)$ (solid line), Burr$(1, 1, 1)$ (dashed line) and Burr$(1, 1, 2)$ (dotted line).

Particular values of the distribution function are available from the basic **R** package – as an illustration we find $\Pr(X \leq 50)$ in the case $X \sim \text{Burr}(6, 20, 0.5)$ by evaluating the distribution function directly. We use the explicit command

```
1-(20/(20+50^0.5))^6
```

which returns the value 0.8373898.

Simulation We can simulate an observation of a $\text{Burr}(\alpha, \lambda, \tau)$ distribution in **R** using the command

```
x = (lambda*(u^(-1/alpha) - 1))^(1/tau)
```

where `lambda`, `alpha` and `tau` contain the values of λ, α and τ, respectively. The **R** object u contains an observation of a uniform distribution on $(0, 1)$, which is equivalent to taking the τth root of an observation simulated from $\text{Pa}(\alpha, \lambda)$ as described earlier (and see the note below).

The histogram in Figure 2.16 displays a sample of claim sizes (the 997 of 1000 which were less than 1000) simulated from $\text{Burr}(6, 20, 0.5)$, which has mean 40 and variance 30 400. The command used for the simulation was

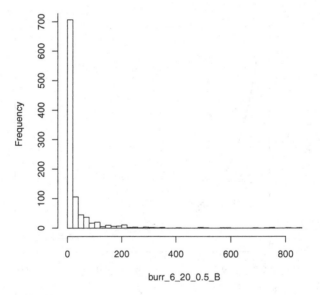

burr_6_20_0.5_B

Figure 2.16. Histogram of observations less than 1000 in a sample of size 1000 simulated from a Burr distribution with parameters $\alpha = 6$, $\lambda = 20$, $\tau = 0.5$. There were three observations above 1000, namely 1337, 3011 and 3051.

```
burr6_20_0.5=(20*(runif(1000)^(-1/6) - 1))^2
```

The sample mean and variance, and the minimum and maximum values that were observed, were as follows:

sample from Burr(6, 20, 0.5): mean 40.49, variance 26779, min 0.000, max 3051.

The probability density function (2.40) can be reparameterised ($\lambda \to \beta^\tau$) and expressed as

$$\frac{\alpha\tau}{\beta}\left(\frac{x}{\beta}\right)^{\tau-1}\frac{1}{[1 + (x/\beta)^\tau]^{\alpha+1}},$$

revealing $\beta = \lambda^{1/\tau}$ to be a scale parameter.

Note Using the facilities of the **R** add-on package `actuar`, we can reproduce the results above. But first note that the parameterisation used by `actuar` is different to that used above – the new parameterisation is Burr 2(a, b, c), where $a = \alpha$, $b = \tau$ and $c = \lambda^{-1/\tau}$. So, for $\Pr(X \le 50)$ in the case $X \sim$ Burr(6, 20, 0.5) in the original parameterisation, we reparameterise and use the command `pburr(50,6,0.5,0.0025)`, which returns 0.8373898; a sample of 1000 simulated claim sizes from Burr(6, 20, 0.5) is likewise obtainable using the command `rburr(1000,6,0.5,0.0025)`.

2.2.10 Loggamma distribution

The loggamma family has two parameters, usually denoted α (> 0) and λ (> 0).

Notation $X \sim$ loggamma(α, λ).

The probability density function is given by

$$f(x) = \frac{\lambda^\alpha}{\Gamma(\alpha)}x^{-(\lambda+1)}(\log x)^{\alpha-1}, \quad x > 1. \tag{2.43}$$

It is important to note that the loggamma variable only assumes values greater than 1 – this can give rise to complications and difficulties with the chosen units in use. There is no scale parameter.

The name "loggamma distribution" arises from the fact that

$$X \sim \text{loggamma}(\alpha, \lambda) \Leftrightarrow Y = \log X \sim \text{gamma}(\alpha, \lambda). \tag{2.44}$$

So, if $X \sim$ loggamma(α, λ) then $\log X$ has a gamma distribution with parameters α and λ. The loggamma parameters α and λ are sometimes referred to as the "shapelog" and "ratelog" parameters.

The distribution function $F(x)$ is given by

$$F(x) = \Pr(X \le x) = \Pr(e^Y \le x) = \Pr(Y \le \log x), \qquad (2.45)$$

where $Y \sim \text{gamma}(\alpha, \lambda)$, and hence is expressible as an incomplete gamma function.

The general moment $\mathbb{E}[X^r]$ exists for $r < \lambda$, and can be found easily using the transformation to gamma, on noting $X = e^Y$, where $Y \sim \text{gamma}(\alpha, \lambda)$:

$$\mathbb{E}[X^r] = \mathbb{E}[e^{rY}] = M_Y(r),$$

where M_Y is the moment generating function of Y. Hence, using the relevant result (2.21) for the gamma distribution, we have

$$\mathbb{E}[X^r] = \left(1 - \frac{r}{\lambda}\right)^{-\alpha}, \quad r < \lambda. \qquad (2.46)$$

Figure 2.17 shows the probability density functions for three loggamma distributions.

Simulation A random sample of size n from a loggamma(α, λ) distribution may be simulated in **R** using the command

```
sample = exp(rgamma(n, alpha, lambda))
```

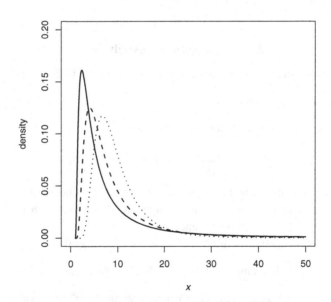

Figure 2.17. Probability density functions of loggamma(3.5, 1.9) (solid line), loggamma(8, 4) (dashed line), loggamma(22, 10) (dotted line).

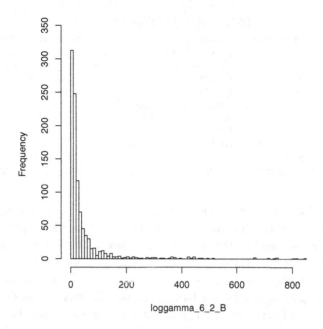

Figure 2.18. Histogram of observations less than 1000 in a sample of size 1000 simulated from a loggamma distribution with parameters $\alpha = 6$, $\lambda = 2$. There were three observations above 1000, namely 1197, 1506 and 42272.

where n, alpha and lambda contain the values of n, α and λ, respectively.

The histogram in Figure 2.18 displays a sample of claim sizes (the 997 of 1000 which were less than 1000) simulated from loggamma(6, 2), which has mean 64 but does not have finite moments of higher orders. The sample mean and variance, and the minimum and maximum values that were observed, were as follows:

sample from loggamma(6, 2): mean 91.45, variance 1795530, min 1.69, max 42272.

Note Particular values of the distribution function are available from the basic *R* package – as an illustration, when $X \sim$ loggamma(3.5, 1.9) we can find $\Pr(X \le 10)$ using the command pgamma(log(10),3.5,1.9), which returns the value 0.7288797. Using the facilities of the *R* add-on package actuar this result can be reproduced using the command plgamma(10,3.5,1.9). Simulations from loggamma(α, λ) can also be carried out using the command rlgamma(n, alpha, lambda).

Table 2.3. *Further tail probabilities*

	Pr($X > 5000$)	Pr($X > 10\,000$)	Pr($X > 15\,000$)
Wei2(0.5, 500)	0.0423	0.0114	0.00418
Wei2(0.3, 107.985)	0.0424	0.0204	0.0124
Burr(2, 43691, 1.5)	0.0121	0.00175	0.000540
Burr(1.5, 500, 1)	0.0274	0.0104	0.00579
Burr(1, 5000, 1)	0.500	0.333	0.25
loggamma(5, 1.33545)	0.0117	0.00616	0.00419
loggamma(20, 3.42402)	0.0306	0.0114	0.00617

Example 2.14   In Table 2.3 we compare some more tail probabilities for selected distributions. The two Weibull distributions have mean $\mathbb{E}[X] = 1000$ and standard deviations 2240 and 5410, respectively. The first Burr has mean 1000 and standard deviation 1217. The second Burr has mean 1000 but does not have a finite standard deviation. The third Burr has no finite moments. The two loggamma distributions have mean 1000; the first one does not have a finite standard deviation; the second one has approximate standard deviation 6380 (most values given are rounded). The table again illustrates the wide range of behaviours we can capture with these distributions.

2.3 Mixture distributions

Mixture distributions

(a) enable us to include in models for claim amounts the variability amongst risks in a portfolio (that is, they allow us to model heterogeneity of risks);
(b) provide a source of further fat-tailed loss distributions; and
(c) shed further light on some distributions we have already met.

Suppose we model a policyholder's claim sizes X using a conditional distribution $X \mid \theta$, where θ can be thought of as a "risk parameter" for that policyholder. Policyholders represent a variety of risks and have different risk parameters, and we model the variation across policyholders by regarding the various θs as being independent realisations of a random variable with known probability distribution. This gives the joint density, which we can write as $f_{X,\theta}(x, \theta) = f_\theta(\theta) f_{X|\theta}(x \mid \theta)$.

In Bayesian work, the distribution with density $f_\theta(\theta)$ is called the "prior distribution", and, given data $\underline{x} = (x_1, \ldots, x_n)$, we are mostly interested in estimating quantities such as $\mathbb{E}[\theta \mid \underline{x}]$.

Here we are interested in the overall distribution of X – the distribution of claim amounts averaged over the portfolio with respect to the chosen "prior". The probability density function of this distribution is given as

$$f_X(x) = \int f_{X,\theta}(x,\theta)d\theta = \int f_\theta(\theta)f_{X|\theta}(x \mid \theta)d\theta, \qquad (2.47)$$

and the distribution is usually called the *marginal distribution* of X. Here we call it a "mixture distribution" with respect to the conditional claim size distribution and the "prior" distribution of θ, here called the "mixing distribution".

To illustrate this, consider claim sizes which have exponential distributions, that is suppose $X \mid \lambda \sim \text{Exp}(\lambda)$. Let us suppose that the variation in λ across the portfolio of risks can be modelled using a gamma(α, β) distribution with known parameters, and let us use this to average across the risks. We are considering a "mixture of exponentials with a gamma mixing distributon". By (2.47) the density function of the mixture distribution $f_X(x)$ is found as follows:

$$
\begin{aligned}
f_X(x) &= \int f_{X,\lambda}(x,\lambda)\,d\lambda \\
&= \int f_\lambda(\lambda)\,f_{X|\lambda}(x \mid \lambda)\,d\lambda \\
&= \int_0^\infty \frac{\beta^\alpha}{\Gamma(\alpha)}\lambda^{\alpha-1}\,e^{-\beta\lambda}\lambda e^{-\lambda x}\,d\lambda \\
&= \frac{\beta^\alpha}{\Gamma(\alpha)}\frac{\Gamma(\alpha+1)}{(\beta+x)^{\alpha+1}} \times \int_0^\infty h(\lambda)d\lambda,
\end{aligned}
$$

where $h(\lambda)$ is the density function of $\lambda \sim \text{gamma}(\alpha+1, \beta+x)$. The integral has value 1, and so

$$f_X(x) = \frac{\alpha\beta^\alpha}{(\beta+x)^{\alpha+1}},$$

which, from (2.30), is the density of a Pa(α, β) distribution.

This provides an illuminating view of the Pareto distribution – it arises as a mixture of exponentials where the mixing distribution is gamma. The resulting Pareto distribution has fatter tails than the original exponential distribution.

Certain properties of the $X \sim \text{Pa}(\alpha, \beta)$ distribution can be established (or verified) using the mixture structure. For example, we can verify the expressions for $\mathbb{E}[X]$ and $\mathbb{E}[X^2]$ (see (2.32)) as follows. Using the conditional expectation result $\mathbb{E}[\mathbb{E}[X \mid Y]] = \mathbb{E}[X]$, and noting that $X \mid \lambda \sim \text{Exp}(\lambda)$ with $\mathbb{E}[X \mid \lambda] = 1/\lambda$, we have, for $\alpha > 1$,

$$\mathbb{E}[X] = \mathbb{E}[\mathbb{E}[X \mid \lambda]] = \mathbb{E}[1/\lambda] = \int_0^\infty \frac{\beta^\alpha}{\Gamma(\alpha)} \frac{1}{\lambda} \lambda^{\alpha-1} e^{-\beta\lambda} d\lambda$$

$$= \frac{\beta}{\alpha - 1} \times \int_0^\infty h(\lambda) d\lambda,$$

where $h(\lambda)$ is the density function of $\lambda \sim$ gamma$(\alpha - 1, \beta)$. Hence

$$\mathbb{E}[X] = \frac{\beta}{\alpha - 1}.$$

Similarly, noting that $\mathbb{E}[X^2 \mid \lambda] = 2/\lambda^2$, we have, for $\alpha > 2$,

$$\mathbb{E}[X^2] = \mathbb{E}[\mathbb{E}[X^2 \mid \lambda]] = \mathbb{E}[2/\lambda^2] = \int_0^\infty \frac{\beta^\alpha}{\Gamma(\alpha)} \frac{2}{\lambda^2} \lambda^{\alpha-1} e^{-\beta\lambda} d\lambda$$

$$= \frac{2\beta^2}{(\alpha - 1)(\alpha - 2)} \times \int_0^\infty h(\lambda) d\lambda,$$

where $h(\lambda)$ is the density function of $\lambda \sim$ gamma$(\alpha - 2, \beta)$. Hence

$$\mathbb{E}[X^2] = \frac{2\beta^2}{(\alpha - 1)(\alpha - 2)}.$$

Notes

(i) The "mixture of exponentials with a gamma mixing distribution" can be generalised by adopting a gamma model for the claim sizes instead of an exponential one, that is taking $X \mid \lambda \sim$ gamma(δ, λ), with δ known, as the claim size model, and with a gamma(α, β) mixing distribution for λ as before. This gives a mixture distribution called a *generalised Pareto distribution*, a three-parameter distribution with density function

$$f(x) = \frac{\Gamma(\alpha + \delta)}{\Gamma(\alpha)\Gamma(\delta)} \frac{\beta^\alpha x^{\delta-1}}{(\beta + x)^{\alpha+\delta}}, \quad x > 0.$$

The Burr distribution also arises as a mixture distribution for which the mixing distribution is a gamma distribution – see Exercise 2.20.

(ii) The Pareto, generalised Pareto and Burr distributions are all special cases of a family of distributions known as the *transformed beta family*.

It is worth noting here that the concept of mixture distributions applies to models for *claim numbers* as well. This again enables us to allow for variability in the risks across a portfolio; that is, to model the heterogeneity of the numbers of claims occurring for different risks. Two situations stand out as being especially interesting.

(1) $N \mid \lambda \sim \text{Poi}(\lambda)$ with mixing distribution gamma(α, β).

The probability mass function of the mixture distribution is found as follows. For $k = 0, 1, 2, \ldots$, we have

$$
\begin{aligned}
\Pr(N = k) &= \int f_\lambda(\lambda) \Pr(N = k \mid \lambda) \, d\lambda \\
&= \int_0^\infty \frac{\beta^\alpha}{\Gamma(\alpha)} \lambda^{\alpha-1} e^{-\beta\lambda} e^{-\lambda} \frac{\lambda^k}{k!} \, d\lambda \\
&= \frac{\Gamma(\alpha + k)}{\Gamma(\alpha)\Gamma(1 + k)} \frac{\beta^\alpha}{(\beta + 1)^{\alpha+k}} \times \int_0^\infty h(\lambda) d\lambda,
\end{aligned}
$$

where $h(\lambda)$ is the probability density function of $\lambda \sim \text{gamma}(\alpha + k, \beta + 1)$. Hence

$$
\Pr(N = k) = \frac{\Gamma(\alpha + k)}{\Gamma(1 + k)\Gamma(\alpha)} \left(\frac{\beta}{\beta + 1} \right)^\alpha \left(\frac{1}{\beta + 1} \right)^k, \quad k = 0, 1, 2, \ldots,
$$

which is the probability mass function of a nb$(\alpha, \beta/(\beta + 1))$ distribution. This provides an illuminating view of the negative binomial distribution – it arises as a mixture of Poissons where the mixing distribution is gamma. We will see this again in the context of the risk models in Chapter 3.

Noting $\mathbb{E}[N \mid \lambda] = \lambda$ we have $\mathbb{E}[N] = \mathbb{E}[\lambda] = \alpha/\beta$ $(= \alpha q/p$, where $p = \beta/(\beta + 1))$.

(2) $N \mid p \sim \text{bi}(n, p)$ with mixing distribution with density function

$$
f_p(p) = \frac{\Gamma(\alpha + \beta)}{\Gamma(\alpha)\Gamma(\beta)} p^{\alpha-1}(1 - p)^{\beta-1}, \, 0 < p < 1, \, (\alpha > 0, \beta > 0).
$$

This mixing distribution is a widely used model for proportions, probabilities and some ratios. It is called the *beta distribution* with parameters α and β – we denote the distribution beta(α, β). The distribution has mean $\alpha/(\alpha + \beta)$ and variance $\alpha\beta/[(\alpha + \beta)^2(\alpha + \beta + 1)]$.

The probability mass function of the mixture distribution is found as follows. For $k = 0, 1, \ldots, n$ we have

$$
\begin{aligned}
\Pr(N = k) &= \int f_p(p) \Pr(N = k \mid p) \, dp \\
&= \int_0^1 \frac{\Gamma(\alpha + \beta)}{\Gamma(\alpha)\Gamma(\beta)} p^{\alpha-1}(1 - p)^{\beta-1} \frac{n!}{k!(n - k)!} p^k (1 - p)^{n-k} \, dp \\
&= \frac{n! \, \Gamma(\alpha + \beta) \, \Gamma(\alpha + k) \, \Gamma(\beta + n - k)}{k! \, (n - k)! \, \Gamma(\alpha) \, \Gamma(\beta) \, \Gamma(\alpha + \beta + n)} \times \int_0^1 h(p) dp,
\end{aligned}
$$

where $h(p)$ is the density function of $P \sim$ beta$(\alpha + k, \beta + n - k)$. Hence we have

$$\Pr(N = k) = \frac{n! \, \Gamma(\alpha + \beta) \, \Gamma(\alpha + k) \, \Gamma(\beta + n - k)}{k! \, (n - k)! \, \Gamma(\alpha) \Gamma(\beta) \Gamma(\alpha + \beta + n)}, \; k = 0, 1, 2, \ldots, n.$$

This mixture distribution is called the *beta-binomial distribution*.

Noting $\mathbb{E}[N \mid p] = np$, we have $\mathbb{E}[N] = \mathbb{E}[np] = n\alpha/(\alpha + \beta)$.

2.4 Fitting models to claim-number and claim-size data

This section is concerned with modelling claim numbers and claim sizes; that is, fitting probability distributions from selected families to sets of data consisting of observed claim numbers or claim sizes. The family may be chosen after an exploratory analysis of the data set – looking at numerical summaries such as mean, median, mode, standard deviation (or variance), skewness, kurtosis and plots such as the empirical distribution function. Of course, one may want to fit a distribution from each of several families to provide comparisons among the fitted models, comparisons with previous work and choice.

To fit a parametric model, we have to calculate estimates of the unknown parameters of the probability distribution. Various criteria are available, including the *method of moments*, the *method of maximum likelihood*, the *method of percentiles* and the *method of minimum distance*.

The *method of moments* leads to parameter estimates by simply matching the moments of the model, $\mathbb{E}[X], \mathbb{E}[X^2], \mathbb{E}[X^3], \ldots$, in turn to the required number of corresponding sample moments calculated from the data $x_1, x_2, \ldots, x_n$, where n is the number of observations available. The sample moments are simply

$$\frac{1}{n} \sum_{i=1}^{n} x_i, \quad \frac{1}{n} \sum_{i=1}^{n} x_i^2, \quad \frac{1}{n} \sum_{i=1}^{n} x_i^3, \quad \ldots.$$

It is often more convenient to match the mean and central moments, in particular matching $\mathbb{E}[X]$ to the sample mean $\bar{x}$ and Var$[X]$ to the sample variance

$$s^2 = \frac{1}{n - 1} \sum_{i=1}^{n} (x_i - \bar{x})^2.$$

An estimate produced using the method of moments is called an MME, and the MME of a parameter θ, say, is usually denoted $\tilde{\theta}$. The approach is usually relatively easy to implement, but the estimates it produces tend to have high standard errors (that is, they are imprecise) – in some cases, MMEs can be very poor and unreliable. However, the method can be used if other methods are unavailable, or to provide starting values for other methods which require them (or at least benefit from having them).

The *method of maximum likelihood* is the most widely used method for parameter estimation. The estimates it produces are those values of the parameters which give the maximum value attainable by the *likelihood function*, denoted L, which is the joint probability mass or density function for the data we have (under the chosen parametric distribution), regarded as a function of the unknown parameters. In practice, it is often easier to maximise the log-likelihood function, which is the logarithm of the likelihood function, rather than the likelihood itself. An estimate produced using the method of maximum likelihood is called an MLE, and the MLE of a parameter θ, say, is denoted $\widehat{\theta}$. MLEs have many desirable theoretical properties, especially in the case of large samples.

In some simple cases we can derive MLE(s) analytically as explicit functions of summaries of the data. Thus, suppose our data consist of a random sample $x_1, x_2, \ldots, x_n$ from a parametric distribution whose parameter(s) we want to estimate. Some straightforward cases include the following:

- the MLE of λ for a Poi(λ) distribution is the sample mean, that is $\widehat{\lambda} = \bar{x}$;
- the MLE of λ for an Exp(λ) distribution is the reciprocal of the sample mean, that is $\widehat{\lambda} = 1/\bar{x}$;
- the MLEs of μ and σ^2 for a N(μ, σ^2) distribution are $\widehat{\mu} = \bar{x}$ and

$$\widehat{\sigma^2} = \frac{1}{n} \sum_{i=1}^{n} (x_i - \bar{x})^2;$$

- the MLEs μ and σ for a lognormal(μ, σ) distribution are obtained from the $y_i = \log x_i$ data as $\widehat{\mu} = \bar{y}$ and

$$\widehat{\sigma} = \left(\frac{1}{n} \sum_{i=1}^{n} (y_i - \bar{y})^2 \right)^{1/2};$$

- the MLE of α for a Pa(α, λ^*) distribution, with the parameter value λ^* known, is given by

$$\widehat{\alpha} = \frac{n}{\sum_{i=1}^{n} \log(1 + x_i/\lambda^*)};$$

- the MLE of β for a Wei2(α^*, β) distribution, with the parameter value α^* known, is given by

$$\widehat{\beta} = \left(\sum_{i=1}^{n} x_i^{\alpha^*} / n \right)^{1/\alpha^*}.$$

In other cases we find that the equations satisfied by the MLEs cannot be solved explicitly – in such cases we find the MLEs by solving equations or maximising likelihoods (or log-likelihoods) by numerical methods.

The *R* command `fitdistr` is available for fitting the following distributions (among others) using the method maximum likelihood: Poisson, geometric, negative binomial, exponential, gamma, lognormal, normal and Weibull. The command is implemented in a package called "MASS" – this package should be attached to the workspace at the start of a session using the command `library(MASS)`. The command does not support the binomial, Pareto, Burr or loggamma distributions. We will see examples of the use of `fitdistr` in the following sections.

The *method of percentiles* is based on matching a set of particular percentiles of the distribution to be fitted with those of the sample. For example, with three parameters to be estimated, we could match the lower quartile, the median and the upper quartile. To use the method, the distribution function of the distribution to be fitted must have an explicit, tractable form.

The *method of minimum distance* minimises a particular "distance function" between the empirical distribution and the chosen parametric distribution. One such function is the "Cramér–von Mises distance function", which is the sum of the squared differences between the empirical distribution function of the data (at the *n* data values) and the theoretical distribution function of the chosen distribution. We will use this approach (which is implemented in the *R* add-on package `actuar` using the command `mde`) in one case, for illustration.

Unless otherwise stated, we will estimate parameters of fitted distributions using maximum likelihood estimation.

2.4.1 Fitting models to claim numbers

We illustrate using a data set consisting of the numbers of claims in one year on a sample of 10 000 policies from a general insurance portfolio. The numbers of claims range from 0 to 5 and are summarised in the frequency distribution in Table 2.4. The table shows that, for example, 115 policies had exactly two claims each.

Summary measures for the sample of 10 000 observations of claim numbers are as follows:

mean 0.1161, variance 0.1420, standard deviation 0.3769, min 0, max 5.

Note that the sample variance is larger than the mean.

Table 2.4. *Frequency distribution of claim numbers*

No. of claims r	0	1	2	3	4	5	≥ 6
No. of policies f_r	9002	862	115	16	4	1	0

We will fit a Poisson distribution (Poi(λ)), a geometric distribution (geo(p)) and a negative binomial distribution (nb(k, p)) to the data and compare the results.

Poisson fit

$$\text{MLE } \widehat{\lambda} = \bar{r} = \frac{\sum_{j=0}^{5} r f_r}{10000} = 0.1161 \ (= \text{MME } \tilde{\lambda}),$$

where r and f_r are as in Table 2.4.

The fitted distribution can then be calculated using the fitted frequency for r claims, $\widehat{f_r} = 10\,000 \times \Pr(N = r)$, where $N \sim \text{Poi}(0.1161)$.

The whole operation can be conveniently carried out in **R** as follows:

```
library(MASS)
fitp=fitdistr(nclaims, "poisson")
lam=fitp$estimate[1]
r=0:6
fittedp=10000*dpois(r, lam)
```

The first line enables the command fitdistr, which uses the method of maximum likelihood to fit a distribution from the specified family to the data set of 10 000 claim numbers, here called nclaims. The object fitp holds the estimate $\hat{\lambda}$ (and its standard error). The estimate is made available as object lam using the command fitp$estimate[1]. The distribution is tabulated for $r = 0, 1, \ldots, 6$ so we set a corresponding vector r. The values of the fitted probability mass function are then produced by the command dpois(r,lam); we calculate the fitted frequencies using 10000*dpois(r,lam). The resulting frequencies are shown in Table 2.5.

Note Although not required here, one may wish to define fittedp[7] separately to cover the tail $\Pr(N \geq 6)$ and ensure that the sum of fitted frequencies is exactly 10 000.

Geometric fit

$$\text{MLE } \widehat{p} = 1/(1 + \bar{r}) = 0.8960 \ (= \text{MME } \tilde{p}).$$

Table 2.5. *Frequency distributions of observed and fitted claim numbers*

No. of claims	Frequency	Poisson	Geometric	Neg. binomial
0	9002	8903.9	8959.8	9002.6
1	862	1033.7	932.0	858.8
2	115	60.0	97.0	117.5
3	16	2.3	10.1	17.7
4	4	0.1	1.0	2.8
5	1	0.0	0.1	0.5
≥ 6	0	0.0	0.0	0.1
	10 000	10 000	10 000	10 000

The fitting was performed using **R** as follows:

```
fitg=fitdistr(nclaims,"geometric")
p=fitg$estimate[1]
r=0:6
fittedg=10000*dgeom(r,p)
```

The resulting frequencies are shown in Table 2.5.

Negative binomial fit

$$\text{MLEs } \widehat{k} = 0.5349 \text{ , } \widehat{p} = 0.8217.$$

The fitting was performed using **R** as follows:

```
fitnb=fitdistr(nclaims,"negative binomial")
p1=fitnb$estimate[1]
p2=fitnb$estimate[2]
p3=p1/(p1+p2)
r=0:6
fittednb=10000*dnbinom(r,p1,p3)
```

The command `fitdistr` applied to the negative binomial uses the parameterisation k and μ, where $\mu = k(1 - p)/p$. Line 4 in the sequence of commands above is the calculation of p using $p = k/(k + \mu)$, which we require for the use of `dnbinom(n,k,p)`. The resulting frequencies are shown in Table 2.5.

Note The MMEs in this case are the solutions of $\mathbb{E}[N] = (1 - p)/p = \bar{r} = 0.1161$ and $\text{Var}[N] = (1 - p)/p^2 = $ sample variance $= 0.142035$, which immediately give $\tilde{p} = 0.1161/0.142035 = 0.8174$ and then $\tilde{k} = 0.5197$. In this case, the MMEs are very close to the MLEs.

We can assess how well the fitted distributions reflect the distribution of the data in various ways. We should, of course, examine and compare the tables of frequencies and, if appropriate, plot and compare empirical distribution functions. More formally, we can perform certain statistical tests. Here we will use the Pearson *chi-square goodness-of-fit* criterion. In this approach we construct the test statistic $\chi^2 = \sum (O - E)^2/E$, where O is the observed frequency in a cell in the frequency table and E is the fitted or *expected* frequency (the frequency expected in that cell under the fitted model), and where we sum over all usable cells. A cell is deemed to be usable provided the expected frequency in it is not too small – rules of thumb for deciding this include the over-conservative "all cells must have $E \geq 5$" and "all cells must have $E \geq 1$, and not more than 20% of the cells should have $E < 5$". Neighbouring cells are combined if necessary, which is often the case in the right tail of a positively skewed distribution.

The value of the test statistic is then evaluated in one of two ways.

(1) We convert it to a *P-value*, which is a measure of the strength of the evidence against the hypothesis that the data do follow the fitted distribution. If the *P-value* is small enough, we conclude that the data do not follow the fitted distribution – we say "the fitted distribution does not provide a good fit to the data" (and quote the *P-value* in support of this conclusion).

(2) We compare it with values in published tables of the distribution function of the appropriate χ^2 distribution, and if the value of the statistic is high enough to be in a tail of specified size of this reference distribution, we conclude that the fitted distribution does not provide a good fit to the data.

The appropriate χ^2 reference distribution depends on the number of cells we used when we summed terms to calculate the test statistic. For example, if we estimate one parameter in the fitting process and then use four cells in our calculations, the appropriate χ^2 reference distribution has parameter ("*degrees of freedom*") (df) 2; if we estimate two parameters in the fitting process and then use six cells in our calculations, the appropriate χ^2 reference distribution has three degrees of freedom. The *P-values* can be found in **R** by using the command 1-pchisq(c, d), where c and d contain the values of the χ^2 statistic and the appropriate degrees of freedom, respectively.

Examining Table 2.5, we see *empirically* that the fit of the Poisson distribution is poor – apart from the dominating cell for $r = 0$, the frequencies expected under the fitted model are very far away from the observed frequencies; in particular, the Poisson model totally fails to reproduce the tail of the observed data. The observed data include 21 policies with three or more claims, while the Poisson fit manages an expected frequency of only 2.4 – the tail of the fitted Poisson is much too light. The geometric fit is better, but, like the Poisson, the geometric is a member of a single-parameter family, so its distribution is not very flexible, and its ability to fit an observed frequency distribution is restricted. The negative binomial fit is much better – in fact, it is very good indeed for these particular data – even the tail has been picked up very successfully. The negative binomial distribution is a member of a two-parameter family, and this additional parameter gives the family much more flexibility than the others considered. The distribution also allows for "variance > mean". The fact that it is a generalistion of the geometric distribution is reflected in its much better fit compared to the geometric.

Table 2.6 shows various quantities for each of the distributions fitted to the data. For each distribution we present the value of the log-likelihood of the data achieved using the maximum likelihood estimates, say $\max(\log L)$; for convenience we actually present $-\max(\log L)$, for which small values are desirable. Since the geometric family is a (one-parameter) sub-family of the (two-parameter) negative binomial family, the models "$N \sim \text{geo}(p)$" and "$N \sim \text{nb}(k, p)$" are described as being *nested*, and the latter will fit the data better than the former. We note that the value of $-\max(\log L)$ for the negative binomial fit is lower than that for the geometric fit. A standard result for nested models states that the reduction in $-2 \log L$ in this case (where we are comparing one nested model with a model with one additional parameter) has a χ^2 distribution with one degree of freedom; such a distribution has an upper 5% point of approximately 3.84. In consequence, many statisticians regard a reduction of at least 2 in $-\log L$ as the minimum improvement required to justify the inclusion of an extra parameter in the preferred model.

We examine the evidence – the values of the log-likelihoods, the chi-square test statistics and the *P-values* for the three fits – and make a judgement as to our preferred model. Here the evidence supports the conclusions that the negative binomial distribution fits the data well, whereas the Poisson and geometric distributions do not.

Table 2.6. *Summary of Poisson, geometric and negative binomial fits*

	Poisson	Geometric	Neg. binomial
No. of parameters	1	1	2
$-\max(\log(L))$	3786.9	3725.9	3717.0
No. cells used	4	4	5
	$(r = 0, 1, 2, \geq 3)$	$(r = 0, 1, 2, \geq 3)$	$(r = 0, 1, 2, 3, \geq 4)$
χ^2 (df)	224.0 (2)	17.4 (2)	0.981 (2)
P-value from **R**	0.00	0.00017 ($< 0.02\%$)	0.61

In a fuller analysis we would also examine the estimated standard errors of the parameter estimates; in particular, we would note whether the estimate of the additional parameter in the negative binomial fit is significantly different from zero – in this case it is, and this supports the inclusion of the extra parameter in the model. The reader may want to extend the account given above by finding the standard errors of fitted parameters – by exploring further the fitting procedures and output from *R* and/or analytically by explicitly evaluating the second derivatives of log-likelihoods at the MLEs and using asymptotic theory (see, for example, sect. 7.8 of DeGroot and Schervish (2002) or chap. 4 of Morgan (2000)).

Formally, we reject the hypothesis that the number of claims has a Poisson (or geometric) distribution. We conclude that these distributions do not provide an adequate description of the variation in the claim numbers that we have observed.

We conclude that it is certainly worth including the additional parameter and fitting the negative binomial distribution – our preferred model for the claim numbers is the negative binomial.

2.4.2 Fitting models to claim sizes

We illustrate using a data set consisting of a sample of 140 claim sizes (settled amounts paid, after excess, in pounds sterling) arising from claims on a general insurance portfolio. The claims range from £8 to £34 975, and are given in Table 2.7.

Half of the claim sizes are in the range (8, 1114), but there are many quite large claims and a few very large ones indeed. The histograms in Figure 2.19 give a good visual summary of the data; Figure 2.19(b) shows the data set after removing the four largest claims.

Summary measures for the sample of claims are as follows:

mean 2939.3, standard deviation 5000.2, min 8, max 34,975, sample coefficient of skewness 3.74, sample coefficient of kurtosis 19.67.

The high positive skewness of the sample reflects the fact that the standard deviation is large compared to the mean, and this suggests in turn that the exponential distribution (for which mean = standard deviation) may not fit the data well. We note also the very high value of the kurtosis coefficient.

Examining the tails, we note that nine claims (6.4%) are greater than £10 000 and three (2.1%) are greater than £20 000.

We fit the following loss distributions one by one (using various computational methods) to the set of 140 claim sizes: exponential, gamma, lognormal,

Table 2.7. *Sample of 140 claim sizes*

8	42	72	103	108	117	120	122
134	150	178	191	207	215	228	231
266	275	286	311	320	321	323	350
358	380	411	422	448	486	511	514
532	559	560	564	582	593	605	612
622	634	649	652	666	678	701	722
730	765	770	778	796	815	823	844
865	876	889	902	911	935	960	993
1001	1023	1087	1092	1095	1114	1148	1196
1228	1288	1351	1396	1432	1476	1504	1531
1577	1599	1609	1656	1687	1742	1772	1804
1832	1919	1976	2030	2077	2114	2162	2215
2265	2289	2390	2426	2443	2500	2572	2604
2630	2723	2827	2969	3116	3332	3706	4103
4161	4414	4673	4870	5161	5253	5447	5410
5622	5801	5989	6184	6256	6742	7111	7555
8319	8633	9413	10216	11899	12829	13821	14433
19832	22421	27641	34975				

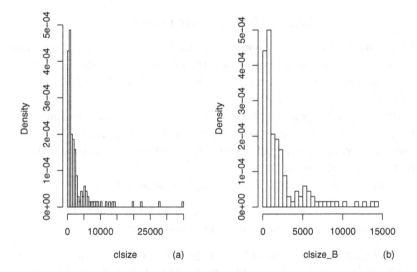

Figure 2.19. Histograms of claim size data: (a) histogram of the full data set; (b) histogram of all observations less than 15 000 (that is, omitting four observations, namely 19832, 22421, 27641 and 34975).

Pareto, Weibull, Burr and loggamma. In each case we will then construct a grouped frequency distribution with the same ten cells, basing this on a set of equi-probable groups determined by the fitted exponential distribution. For each fit we present two informative displays: (1) a histogram with the fitted density function superimposed and (2) a graph of the "ecdf", the empirical (cumulative) distribution function of the sample, with the "cdf", the (cumulative) distribution function of the fitted distribution superimposed. Recall that the empirical distribution function $\widehat{F}_n$ based on a sample $x_1, \ldots, x_n$ is given by

$$\widehat{F}_n(x) = \frac{1}{n} \sum_{i=1}^{n} 1(x_i \leq x) = \frac{\#\{i : x_i \leq x\}}{n}, \quad x \in \mathbb{R},$$

where $1(x_i \leq x)$ is unity if $x_i \leq x$ and is zero otherwise. At the end of this chapter we present a set of further revealing displays in the form of quantile–quantile plots (Q–Q plots); in each plot the quantiles of the sample are plotted against the corresponding values calculated for the fitted distribution. All of these displays assist us in exploring the data – we can compare the distribution of the data with each of the fitted distributions in turn and come to an informal view on the goodness-of-fit of each one.

In each case we again present the value of $-\max(\log L)$, the negative of the log likelihood of the data achieved using the maximum likelihood estimates, for which small values are desirable.

We compare the observed frequencies with each of the seven sets of expected frequencies formally by performing two distribution-free tests: (a) a chi-square goodness-of-fit test and (b) a Kolmogorov–Smirnov (K–S) test. In both cases the null hypothesis that the sample comes from a specified distribution is contrasted with a general ("not so") alternative. The K–S test statistic is the maximum difference between the values of the ecdf of the sample and the cdf of the fully specified fitted distribution.

Note   The **R** command used for a K–S test of the null hypothesis that the sample `clsize` comes from a distribution from the family `dist` is

```
ks.test(clsize, "pdist", p1, p2, ...)
```

where `p1, p2, ...` are the parameters of the fitted distribution from the family `dist`. The command returns the value of the test statistic and the *P-value* for the sample.

Exponential: $X \sim \text{Exp}(\lambda)$

MLE : $\widehat{\lambda} = 1/\bar{x} = 1/2939.286 = 0.0003402187.$

We will use ten groups whose boundaries $\{\text{upbd}_j\}$ are determined using $\Pr(X \leq \text{upbd}_j) = j/10, \quad j = 1, 2, \ldots, 9.$ By (2.13) we have $\Pr(X \leq x) = 1 - \exp(-\lambda x)$, which we estimate by $1 - \exp(-\widehat{\lambda} x)$, giving

$$\text{upbd}_j = -\frac{1}{\widehat{\lambda}} \log(1 - j/10).$$

Table 2.8. *Frequency distributions for observed and fitted*
claim sizes for exponential, gamma and lognormal fits

Class interval	O	E (Exp)	E (gamma)	E (lognormal)
0–310	19	14	24.6	22.3
310–656	25	14	15.3	23.3
656–1048	22	14	13.3	18.2
1048–1501	12	14	12.2	14.6
1501–2037	14	14	11.7	12.0
2037–2693	13	14	11.3	10.1
2693–3539	5	14	11.3	8.8
3539–4731	5	14	11.5	8.0
4731–6768	11	14	12.2	7.8
>6768	14	14	16.6	14.9

Table 2.9. *Summary of exponential, gamma and lognormal fits*

	Exponential	Gamma	Lognormal
$-\max(\log(L))$	1258.0	1250.6	1237.7
χ^2 (df)	27.6 (8)	21.54 (7)	7.17 (7)
P-value	0.00056 ($< 0.1\%$)	0.0030 (0.3%)	0.411
K–S statistic	0.185	0.125	0.049
P-value	0.000144 (<0.02%)	0.025 (2.5%)	0.897

With the claim sizes held in the R object clsize, we can find the cell boundaries easily and in a general way, which we will use later, with the quantile
command as follows:

```
j=1:9
upbd=qexp(j/10, 1/mean(clsize))
```

This produces the (rounded) cell boundaries as 310, 656, 1048, 1501, 2037,
2693, 3539, 4731, 6768, and the resulting observed frequencies are given in
the second column of Table 2.8.

The expected frequencies under the fitted exponential distribution are given
in the third column of Table 2.8, and the summary values of the exponential fit
are given in the second column of Table 2.9. The fit is poor – we see that the
model under-fits the data for claims up to about £1000 (that is, gives expected
frequencies which are lower than the observed frequencies) and over-fits for
claims in the approximate range £2500 to £5000. In addition, using the fitted
model $X \sim \text{Exp}(0.00034022)$ we find $\Pr(X > 10000) = \exp(-3.4022) = 0.033$

and $\Pr(X > 20000) = \exp(-6.8044) = 0.0011$. The fitted exponential model gives only 3.3% of claims greater than £10 000 and only 0.11% greater than £20 000 – the sample contains 6.4% and 2.1%, respectively.

Formally we reject the hypothesis that the number of claims has an exponential distribution – we conclude that the exponential distribution does not provide an adequate description of the variation in the claim sizes that we have observed.

Figure 2.20 shows a histogram of the claim sizes (excluding the four largest and with more smoothing (combined cells) above the value 7000) with the fitted exponential density superimposed – the unsatisfactory under-fitting and over-fitting mentioned above is clear. The figure also shows the empirical distribution function (ecdf) of the whole sample of 140 claim sizes with the distribution function (cdf) of the fitted exponential superimposed. This display

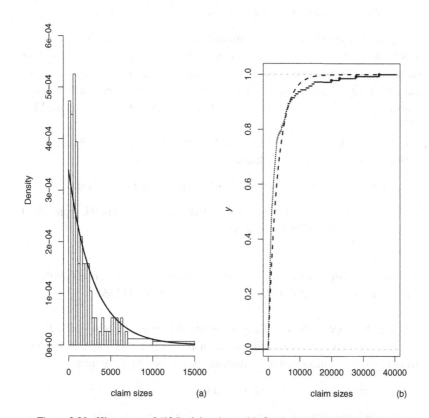

Figure 2.20. Histogram of (136) claim sizes with fitted exponential distribution added (a); ecdf of (all 140) claim sizes with cdf of fitted exponential added (dashed line) (b).

is very effective in highlighting the poor fit in the tail of the distribution – the fitted exponential reaches values close to unity at lower claim sizes than is the case for the observed claim sizes, that is the fitted tail is too light/thin.

Gamma: $X \sim$ gamma(α, λ)

The maximum likelihood estimation for this distribution is most conveniently performed analytically under a simple reparameterisation, namely using parameters α and μ, where $\mu = \mathbb{E}[X] = \alpha/\lambda$. The MLEs of α and μ can be shown to satisfy the following:

$$\widehat{\mu} = \overline{x},$$

and $\widehat{\alpha}$ is the value of α which maximises the following log-likelihood expression:

$$\ell_n(\alpha) = n\alpha(\log \alpha - \log \overline{x} - 1) + (\alpha - 1) \sum \log x_i - n \log \Gamma(\alpha),$$

where n is the sample size. The optimisation can be carried out in many ways: in **R** using the non-linear optimisation procedures nlm or optim, by computer search, and in mathematics computer packages. When $\widehat{\alpha}$ has been found, we can find $\widehat{\lambda}$ from the relation $\widehat{\lambda} = \widehat{\alpha}/\widehat{\mu}$, this being justified using the "invariance property" of MLEs (which states that, if $\widehat{\theta}$ is the MLE of θ then $h(\widehat{\theta})$ is the MLE of $h(\theta)$).

Here we have $\widehat{\mu} = \overline{x} = 2939.286$, and, noting that $\sum \log x_i = 995.0292$, we conduct a computer search to find the value of α that maximises

$$\ell_n(\alpha) = 140\alpha(\log \alpha - 8.985922) + 995.0292(\alpha - 1) - 140 \log \Gamma(\alpha).$$

The maximising value of this function is found to be $\widehat{\alpha} = 0.6893$, from which we find $\widehat{\lambda} = 0.6893/2939.286 = 0.0002345$.

Notes

(1) Using nlm in **R** with any reasonable starting value returns $\widehat{\alpha} = 0.68935$.
(2) For interest, the MMEs are $\tilde{\alpha} = 0.34554$, $\tilde{\lambda} = 0.00011756$.

Using the upper cell boundaries determined above and held in the **R** vector upbd, we can find the frequencies in the cells under the fitted gamma distribution from the cumulative frequencies returned by the command

```
140*pgamma(upbd,0.6893,0.0002345)
```

The cell frequencies are shown in the fourth column of Table 2.8, and the summary values of the gamma fit are given in Table 2.9. The (two-parameter) gamma family includes the (one-parameter) exponential family as a special

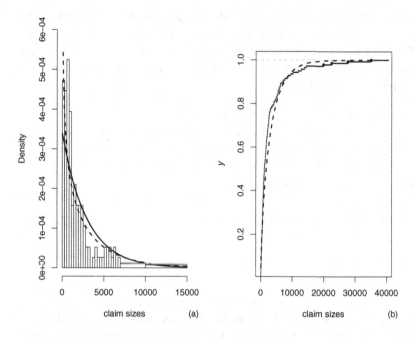

Figure 2.21. Histogram of (136) claim sizes with fitted exponential (solid line) and gamma (dashed line) distributions added (a); ecdf of (all 140) claim sizes with cdf of fitted gamma added (dashed line) (b).

case (the models are nested), but the gamma fit is again poor – we see that the model under-fits the data for claims between about £300 and £1000 and over-fits for claims between about £2500 and £5000.

Formally, we reject the hypothesis that the claim sizes have a gamma distribution – we conclude that the gamma distribution does not provide an adequate description of the variation in the claim sizes that we have observed.

Figure 2.21 shows the histogram for the claim sizes (excluding the four largest) with the fitted gamma density added to that of the fitted exponential – the unsatisfactory nature of the fit is clear. The figure also shows the ecdf of the whole sample of 140 claim sizes with the cdf of the fitted gamma superimposed. This display highlights the poor fit in the tails of the distributions.

Note   For illustration, we also fit a gamma distribution using another method of parameter estimation, namely the *method of minimum distance*, which can be implemented using the `actuar` command `mde`. To avoid problems with the workings of the optimisation routine, it has been found that using the logs of the parameters is more reliable in some cases, and we adopt this approach.

First we define a function based on the logged parameters which will produce a gamma cdf and be acceptable to `mde`, and then specify a set of starting values. Then we carry out the minimisation of the Cramér–von Mises distance between the ecdf and the gamma cdf. (Note that + at the beginning of a line is the continuation prompt in **R**.)

```
pgammalog=function(x,logshape,lograte)
+ pgamma(x,shape=exp(logshape),rate=exp(lograte))
mdefit=mde(clsize,pgammalog,start=list(logshape=
+ log(0.6),lograte=log(0.0002)),measure="CvM")
exp(mdefit$estimate[1])
0.8377546
exp(mdefit$estimate[2])
0.0004142281
```

The fitted distribution is gamma(0.8378, 0.0004142).

Lognormal: $X \sim$ lognormal(μ, σ)

Before fitting a lognormal distribution, it is instructive to view a display of the logged data – see Figure 2.22. The display shows that the logged claim sizes have a distribution which may be similar to a normal distribution but

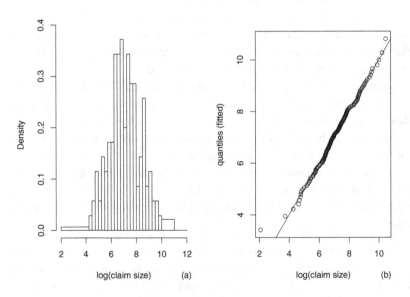

Figure 2.22. Histogram (a) and normal quantile–quantile plot (b) of the log(claim sizes).

with a modest negative skew, and hence suggests that the claim sizes may be modelled quite well by a lognormal distribution.

Noting that $Y = \log X \sim N(\mu, \sigma^2)$ we can find the MLEs of μ and σ directly from the logged data as the sample mean and standard deviation of the $\log(x_i)$ values: we find $\widehat{\mu} = 7.1074$ and $\widehat{\sigma} = 1.3748$.

Notes

(1) For interest, we also fit the distribution using the **R** command `fitdistr` as follows:

```
fitln=fitdistr(clsize,"lognormal")
p1=fitln$estimate[1]
p2=fitln$estimate[2]
```

This gives $\widehat{\mu} = $ `p1` $= 7.1074$ and $\widehat{\sigma} = $ `p2` $= 1.3699$.

(2) The MMEs are $\tilde{\mu} = 7.3062$ and $\tilde{\sigma} = 1.1659$.

The fitted cell frequencies come from the cumulative frequencies returned by the command

```
140*plnorm(upbd,mean(log(clsize)), sd(log(clsize)))
```

and are shown in Table 2.8. The summary of the lognormal fit is given in Table 2.9. The fit is very good indeed – we accept the hypothesis that the lognormal distribution provides an adequate description of the variation in the claim sizes that we have observed.

Figure 2.23 shows the histogram for the claim sizes (excluding the four largest) with the fitted lognormal density superimposed. The figure also shows the ecdf of the whole sample of 140 claim sizes with the cdf of the fitted lognormal superimposed. This display highlights the good fit of the lognormal distribution.

Pareto: $X \sim \text{Pa}(\alpha, \lambda)$

The log-likelihood for this distribution is given by

$$\ell_n(\alpha, \lambda) = n \log \alpha + n\alpha \log \lambda - (\alpha + 1) \sum \log(\lambda + x_i),$$

and the MLEs therefore satisfy the equations

$$\widehat{\alpha} = \frac{n}{\sum \log\left(1 + x_i/\widehat{\lambda}\right)} \quad \text{and} \quad \frac{n\widehat{\alpha}}{\widehat{\lambda}} - (\widehat{\alpha} + 1) \sum \left(\frac{1}{\widehat{\lambda} + x_i}\right) = 0,$$

which must be solved by numerical methods (and see Exercise 2.21).

We will find the MLEs by maximising the log-likelihood in **R**; we need starting values, which we take from the MMEs, which are $\tilde{\alpha} = 3.0560$ and

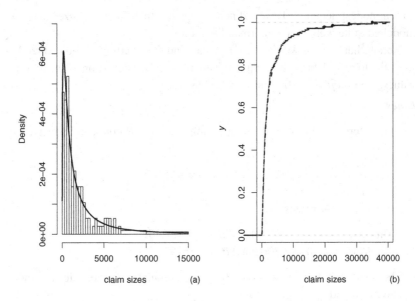

Figure 2.23. Histogram of (136) claim sizes with fitted lognormal distribution added (a); ecdf of (all 140) claim sizes with cdf of fitted lognormal added (dashed line) (b).

$\tilde{\lambda} = 6043.1$. The R function nlm is a minimisation procedure, so the negative of the log-likelihood function is declared (as a function of a vector x of length 2) as follows:

```
fp = function(x){
+ -(140*log(x[1]) + 140*x[1]*log(x[2])
+ - (x[1]+1)*sum(log(x[2]+clsize))) }
```

(Note that + at the beginning of a line is the continuation prompt in R.)

With starting values $\alpha = 3$, $\lambda = 6000$, we achieve the maximisation of the log-likelihood using the command nlm(fp,c(3,6000)), which returns the MLEs $\widehat{\alpha} = 1.9870$ and $\widehat{\lambda} = 3074.5$.

The fitted cell frequencies come from the cumulative frequencies returned by the actuar command 140*ppareto(upbd,1.987,3075) and are shown in Table 2.10. The summary of the Pareto fit is given in Table 2.11. The fit is again very good – we accept the hypothesis that the Pareto distribution provides an adequate description of the variation in the claim sizes that we have observed.

Figure 2.24 shows the histogram for the claim sizes (excluding the four largest) with the fitted Pareto density superimposed. The figure also shows

Table 2.10. *Frequency distributions for observed and fitted claim sizes for Pareto, Weibull, Burr and loggamma fits*

Class interval	O	E (Pareto)	E (Weibull)	E (Burr)	E (loggamma)
0–310	19	24.3	26.8	21.8	24.2
310–656	25	20.3	16.9	22.2	25.5
656–1048	22	17.2	14.2	19.0	18.2
1048–1501	12	14.6	12.6	15.7	13.7
1501–2037	14	12.6	11.6	13.0	10.8
2037–2693	13	10.9	10.9	10.7	8.9
2693–3539	5	9.6	10.4	9.0	7.7
3539–4731	5	8.6	10.3	7.8	7.0
4731–6768	11	8.1	10.7	7.2	6.9
>6768	14	13.9	15.7	13.6	17.1

Table 2.11. *Summary of Pareto, Weibull, Burr and loggamma fits*

	Pareto	Weibull	Burr	Loggamma
$-\max(\log(L))$	1238.7	1245.8	1237.4	1243.9
χ^2 (df)	9.36 (7)	17.09 (7)	7.43 (6)	9.48 (7)
P-value	0.228	0.0168 (1.7%)	0.283	0.220
K–S statistic	0.049	0.0927	0.0377	0.0762
P-value	0.891	0.180	0.989	0.390

the ecdf of the whole sample of 140 claim sizes with the cdf of the fitted Pareto superimposed. This display highlights the good fit of the Pareto distribution.

Figure 2.25 shows the histogram for the claim sizes (excluding the four largest) with the fitted lognormal and Pareto densities superimposed. Despite the very different nature of the two models (note the behaviours close to the origin), the fits are very close over most of the range of claim sizes.

Weibull: $X \sim \text{Wei}(c, \tau)$ **or** $X \sim \text{Wei2}(\alpha, \beta)$

The moments of the distribution involve gamma functions, and finding MMEs is awkward. In addition, finding MLEs is also complicated unless carried out using a computer package. Other methods of estimating the parameters are sometimes used – the *method of percentiles*, based on matching the quartiles of the distribution and the sample, is easy to implement in this case (see notes below).

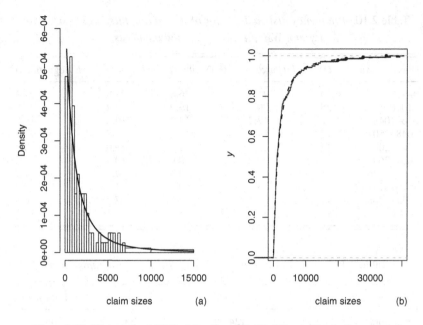

Figure 2.24. Histogram of (136) claim sizes with fitted Pareto distribution added (a); ecdf of (all 140) claim sizes with cdf of fitted Pareto added (dashed line) (b).

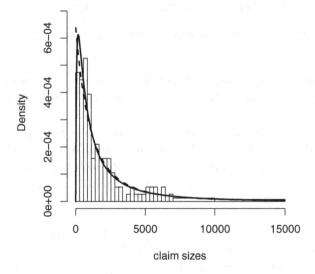

Figure 2.25. Histogram of claim sizes with fitted lognormal (solid line) and Pareto (dashed line) distributions added.

The log-likelihood in the Wei2 parameterisation is given by

$$\ell_n(\alpha,\beta) = n \log \alpha - n \log \beta + (\alpha - 1) \sum \log\left(\frac{x_i}{\beta}\right) - \sum \left(\frac{x_i}{\beta}\right)^\alpha.$$

We will fit the distribution here (in the Wei2 parameterisation) using the **R** command nlm to minimise the negative of the log-likelihood as follows:

```
fw = function(x){
+ -(140*log(x[1]) - 140*log(x[2])
+ + (x[1]- 1)*sum(log((clsize/x[2])))
+ - sum((clsize/x[2])^x[1]))}
```

(Note that + at the beginning of a line is the continuation prompt in **R**.) With starting values $\alpha = 0.7$, $\beta = 2300$, we achieve the maximisation of the log-likelihood using the command nlm(fw,c(0.7,2300)), which returns the MLEs $\widehat{\alpha} = 0.75697$ and $\widehat{\beta} = 2402.7$.

The fitted cell frequencies come from the cumulative frequencies returned by the command 140*pweibull(upbd,0.75697,2402.7). The cell frequencies are shown in Table 2.10. The summary of the Weibull fit is given in Table 2.11.

Notes

(1) The **R** optimising function fitdistr gives slightly different maximising values, but the fitted frequencies are very close using the two pairs of estimated parameters.

(2) The lower and upper quartiles of the claim sizes are 561 and 2699.75, respectively (using a standard definition of the positions of the quartiles). The method of percentiles based on these two quartiles gives $\exp(-561/\beta)^\alpha = 0.75$ and $\exp(-2699.75/\beta)^\alpha = 0.25$, which give estimates $\alpha^* = 1.001$ and $\beta^* = 1948$.

Figure 2.26 shows the histogram for the claim sizes (excluding the four largest) with the fitted Weibull density superimposed. The (two-parameter) Weibull family includes the (one-parameter) exponential family as a special case – the models are nested. The Weibull is only a slightly better fit than the exponential, which is also shown. The figure also shows the ecdf of the whole sample of 140 claim sizes with the cdf of the fitted Weibull superimposed.

Taking into account all the evidence – the histogram and fitted density, the ecdf and cdf of fitted distribution, and the *P-value* of the chi-square goodness-of-fit test (and not ignoring the *P-value* of the K–S test statistic) – it appears that the Weibull model is not a particularly good fit to the data. On the basis of the chi-square test, we can formally reject the hypothesis (at levels

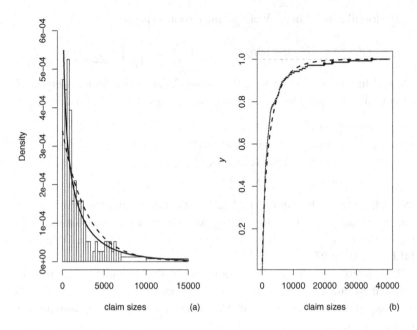

Figure 2.26. Histogram of (136) claim sizes with fitted Weibull (solid line) and
exponential (dashed line) distributions added (a); ecdf of (all 140) claim sizes with
cdf of fitted Weibull added (dashed line) (b).

of testing down to 1.8%) that the Weibull distribution provides an adequate
description of the variation in the claim sizes that we have observed. Several
other fits are better.

Burr: $X \sim \text{Burr}(\alpha, \lambda, \tau)$

The log-likelihood for the Burr(α, λ, τ) distribution is given by

$$\ell_n(\alpha, \lambda, \tau) = n(\log \alpha + \log \tau + \alpha \log \lambda)$$
$$+ (\tau - 1) \sum \log x_i - (\alpha + 1) \sum \log(\lambda + x_i^\tau).$$

We will maximise this using the command $\texttt{nlm}$ in *R*, with starting values found
by evaluating the expressions for the quartiles and the mean $\mathbb{E}[X]$ for various
choices of the parameters until suitable values are found. The MLEs are $\widehat{\alpha} =$
1.2191, $\widehat{\lambda} = 6077.3$, $\widehat{\tau} = 1.1864$.

The fitted cell frequencies come from the cumulative frequencies returned
by the $\texttt{actuar}$ command $\texttt{140*pburr(upbd, p1, p2, p3)}$, where $\texttt{p1}$, $\texttt{p2}$
and $\texttt{p3}$ are the fitted parameters in the form required by *R* (see the note at the
end of §2.2.9). The frequencies are shown in Table 2.10. The summary of the

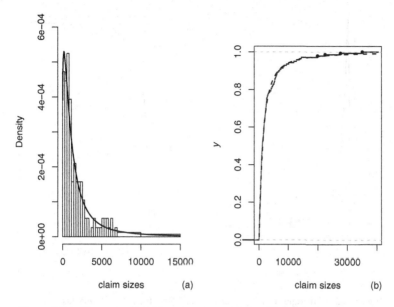

Figure 2.27. Histogram of (136) claim sizes with fitted Burr distribution added (a); ecdf of (all 140) claim sizes with cdf of fitted Burr added (dashed line) (b).

Burr fit is given in Table 2.11. Since the Pareto family is a two-parameter sub-family of the three-parameter Burr family, the models are nested – the latter will fit the data better, and the value of $-\max(\log L)$ for the Burr fit will be lower than that for the Pareto fit. The fit is very good indeed – we accept the hypothesis that the Burr distribution provides an adequate description of the variation in the claim sizes that we have observed.

Figure 2.27 shows the histogram for the claim sizes (excluding the four largest) with the fitted Burr density superimposed. The figure also shows the ecdf of the whole sample of 140 claim sizes with the cdf of the fitted Burr superimposed. This display highlights the very good fit of the Burr distribution.

The reduction in $-\max(\log L)$ for the Burr fit compared to the Pareto fit, however, is small (only 1.3), and, from a statistical point of view, one would question whether this justifies the "cost" of including the extra parameter (see the discussion in §2.4.1).

Loggamma: $X \sim \text{loggamma}(\alpha, \lambda)$

Recall from (2.44) that $X \sim \text{loggamma}(\alpha, \lambda) \Leftrightarrow Y = \log X \sim \text{gamma}(\alpha, \lambda)$. We can find the MLEs of α and λ from the logged data $y_i = \log(x_i)$ in a similar

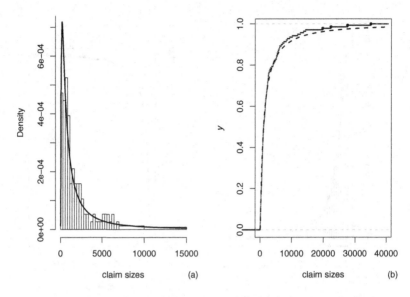

Figure 2.28. Histogram of (136) claim sizes with fitted loggamma distribution added (a); ecdf of (all 140) claim sizes with cdf of fitted loggamma added (dashed line) (b).

way to that used earlier for the gamma distribution. We find $\widehat{\alpha} = 23.959$ and $\widehat{\lambda} = 3.371$.

The fitted cell frequencies come from the cumulative frequencies returned by the command (in `actuar`) `140*plgamma(upbd,23.959,3.371)`. The frequencies are shown in Table 2.10, and the summary of the loggamma fit is given in Table 2.11.

Figure 2.28 shows the histogram for the claim sizes (excluding the four largest) with the fitted loggamma density superimposed (solid line). The figure also shows the ecdf of the whole sample of 140 claim sizes with the cdf of the fitted loggamma superimposed. This display supports the view that the fit is good, but with a slight over-fitting in the tail.

On the basis of the chi-square statistic alone, the fit appears to be satisfactory – but a further visual display (see Figure 2.29 in the following) shows that the fit in the tail is, in fact, poor.

Quantile–quantile plots
As a further set of visual displays, we present in Figure 2.29 quantile–quantile (Q–Q) plots for all seven fitted distributions – the added straight lines help us

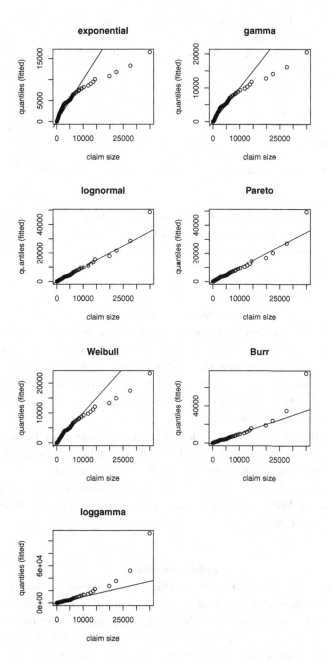

Figure 2.29. Q–Q plots for fitted exponential, gamma, lognormal, Pareto, Weibull, Burr and loggamma distributions.

Table 2.12. *Summary statistics for the seven fits*

	Parameters	$-\max(\log L)$	P-value (χ^2)	P-value (K–S)
Exponential	1	1258.0	0.185	0.00014
Gamma	2	1250.6	0.003	0.025
Lognormal	2	1237.7	0.411	0.897
Pareto	2	1238.7	0.228	0.891
Weibull	2	1245.8	0.017	0.180
Loggamma	2	1243.9	0.220	0.390
Burr	3	1237.4	0.283	0.989

to assess how well the fitted distributions perform. We note that, as is often the case with claim sizes, a fit can be good at lower claim sizes, can be assessed by goodness-of-fit statistics as good overall, but can still be less than satisfactory in the tail.

Summary of fits

In Table 2.12 we present summary statistics for the seven fits. Noting the information in Table 2.12, and taking into account the histograms, ecdfs and Q–Q plots, it appears that the lognormal and Burr distributions provide the best fits to the claim size data, followed closely by the Pareto. The loggamma fit is poor, the Weibull rather worse, and the gamma, and thus the exponential, are totally inadequate as models for the data.

Taking the view that the inclusion of the third parameter for the Burr fit is not justified when compared to the Pareto fit, our judgement is that the most effective fits are provided by the lognormal and Pareto models.

As suggested in §2.4.1, the reader may want to calculate and examine the estimated standard errors of fitted parameters to extend the account here and provide a fuller analysis.

Further distributions

There exist other distributions in use which arise as reciprocals or other functions of distributions already considered here. Further types of distribution are also used for certain purposes, including *threshold distributions* (distributions of $X \mid X > T$), *limited distributions* (distributions of $X \mid X \le T$) and *limited distributions with additional probability mass*, as used, for example, in *excess of loss reinsurance*; see Chapter 5.

Exercises

2.1 Let $N \sim \text{Poi}(\lambda)$.

(a) Verify the recursion formula

$$\Pr(N = n + 1) = \frac{\lambda}{n + 1} \Pr(N = n), \quad n = 0, 1, 2, \ldots,$$

with starting value $\Pr(N = 0) = e^{-\lambda}$.

(b) Consider the case in which λ is not an integer. By using the recursion formula in (a) to compare $\Pr(N = n)$ with $\Pr(N = n + 1)$ and $\Pr(N = n - 1)$, show that the most likely number of claims which arise is given by $[\lambda]$, where $[x]$ denotes the integer part of the number x. What is the corresponding result in the case that λ *is* an integer?

2.2 Let $N \sim \text{nb}(k, p)$ for general k. Verify the recursion formula

$$\Pr(N = n + 1) = \frac{n + k}{n + 1}(1 - p)\Pr(N = n), \quad n = 0, 1, 2, \ldots,$$

with starting value $\Pr(N = 0) = p^k$.

2.3 Let $N \sim \text{geo}(p)$.

(a) For $n = 0, 1, 2, \ldots$ and for $k = 0, 1, 2, \ldots$, show that

$$\Pr(N \geq n + k \mid N \geq k) = \Pr(N \geq n)$$

and comment in the context of the occurrence of additional claims over and above claims which have already arisen.

(b) The number of claims which arise on a group of policies in a year, N, is modelled as having a geo(0.9) distribution. Find the conditional probability $\Pr(N > 9 \mid N \geq 8)$.

2.4 Policies of a certain type are examined one by one until one is found on which a claim arose last year.

Assuming independence from policy to policy, and that for each policy the probability of a claim arising in a year is 0.1, calculate how many policies have to be examined to ensure a probability of at least 0.95 that a policy is found on which a claim arose last year.

2.5 Consider a random variable X with moment generating function $M_X(t)$ and cumulant generating function $K_X(t) = \log M_X(t)$. Let κ_j be the jth cumulant; that is, the coefficient of $t^j / j!$ in the power series expansion of $K_X(t)$. Assume moments are finite as necessary.

By considering the first four derivatives of $K_X(t)$ and $M_X(t)$ evaluated at $t = 0$, and using the facts that $\mathbb{E}[X^j] = M^{(j)}(0)$ and $\kappa_j = K^{(j)}(0)$, verify that

$$\text{Var}[X] = \kappa_2, \quad \mathbb{E}[(X - \mathbb{E}[X])^3] = \kappa_3, \quad \mathbb{E}[(X - \mathbb{E}[X])^4] = \kappa_4 + 3\kappa_2^2.$$

2.6 Claim sizes are modelled as an exponential random variable with parameter λ. Suppose we have a random sample $(X_1, X_2, \ldots, X_n)$ of n such claim sizes. Let

$$X_{\min} = \min(X_1, X_2, \ldots, X_n), \quad X_{\max} = \max(X_1, X_2, \ldots, X_n);$$

that is, $X_{\min}$ is the smallest and $X_{\max}$ is the largest of the n claim sizes.
(a) By considering $\Pr(X_{\min} > x)$, show that $X_{\min} \sim \text{Exp}(n\lambda)$.
(b) By considering $\Pr(X_{\max} \leq x)$, find an expression for the probability density function of $X_{\max}$.

2.7 Let $X \sim \text{gamma}(\alpha, \lambda)$.
(a) Verify that the third and fourth central moments are given by

$$\mathbb{E}[(X - \mathbb{E}[X])^3] = 2\alpha/\lambda^3, \quad \mathbb{E}[(X - \mathbb{E}[X])^4] = (6\alpha + 3\alpha^2)/\lambda^4,$$

and hence that the coefficients of skewness and kurtosis are $2/\sqrt{\alpha}$ and $3 + 6/\alpha$, respectively.
State the corresponding results for $X \sim \text{Exp}(\lambda)$.
(b) Consider the case that 2α is an integer.
Show that $2\lambda X \sim \text{gamma}(\alpha, 1/2)$; that is, $2\lambda X \sim \chi^2$ with 2α degrees of freedom.

2.8 Claims arise on a portfolio of business in a Poisson process at a rate of four per hour.
(a) Calculate the probability that the time between two successive claim arrivals exceeds 45 minutes.
(b) Calculate the probability that the time from any fixed time origin to the third subsequent claim arising exceeds 45 minutes.

2.9 Let $X \sim \text{gamma}(\alpha, \lambda)$ and $Y \sim \text{gamma}(\beta, \theta)$, with X and Y independent. Let $W = X - Y$.
(a) Show that the cumulant generating function of W is given by

$$K_W(t) = -\alpha \log(1 - t/\lambda) - \beta \log(1 + t/\theta),$$

and hence (or otherwise) show that coefficients of skewness and kurtosis of W are given by

$$\frac{2(\alpha\theta^3 - \beta\lambda^3)}{(\alpha\theta^2 + \beta\lambda^2)^{3/2}}$$

and

$$\frac{6(\alpha\theta^4 + \beta\lambda^4)}{(\alpha\theta^2 + \beta\lambda^2)^2} + 3,$$

respectively.

(b) Setting $\theta = \beta\lambda/\alpha$ gives $\mathbb{E}[W] = 0$. Find expressions for the coefficients of skewness and kurtosis of W in this case (they are functions of α and β only).

2.10 The sizes of claims last year on a portfolio of motor insurance policies have a lognormal distribution with parameters $\mu = 6$ and $\sigma = 0.65$. It is estimated that the sizes of all claims next year will increase by 12%.

Calculate the probability that the size of a claim next year will exceed 1000.

2.11 Let $X \sim$ lognormal(μ, σ). Show that, for $a > 0$ and $k = 0, 1, 2, \ldots,$

$$\int_0^a x^k f_X(x)dx = \exp\left(k\mu + \frac{k^2\sigma^2}{2}\right)\Phi\left(\frac{\log a - \mu - k\sigma^2}{\sigma}\right).$$

2.12 Individual losses on an insurance portfolio, X, are modelled as a lognormal random variable with parameters $\mu = 2$ and $\sigma = 0.5$. Let S be the aggregate (total) loss for 100 independent individual losses. The distribution of S can be approximated by that of a normal distribution with appropriate parameters.
   (a) Calculate the probability that an individual loss exceeds 10.
   (b) Calculate the mean and variance of S.
   (c) Calculate the approximate probability that S exceeds 920.
   (d) Calculate the approximate interval, symmetrical and centred on the mean, within which 90% of values of S lie.
   (e) Calculate the 99th percentile of the distribution of S, that is the value $s_{0.99}$ such that $\Pr(S > s_{0.99}) = 0.01$.

2.13 Let $X \sim$ Pa(α, λ).
   (a) By writing x as $(\lambda + x) - \lambda$, or otherwise, show that, in the case $\alpha > 1$, the mean is given by

$$\mathbb{E}[X] = \frac{\lambda}{\alpha - 1}.$$

   (b) By considering $\Pr(X - w > x \mid X > w)$, $x > 0$, $w > 0$, show that $X - w \mid X > w \sim$ Pa$(\alpha, \lambda + w)$.
   (c) Let $\alpha > 2$ and $M > 0$. Find expressions in terms of α, λ and M for
      (1) $\mathbb{E}[X - M \mid X > M]$,
      (2) $\mathbb{E}[(X - M)^2 \mid X > M]$.

(d) Show that, for $\alpha > 3$, the coefficient of skewness is given by

$$\left(\frac{2(\alpha + 1)}{\alpha - 3}\right)\left(\frac{\alpha - 2}{\alpha}\right)^{1/2}$$

and comment on its value.

2.14 Let X have a lognormal distribution with parameters $\mu = 5$ and $\sigma = 1$, and let Y have a Pareto distribution with parameters α and λ. Calculate the values of α and λ such that X and Y have the same mean and variance as each other.

2.15 Suppose that claims occur on a portfolio of general insurance policies independently of one another and are of two types: each claim is classified as being of "type A" or "type B". Type A claim sizes are distributed $Pa(4, 300)$ and type B claim sizes are distributed $Pa(6, 2500)$; 80% of claims are of type A.

Let X denote the size of a randomly chosen claim arising on the portfolio.

(a) Calculate: (i) $\Pr(X > 2000)$; (ii) $\mathbb{E}[X]$ and $\text{Var}[X]$.
(b) Calculate $\Pr(Y > 2000)$, where Y has a Pareto distribution with the same mean and variance as X in part (a)(ii).
(c) Comment on the difference in the answers to parts (a)(i) and (b).

2.16 By considering $\Pr(Y \leq y)$, or otherwise, show that (for positive c, k, α, β, γ, λ, σ, τ)
(a) $X \sim \text{gamma}(\alpha, \lambda) \Rightarrow Y = kX \sim \text{gamma}(\alpha, \lambda/k)$;
(b) $X \sim \text{lognormal}(\mu, \sigma) \Rightarrow Y = kX \sim \text{lognormal}(\mu + \log k, \sigma)$;
(c) $X \sim \text{Pa}(\alpha, \lambda) \Rightarrow Y = kX \sim \text{Pa}(\alpha, k\lambda)$;
(d) $X \sim \text{Weibull}(c, \gamma) \Rightarrow Y = kX \sim \text{Weibull}(ck^{-\gamma}, \gamma)$;
(e) $X \sim \text{Wei2}(\alpha, \beta) \Rightarrow Y = kX \sim \text{Wei2}(\alpha, k\beta)$;
(f) $X \sim \text{Burr}(\alpha, \lambda, \tau) \Rightarrow Y = kX \sim \text{Burr}(\alpha, \lambda k^{\tau}, \tau)$;
(g) $X \sim \text{Exp}(\lambda) \Rightarrow Y = X^{1/\gamma} \sim \text{Wei}(\lambda, \gamma)$;
(h) $X \sim \text{Pa}(\alpha, \lambda) \Rightarrow Y = X^{1/\tau} \sim \text{Burr}(\alpha, \lambda, \tau)$;
(i) $X \sim \text{lognormal}(\mu, \sigma) \Rightarrow Y = \log X \sim \text{N}(\mu, \sigma^2)$;
(j) $X \sim \text{loggamma}(\alpha, \lambda) \Rightarrow Y = \log X \sim \text{gamma}(\alpha, \lambda)$.

2.17 The *failure rate* (or, depending on circumstances, the *hazard rate* or the *force of mortality*) at value x for a non-negative random variable X, denoted $q(x)$, is defined as $q(x) = f(x)/[1 - F(x)]$, where F and f are the cdf and pdf of X, respectively.

For an item whose lifetime distribution has probability density function $f(x)$, $q(x)$ is the probability of failure in $(x, x + dx)$, given survival up to time x, and it follows from the definition that

$$f(x) = q(x) \exp\left(-\int_0^x q(t)dt\right).$$

Let the parameters c, α, γ, λ and τ all be positive.

(a) Show that $X \sim \text{Exp}(\lambda)$ has a constant failure rate $q(x) = \lambda$.

(b) Show that a decreasing failure rate $q(x) = \alpha/(\lambda + x)$ corresponds to $X \sim \text{Pa}(\alpha, \lambda)$.

(c) Show that the failure rate $q(x) = c\gamma x^{\gamma-1}$ corresponds to

$$X \sim \text{Weibull}(c, \gamma).$$

Comment on the cases

(1) $\gamma < 1$,

(2) $\gamma > 1$,

(3) $\gamma = 1$.

(d) Find the failure rate for $X \sim \text{Burr}(\alpha, \lambda, \tau)$, and comment on the case $\tau = 1$.

Note Some readers will recognise these results (in different notation) from a life table perspective.

2.18 Show that using a normal mixing distribution for normal claim amount means (with known variance) gives a normal mixture distribution; that is, show that, if $X \mid \mu \sim \text{N}(\mu, \sigma^2)$ and $\mu \sim \text{N}(\mu_0, \sigma_0^2)$ (with $\mu_0, \sigma^2, \sigma_0^2$ known), then X has a normal distribution.

Note It is sufficient to show that the pdf $f_X(x)$ is proportional to $e^{-g(x)}$, where $g(x)$ is of the form $a(x - b)^2$ with $a > 0$.

2.19 Let N be the number of trials required to get the first success in a series of independent, identical Bernoulli trials with $\text{Pr}(\text{success}) = p$, so N has probability mass function

$$\text{Pr}(N = n) = p(1 - p)^{n-1}, \quad n = 1, 2, 3, \ldots.$$

There is uncertainty about the value of p, which is modelled using a mixing distribution which is beta$(1, \beta)$ (see §2.3). Show that the mixture distribution for N has a probability mass function given by

$$\text{Pr}(N = n) = \frac{\beta}{(\beta + n)(\beta + n - 1)}, \quad n = 1, 2, 3, \ldots.$$

2.20 Show that a mixture of Weibull(c, γ) distributions with a gamma(α, λ) mixing distribution for c is a Burr$(\alpha, \lambda, \gamma)$ distribution.

2.21 Suppose we have a random sample of observations $x_1, x_2, \ldots, x_n$ from $X \sim \text{Pa}(\alpha, \lambda)$, with both parameters unknown. Let $\widehat{\alpha}$ and $\widehat{\lambda}$ be the MLEs of α and λ. Find two expressions for $\widehat{\alpha}$ in terms of $\widehat{\lambda}$, and, by equating them, show that $\widehat{\lambda}$ satisfies the equation

$$\frac{n}{\sum \log \left(1 + x_i/\widehat{\lambda}\right)} - \frac{\sum \left(\frac{1}{1+x_i/\widehat{\lambda}}\right)}{\sum \left(\frac{x_i/\widehat{\lambda}}{1+x_i/\widehat{\lambda}}\right)} = 0.$$

2.22 Suppose we have a random sample of observations $x_1, x_2, \ldots, x_n$ from $X \sim \text{Wei}(c, \gamma)$, with both parameters unknown. Derive an equation satisfied by $\widehat{\gamma}$, the MLE of γ, which does not involve $\widehat{c}$, the MLE of c.

2.23 Given a sample of observations $(x_1, x_2, \ldots, x_n)$, find the method of moments estimator and the maximum likelihood estimator of
   (a) α, in the case $X \sim \text{Pa}(\alpha, 2)$;
   (b) c, in the case $X \sim \text{Weibull}(c, 1/2)$.

2.24 Given a sample of observations $(x_1, x_2, \ldots, x_n)$, find the method of moments estimators of α and λ (obtained by matching the sample mean and sample variance with the mean and variance of the probability distribution) in the cases
   (a) $X \sim \text{gamma}(\alpha, \lambda)$;
   (b) $X \sim \text{Pareto}(\alpha, \lambda)$.

2.25 A random sample of 100 observations $(x_1, x_2, \ldots, x_{100})$ from a distribution modelled as $X \sim \text{Pa}(\alpha, 200)$ gives

$$\sum \log(200 + x_i) = 566.926.$$

   (a) Verify that the MLE of α is $\widehat{\alpha} = 2.696$.
   (b) Find the three quartiles $x_{0.25}$, $x_{0.5}$, $x_{0.75}$ (where x_p is defined by $\Pr(X \leq x_p) = p$) of the fitted distribution.
   (c) The numbers of observations in the sample in the ranges $(0, x_{0.25})$, $(x_{0.25}, x_{0.5})$, $(x_{0.5}, x_{0.75})$ and $(x_{0.75}, \infty)$ are 19, 28, 22 and 31, respectively. Calculate a χ^2 goodness-of-fit statistic, and comment on the extent to which the stated Pareto model provides an adequate description of the variation in the data.

2.26 A random sample of 1000 observations $(x_1, x_2, \ldots, x_{1000})$ from a population gives sample mean $\bar{x} = 754.51$ and standard deviation $s_x = 965.67$. The transformed data $y_i = \log(x_i)$ give sample mean $\bar{y} = 5.9139$ and standard deviation $s_y = 1.4227$.
   (a) Using the method of moments (by matching means and variances), estimate the parameters in the following proposed models for the data:

  (1) gamma(α, λ),
  (2) Pa(α, λ),
  (3) Weibull($c, 1/3$).
(b) Using the method of maximum likelihood, estimate the parameters in the following proposed models for the data:
  (1) Exp(λ),
  (2) lognormal(μ, σ),
  (3) loggamma(α, λ).
(c) Using each of the fitted Pareto, Weibull, exponential and lognormal models
  (1) calculate $\Pr(X > 2000)$;
  (2) find the 95th percentile of the fitted distribution, that is the value $x_{0.95}$, where $\Pr(X > x_{0.95}) = 0.05$.

3

Short term risk models

One of the key quantities of interest to an insurance company is the total amount to be paid out on a particular portfolio of policies over a fixed time interval, such as an accounting period. This quantity may be approached in various ways, and we mention two popular models below. We refer to both these models as examples of *short term risk models* because they model a risk over a fixed time period. This is in contrast to the classical risk model in Chapter 6, where the stochastic evolution of the flow of claim payments and premium income is modelled over time, and properties of this evolution over an infinite time period are derived. As might be expected, the techniques and results of Chapter 6 are deeper and more complex than those in this chapter, but they build on the foundations that we develop here for short term models.

One short term model is the *individual risk model*, where we consider the portfolio to consist of a fixed number, n, of independent policies, and the total amount claimed on the portfolio in a fixed time period is modelled as a random variable T, given by

$$T = Y_1 + \cdots + Y_n,$$

where Y_i is the total amount claimed on policy i, and $Y_1, \ldots, Y_n$ are assumed to be independent, but not necessarily identically distributed. It turns out that it is more difficult than might be expected at first sight to deal with this apparently simple quantity in terms of numerical calculations and in terms of obtaining analytical expressions for the distribution of T. This model is considered in §3.8.

Another short term model is the *collective risk model* (or *aggregate risk model*), where we model successive claims arising from the portfolio as independent, identically distributed (iid) random variables $X_1, X_2, X_3, \ldots$, and we ignore which policy gives rise to which claim. The number of claims in the fixed time period is a random variable, N, say, which is assumed to be independent of the X_i. The total claim amount (or aggregate claims) is modelled as a random variable S given by

$$S = X_1 + \cdots + X_N.$$

The distribution of S is an example of a compound distribution, and the properties and behaviour of compound distributions form the subject matter of §§3.1–3.7.

Both of the above insurance models have been much studied; for early work, see Cramér (1994) and the references to E. F. O. Lundberg (1876–1965) and others given there, while, more recently, issues of journals such as *The ASTIN Bulletin* and *Insurance: Mathematics and Economics* contain papers about these models and modifications of them.

3.1 The mean and variance of a compound distribution

First we define a compound distribution.

Definition 3.1 Let $X_1, X_2, \ldots$ be iid random variables and let N be a random variable taking values in $\{0, 1, 2, 3, \ldots\}$, independent of $\{X_i\}_{i=1}^{\infty}$. Let

$$S = X_1 + \cdots + X_N,$$

with $S = 0$ if $N = 0$. The random variable S is called a *random sum*. The distribution of a random sum is said to be a *compound distribution*.

We sometimes refer to N as "the counting random variable" (even if X_1 is also discrete), and to X_1 as "the step random variable".

Example 3.2 (i) Suppose an experiment consists of throwing a number of fair coins, where the number of coins to be thrown is decided by first throwing a fair die. Suppose we are interested in the total number S of heads. Let N be the number on the die, so that N has a discrete uniform distribution on $\{1, 2, \ldots, 6\}$, with $\Pr(N = n) = 1/6, n = 1, \ldots, 6$. Let X_i take the value 1 or 0 according as to whether the ith coin thrown shows a head or a tail. Then the total number of heads is the random sum $S = X_1 + \cdots + X_N$.

It is easy to see that

$$\mathbb{E}[N] = \frac{1}{6} \sum_{n=1}^{6} n = \frac{7}{2} \text{ and } \mathrm{Var}[N] = \frac{1}{6} \sum_{n=1}^{6} n^2 - \frac{49}{4} = \frac{35}{12}.$$

Conditional on $N = n$, the random variable S has a binomial distribution with parameters n and $1/2$, and so

$$\mathbb{E}[S \mid N = n] = \frac{n}{2} \text{ and } \text{Var}[S \mid N = n] = \frac{n}{4}.$$

Then $\mathbb{E}[S \mid N] = N/2$, and the conditional expectation formula (1.3) gives

$$\mathbb{E}[S] = \mathbb{E}[\mathbb{E}[S \mid N]] = \mathbb{E}\left[\frac{N}{2}\right] = \frac{7}{4}.$$

Similarly we see that $\text{Var}[S \mid N] = N/4$, and, using the conditional variance formula (1.4), we have

$$\begin{aligned}
\text{Var}[S] &= \mathbb{E}[\,\text{Var}[S \mid N]] + \text{Var}[\mathbb{E}[S \mid N]] \\
&= \mathbb{E}[N/4] + \text{Var}[N/2] \\
&= \frac{7}{8} + \frac{35}{48} = \frac{77}{48}.
\end{aligned}$$

(ii) Suppose the number N of claims in one time unit has a Poisson distribution with parameter $\lambda = 10$, and suppose that claims $X_1, X_2, \ldots$ are iid random variables, independent of N, with mean and variance both equal to unity. Then the total amount claimed is $S = X_1 + \cdots + X_N$. Given $N = n$, S is the sum of n iid random variables each with mean and variance 1, so that

$$\mathbb{E}[S \mid N = n] = n \text{ and } \text{Var}[S \mid N = n] = n.$$

Using the conditional expectation and variance formulae, together with properties of the Poisson distribution in §2.1.1, we obtain

$$\mathbb{E}[S] = \mathbb{E}\big[\mathbb{E}[S \mid N]\big] = \mathbb{E}[N] = 10$$

and

$$\text{Var}[S] = \mathbb{E}[\,\text{Var}[S \mid N]] + \text{Var}[\mathbb{E}[S \mid N]] = \mathbb{E}[N] + \text{Var}[N] = 20.$$

From the definition, we see that a random sum S is the sum of a *random number* N of random variables. Since it would be simpler to deal with the sum of a fixed number of iid random variables, we often use conditioning on N, as in Example 3.2, because this conditioning reduces our calculations to those for a fixed number of random variables as an intermediate step. Conditioning will be a major tool when working with compound distributions.

At the beginning of this chapter, we gave the motivation for random sums in insurance for the case where N is the number of claims and the X_i are the claim amounts. However, any random variable that is the sum of a random number N of iid random variables (independent of N) is a random sum. In the following example, the number of claims is itself a random sum.

Example 3.3 Suppose that, in a certain region, the number of floods in a fixed time period is a random variable N, and that the ith flood gives rise to M_i claims, where $M_1, M_2, \ldots$ are iid random variables, independent of N. Then the total number of claims in the time period is $T = M_1 + \cdots + M_N$, a random sum.

In general, compound distributions are determined once we know the two relevant input distributions, i.e. the distributions of N and X_1. Typically we are given the distributions of N and X_1 and we are interested in the behaviour of the resulting random sum S. A first step might well be to determine the mean and variance of S, and in the theorem below we give formulae for these in terms of the means and variances of the input distributions.

Theorem 3.4 *For a random sum $S = X_1 + \cdots + X_N$, we have*

$$\mathbb{E}[S] = \mathbb{E}[N]\,\mathbb{E}[X_1] \tag{3.1}$$

and

$$\mathrm{Var}[S] = \mathbb{E}[N]\,\mathrm{Var}[X_1] + \mathrm{Var}[N]\,(\mathbb{E}[X_1])^2. \tag{3.2}$$

Proof We know that $\mathbb{E}[S \mid N = n] = n\mathbb{E}[X_1]$, and so $\mathbb{E}[S \mid N] = N\mathbb{E}[X_1]$. Using the conditional expectation formula (1.3), we have

$$\mathbb{E}[S] = \mathbb{E}[\mathbb{E}[S \mid N]] = \mathbb{E}[N\mathbb{E}[X_1]] = \mathbb{E}[N]\,\mathbb{E}[X_1].$$

By our independence assumptions in the definition of a random sum, we have $\mathrm{Var}[S \mid N = n] = n\,\mathrm{Var}[X_1]$, and so $\mathrm{Var}[S \mid N] = N\,\mathrm{Var}[X_1]$. Hence, using the conditional variance formula (1.4), we obtain

$$
\begin{aligned}
\mathrm{Var}[S] &= \mathbb{E}[\,\mathrm{Var}[S \mid N]] + \mathrm{Var}\,[\mathbb{E}[S \mid N]] \\
&= \mathbb{E}[N\,\mathrm{Var}[X_1]] + \mathrm{Var}\,[N\mathbb{E}[X_1]] \\
&= \mathbb{E}[N]\,\mathrm{Var}[X_1] + \mathrm{Var}[N]\,(\mathbb{E}[X_1])^2. \qquad \square
\end{aligned}
$$

You might like to check for yourself that the mean and variance of S in Example 3.2(i) and (ii) agree with the results of applying (3.1) and (3.2).

3.2 The distribution of a random sum

We often need more than the mean and the variance of a compound distribution. This is demonstrated in the following example, which shows that two compound distributions with the same mean and variance can behave very differently in their tails.

Example 3.5 Suppose that the number of claims in one time period has a Poisson distribution with mean 10, and consider the total amount S claimed in one time period. We consider two cases as follows.

- Claim sizes have a gamma distribution with parameters $\alpha = 3$, $\lambda = 2$.
- Claim sizes have a translated Pareto distribution with density

$$f(x) = \frac{3}{x^4} \quad \text{for } x > 1. \tag{3.3}$$

As an aside, we show how this translated Pareto distribution relates to the Pareto distribution in Chapter 2; note that, if $Y \sim \text{Pa}(3, 1)$ as defined in §2.2.7, then

$$\Pr(Y > y) = \left(\frac{1}{1+y}\right)^3, \quad y \geq 0.$$

We then find that the random variable $Y + 1$ (obtained from Y by translating one unit to the right) has tail probability

$$\Pr(Y + 1 \geq y) = \Pr(Y \geq y - 1) = \left(\frac{1}{y}\right)^3, \quad y > 1,$$

so that $Y + 1$ has a probability density function as in (3.3).

Both the gamma and the translated Pareto claim-size distributions have mean 3/2 and variance 3/4, so that resulting compound distributions have the same mean, 15, and the same variance, 30.

Table 3.1 shows the tail probability $\Pr(S > x)$ of the total claim amount for various x-values. These tail probabilities were obtained using the fast Fourier transform algorithm in R (see §3.5.2).

Table 3.1. *Tail probabilities for two compound Poisson distributions with the same mean and variance*

x	Pr($S > x$)	
	Gamma	Pareto
10	0.8134	0.8226
20	0.1749	0.1634
30	0.008139	0.01014
40	0.0001196	0.001018
50	7.404×10^{-7}	0.0002769
60	2.330×10^{-9}	0.0001158

While these two compound distributions have similar tail probabilities in the region of the mean, $\mathbb{E}[S] = 15$, they have very different tail probabilities for large x, with the tail probability for a very large total claim amount being much higher for Pareto-distributed claims.

The above example shows that we may need to look beyond the mean and variance of the random sum S if we are concerned, for example, with its tail behaviour, and so we turn now to consider the whole distribution of S. Since S involves sums of independent random variables, we shall use either convolutions or transforms in finding expressions for its distribution. The word "transforms" here is a generic term including probability generating functions, moment generating functions (which are essentially Laplace transforms), characteristic functions (essentially Fourier transforms), cumulant generating functions, etc. The convolution approach is in §3.2.1, and §3.2.2 gives an approach via moment generating functions. Examples and illustrations of these two approaches follow in §3.4.

3.2.1 Convolution series formula for a compound distribution

Convolutions are used to find the distributions of sums of independent random variables.

Definition 3.6 Let $X_1, \ldots, X_n$ be iid with distribution function F. For fixed $n \geq 1$, the *n-fold convolution* or *nth convolution power* $F^{\star n}$ of F is the distribution function of $X_1 + \cdots + X_n$. This means that, for $n \geq 1$,

$$F^{\star n}(x) = \Pr(X_1 + \cdots + X_n \leq x).$$

For $n = 0$, $F^{\star 0}$ is defined by

$$F^{\star 0}(x) = 1_{[0,\infty)}(x),$$

where $1_A(x)$ is the indicator function of the set A, which takes the value 1 if $x \in A$ and is zero otherwise. This means that

$$F^{\star 0}(x) = \begin{cases} 1 & \text{if } x \geq 0 \\ 0 & \text{otherwise.} \end{cases}$$

Note that $F^{\star 0}$ is the distribution function of a random variable that takes the value 0 with probability 1, and that $F^{\star 1}$ is just the distribution function F itself.

Example 3.7   Suppose that $X_1, X_2, \ldots$ are independent exponentially distributed random variables with mean μ. We know from §2.2.4 that $X_1 + \cdots + X_n$ has a gamma$(n, 1/\mu)$ distribution, so that

$$F^{\star n}(x) = \int_0^x \frac{t^{n-1} e^{-t/\mu}}{\mu^n (n-1)!} \, dt.$$

Check that this means that

$$F^{\star 2}(x) = 1 - \left(1 + \frac{x}{\mu}\right) e^{-x/\mu},$$

which completes this example.

We shall mostly deal with the case where F has a density f, say, as in the example above. In this case, we can obtain a formula for $F^{\star n}$ in terms of $F^{\star(n-1)}$ and f by conditioning on X_1. For $n \geq 2$, we obtain

$$
\begin{aligned}
F^{\star n}(x) &= \Pr(X_1 + \cdots + X_n \leq x) \\
&= \int \Pr(X_1 + \cdots + X_n \leq x \mid X_1 = t) f(t) dt \\
&= \int \Pr(X_2 + \cdots + X_n \leq x - t) f(t) dt \\
&= \int F^{\star(n-1)}(x - t) f(t) dt. \qquad (3.4)
\end{aligned}
$$

We can check that (3.4) works when $n = 1$ as follows. Putting $n = 1$ into (3.4) gives $F(x)$ on the left-hand side. On the right-hand side we obtain

$$\int F^{\star 0}(x - t) f(t) dt = \int 1_{[0,\infty)}(x - t) f(t) dt = \int_{-\infty}^x f(t) dt = F(x),$$

so that the left-hand side is equal to the right-hand side in this case.

Further, for the X_i as in Example 3.7, if we put $n = 2$ into (3.4) and write f for the density of X_i, then we obtain

$$
\begin{aligned}
F^{\star 2}(x) &= \int_0^x F(x - t) f(t) dt \\
&= \int_0^x (1 - e^{-(x-t)/\mu}) \frac{1}{\mu} e^{-t/\mu} \, dt \\
&= 1 - e^{-x/\mu} - \frac{x}{\mu} e^{-x/\mu},
\end{aligned}
$$

as given in Example 3.7.

When the X_i are non-negative random variables with density f, then (3.4) becomes

$$F^{\star n}(x) = \int_0^x F_X^{\star(n-1)}(x - t) f(t) dt,$$

and this is the case that arises most often in the rest of the book.

Finally, we note that, for a general distribution function F, (3.4) becomes

$$F^{\star n}(x) = \int F^{\star(n-1)}(x - t) F(dt),$$

using the notation $\int \ldots F(dt)$ as in §1.2.1.

The following theorem gives an expression for the distribution function F_S of a random sum S in terms of convolution powers of the step distribution function.

Theorem 3.8 *The distribution function of the random sum $S = X_1 + \cdots + X_N$ is*

$$F_S(x) = \sum_{n=0}^{\infty} F_X^{\star n}(x) \Pr(N = n), \tag{3.5}$$

where F_X is the distribution function of X_1.

Proof Conditioning on N, we have

$$F_S(x) = \Pr(S \le x) = \sum_{n=0}^{\infty} \Pr(S \le x \mid N = n) \Pr(N = n).$$

Observe that, since $N = 0$ implies that $S = 0$, we have

$$\Pr(S \le x \mid N = 0) = 1_{[0,\infty)}(x) = F_X^{\star 0}(x).$$

For $n \ge 1$, we have

$$\Pr(S \le x \mid N = n) = \Pr(X_1 + \cdots + X_n \le x) = F_X^{\star n}(x),$$

so that

$$F_S(x) = \sum_{n=0}^{\infty} F_X^{\star n}(x) \Pr(N = n). \qquad \square$$

It is sometimes helpful to separate out the term for $n = 0$, and then we have

$$F_S(x) = \Pr(N = 0) 1_{[0,\infty)}(x) + \sum_{n=1}^{\infty} F_X^{\star n}(x) \Pr(N = n). \tag{3.6}$$

When X_1 is a strictly positive random variable, $X_1 + \cdots + X_n$ is also strictly positive, and so $F_X^{\star n}(0) = \Pr(X_1 + \cdots + X_n \le 0) = 0$ for $n \ge 1$. This means that, in this case, $F_S(0) = \Pr(N = 0)$. Hence when $\Pr(N = 0) > 0$ and X_1

is strictly positive, the distribution function F_S has a positive jump of size $\Pr(N = 0)$ at zero.

If F_X has a density f_X, then, for $n \geq 1$, $F_X^{\star n}$ has density f_X^{*n}, where $f_X^{*1} = f_X$ and, for $n > 1$, $f_X^{*n}(x) = \int f_X^{*(n-1)}(x - t) f_X(t) dt$ (see Exercise 3.2). Note the different notation $\star$ and $*$ for convolution of distribution functions and convolution of densities, respectively. Thus (3.6) becomes

$$F_S(x) = \Pr(N = 0) 1_{[0,\infty)}(x) + \int_{-\infty}^{x} \sum_{n=1}^{\infty} f_X^{*n}(t) \Pr(N = n) dt. \tag{3.7}$$

This shows that the distribution of S consists of an atom at zero and a part with a density. For some particular choices for the distributions of N and X_1, there is a simple explicit expression for $F_S(x)$, as we will see in Example 3.17.

3.2.2 Moment generating function of a compound distribution

An alternative approach to the distribution of sums of independent random variables is via moment generating functions. In this subsection, we find the moment generating function of $S = X_1 + \cdots + X_N$ in terms of the probability generating function of N, given by (1.7), that is

$$G_N(z) = \mathbb{E}[z^N] = \sum_{n=0}^{\infty} z^n \Pr(N = n),$$

and of the moment generating function of X_1, given by (1.15), that is

$$M_X(r) = \mathbb{E}[e^{rX_1}] = \int e^{rx} F_X(dx),$$

where F_X is the distribution function of X_1. Recall from §1.2.1 that the generating functions $G_N(z)$ and $M_X(r)$ are not necessarily finite for all real values of z and r. In statements below that relate various generating functions, we tacitly assume that r is such that both sides are defined.

The following theorem gives the moment generating function of S in terms of G_N and M_X.

Theorem 3.9 *The moment generating function of the random sum $S = X_1 + \cdots + X_N$ is*

$$M_S(r) = G_N[M_X(r)]. \tag{3.8}$$

Proof It is easy to see that

$$\mathbb{E}[e^{rS} \mid N = n] = \mathbb{E}[e^{r(X_1 + \cdots + X_n)}] = (M_X(r))^n,$$

so that $\mathbb{E}[e^{rS} \mid N] = (M_X(r))^N$. Using the conditional expectation formula and recalling that the $G_N(z) = \mathbb{E}[z^N]$, we find

$$M_S(r) = \mathbb{E}[\mathbb{E}[e^{rS} \mid N]] = \mathbb{E}[(M_X(r))^N] = G_N[M_X(r)]. \qquad \square$$

Given a formula for $M_S(r)$, we can use it to find moments of S because $\mathbb{E}[S^k] = M_S^{(k)}(0)$ (see (1.6)). For example, recall that $G_N'(1) = \mathbb{E}[N]$ and $M_X(0) = 1$, so that

$$\mathbb{E}[S] = M_S'(0) = G_N'(M_X(0))M_X'(0) = \mathbb{E}[N]\mathbb{E}[X_1],$$

and we have recovered (3.1).

Sometimes it is helpful to have corresponding formulae for other transforms of S. For example, if X_1 is a discrete random variable concentrated on $\{0, 1, 2, \ldots\}$ with probability generating function G_X, then $S = X_1 + \cdots + X_N$ is also discrete and concentrated on $\{0, 1, 2, \ldots\}$. The probability generating function of S is found using the same conditioning argument as in Theorem 3.9, and this gives

$$G_S(z) = G_N[G_X(z)]. \tag{3.9}$$

If we are interested in the characteristic function $\phi_S(r) = \mathbb{E}[e^{iSr}]$, where i is $\sqrt{-1}$, then we find similarly that

$$\phi_S(r) = G_N[\phi_X(r)], \tag{3.10}$$

where $\phi_X(r)$ is the characteristic function of X_1.

Recall that cumulant generating functions were defined in §2.2.5. The cumulant generating function of S is

$$K_S(r) = \log(M_S(r)) = \log\left(G_N[M_X(r)]\right).$$

We use the fact that $G_N(z) = \mathbb{E}[z^N] = \mathbb{E}[e^{N \log z}] = M_N(\log z)$ to see that $K_S(r)$ can also be written

$$K_S(r) = \log\left(M_N[\log(M_X(r))]\right) = K_N[K_X(r)], \tag{3.11}$$

where K_N and K_X are the cumulant generating functions of N and X_1, respectively. This provides us with an easy way to find the cumulants and moments of S. For example, differentiating (3.11), we find that

$$K_S'(r) = K_N'(K_X(r))K_X'(r).$$

It is clear that $K_X(0) = \log(M_X(0)) = \log(1) = 0$, and so we have

$$\mathbb{E}[S] = K_S'(0) = K_N'(0)K_X'(0) = \mathbb{E}[N]\mathbb{E}[X_1].$$

Differentiating again, we have

$$K_S''(r) = K_N''(K_X(r))(K_X'(r))^2 + K_N'(K_X(r))K_X''(r),$$

so that, putting $r = 0$, we obtain

$$\text{Var}[S] = K_S''(0) = K_N''(0)(K_X'(0))^2 + K_N'(0)K_X''(0)$$
$$= \text{Var}[N](\mathbb{E}[X_1])^2 + \mathbb{E}[N]\,\text{Var}[X_1].$$

Thus we have obtained (3.1) and (3.2) again.

Sometimes we can use (3.8) to find an explicit expression for the distribution of S – we will see this in action in Example 3.18.

So far in this chapter we have seen various general properties of compound distributions. In §3.4 we will look at particular cases for the counting random variable in a random sum. In order to do this, we need the notion and properties of a finite mixture distribution, which we consider in §3.3.

3.3 Finite mixture distributions

Recall that mixture distributions were introduced in §2.3, where the mixing distributions had densities. A *finite mixture distribution* is an example of a mixture distribution where the mixing random variable is discrete with a finite number of possible values.

Example 3.10 Suppose that the number of car accidents in a day on a particular road has a $\text{Poi}(\lambda_1)$ distribution if the day is classified as "rainy", a $\text{Poi}(\lambda_2)$ distribution if the day is classified as "wintry", and a $\text{Poi}(\lambda_3)$ distribution otherwise (where, in this simple model, we assume that there are clear criteria by which each day is classified as belonging to exactly one of the categories "rainy", "wintry" and "other"). Suppose further that 40%, 10% and 50% of days are "rainy", "wintry" and "other", respectively. Let N be the number of accidents on a randomly chosen day. Let θ be a random variable that takes the values 1, 2 or 3 according to whether the day is "rainy", "wintry" or "other", respectively. Then

$$\Pr(\theta = i) = \begin{cases} 0.4 & \text{if } i = 1 \\ 0.1 & \text{if } i = 2 \\ 0.5 & \text{if } i = 3. \end{cases}$$

Then the distribution of $N \mid (\theta = i)$ is $\text{Poi}(\lambda_i)$, $i = 1, 2, 3$, and, for $x = 0, 1, 2, \ldots$,

$$\Pr(N \le x) = \sum_{i=1}^{3} \Pr(N \le x \mid \theta = i)\,\Pr(\theta = i)$$
$$= 0.4F_1(x) + 0.1F_2(x) + 0.5F_3(x),$$

where F_i is the distribution function of a $\text{Poi}(\lambda_i)$ distribution, $i = 1, 2, 3$.

This motivates the following definition.

Definition 3.11   A random variable Y has a *finite mixture distribution* with *mixing proportions* $p_1, \ldots, p_n$ if it has distribution function

$$F_Y = p_1 F_1 + \cdots + p_n F_n$$

for some $n \in \{1, 2, \ldots\}$, where $F_1, \ldots, F_n$ are distribution functions, $p_i \in [0, 1]$, $i = 1, \ldots, n$, and $\sum_{i=1}^n p_i = 1$.

Most often, the p_i will be in $(0, 1)$, but it is helpful to allow the values 0 and 1 in the definition. Observe that F_Y is indeed a distribution function (because the F_i are distribution functions). Further, if F_i has density $f_i, i = 1, \ldots, n$, then F_Y has density

$$f_Y = p_1 f_1 + \cdots + p_n f_n.$$

Suppose that random variable W_i has distribution function F_i. Then the mixture random variable Y with distribution function $F_Y = \sum_{i=1}^n p_i F_i$ may be interpreted as

$$Y = \begin{cases} W_1 & \text{with probability } p_1 \\ W_2 & \text{with probability } p_2 \\ \vdots \\ W_n & \text{with probability } p_n. \end{cases}$$

We can check that, with this representation, we do indeed have that $F_Y = \Pr(Y \le y)$ is equal to $\sum_{i=1}^n \Pr(W_i \le y) p_i = \sum_{i=1}^n p_i F_i$.

If we wish to simulate an observation from the mixture distribution F_Y, then one way to do this is first to simulate an observation i from a discrete random variable I that takes values $1, 2, \ldots, n$ with probabilities $p_1, \ldots, p_n$, respectively. Once the value of i is simulated, we would simulate an observation from the distribution F_i of W_i. (Note that this bears no relationship whatsoever to the operation of adding up the random variables $W_1, \ldots, W_n$.)

The moments of Y are given by

$$\mathbb{E}[Y^r] = \int y^r F_Y(dy) = \int y^r \sum_{i=1}^n p_i F_i(dy)$$

$$= \sum_{i=1}^n p_i \int y^r F_i(dy) = \sum_{i=1}^n p_i \mathbb{E}[W_i^r].$$

Similarly, the moment generating function of Y is given by

$$M_Y(r) = \int e^{ry} F_Y(dy) = \sum_{i=1}^n p_i \int e^{ry} F_i(dy) = \sum_{i=1}^n p_i M_i(r), \tag{3.12}$$

where M_i is the moment generating function of W_i.

Example 3.12   (i) Suppose that Y has density

$$f_Y(y) = \frac{5e^{-5y}}{2}\left(1 + 2e^{-5y}\right).$$

This may be written

$$f_Y(y) = \frac{1}{2} \times 5e^{-5y} + \frac{1}{2} \times 10e^{-10y}, \tag{3.13}$$

and we see that the distribution of Y is a finite mixture of an exponential distribution with mean $1/5$ and exponential distribution with mean $1/10$, with both mixing proportions equal to $1/2$. We call this an *equal mixture* of the two exponential distributions. Its moment generating function is given by

$$M_Y(r) = \frac{1}{2} \times \frac{5}{5-r} + \frac{1}{2} \times \frac{10}{10-r} = \frac{5(20-3r)}{2(5-r)(10-r)}$$

for $r < 5$.

(ii) Suppose a policyholder makes at most one claim in a given year, and that the probability of making a claim is p. If a claim is made, then it has an $\text{Exp}(\lambda)$ distribution. Let Y be the amount claimed by the policyholder in one year. Then

$$Y = \begin{cases} 0 & \text{with probability } 1 - p \\ W & \text{with probability } p, \end{cases}$$

where W has an $\text{Exp}(\lambda)$ distribution. Then, for $y \geq 0$, the amount claimed can only be greater than y if a claim is actually made, and so we have

$$1 - F_Y(y) = \Pr(Y > y) = p\Pr(W > y) = pe^{-\lambda y}.$$

This means that Y has distribution function

$$F_Y(y) = \Pr(Y \leq y) = \begin{cases} 0 & \text{if } y < 0 \\ 1 - pe^{-\lambda y} & \text{if } y \geq 0 \end{cases}$$

$$= \begin{cases} 0 & \text{if } y < 0 \\ 1 - p + p(1 - e^{-\lambda y}) & \text{if } y \geq 0. \end{cases}$$

This may be rewritten as

$$F_Y(y) = (1 - p)1_{[0,\infty)}(y) + pF_W(y).$$

Recall from Definition 3.6 that $1_{[0,\infty)}$ is the distribution function of a random variable, Z say, that takes the value 0 with probability 1, i.e. such that $\Pr(Z = 0) = 1$. We sometimes call this distribution "a unit mass at zero". Thus the distribution of Y is a finite mixture of a unit mass at zero and an exponential distribution with mean $1/\lambda$, with mixing proportions $1 - p$ and p, respectively. Note that the moment generating function of Z is

$$M_Z(r) = \mathbb{E}[e^{Zr}] = 1, \tag{3.14}$$

so that the moment generating function of Y is

$$M_Y(r) = (1 - p) + p\frac{\lambda}{\lambda - r} \quad \text{for } r < \lambda.$$

The definition of a finite mixture distribution extends easily to a countable mixture distribution $\sum_{i=1}^{\infty} p_i F_i$, where the F_i are distribution functions and the p_i are in the interval $[0, 1]$ and sum to unity. As a final observation we note that a compound distribution,

$$F_S = \sum_{n=0}^{\infty} \Pr(N = n)F_X^{\star n},$$

is a countable mixture of $F_X^{\star 0}$, $F_X^{\star 1}$, $F_X^{\star 2}, \ldots$ with mixing proportions $\Pr(N = 0)$, $\Pr(N = 1)$, $\Pr(N = 2), \ldots$, respectively.

3.4 Special compound distributions

In this section, we return to a consideration of compound distributions, and we study the properties of various special compound distributions with particular choices for the distribution of the counting random variable.

3.4.1 Compound Poisson distributions

A Poisson distribution (with mean λ, say, see §2.1.1) is a common choice for the distribution of the counting random variable N, and the resulting random sum $S = X_1 + \cdots + X_N$ is said to have a *compound Poisson* distribution. Let the step random variables $X_1, X_2, \ldots$ be iid with distribution function F_X. We use the notation $CP(\lambda, F_X)$ for a compound Poisson distribution with Poisson parameter λ and step distribution function F_X. From the convolution series formula (3.5) for F_S, the compound Poisson distribution function is

$$F_S(x) = \sum_{n=0}^{\infty} \frac{e^{-\lambda}\lambda^n}{n!} F_X^{\star n}(x). \tag{3.15}$$

From the moment generating function (3.8) for a compound distribution, and from the probability generating function (2.1) for the Poisson distribution, we find that the moment generating function of S is

$$M_S(r) = \exp\left[\lambda(M_X(r) - 1)\right], \tag{3.16}$$

where M_X is the moment generating function corresponding to F_X. The cumulant generating function of S is given by

$$K_S(r) = \log(M_S(r)) = \lambda(M_X(r) - 1).$$

From this, it is particularly simple to find compound Poisson cumulants in terms of λ and moments of X_1, since, for $j = 1, 2, \ldots,$

$$\kappa_{S,j} = K_S^{(j)}(0) = \lambda M_X^{(j)}(0) = \lambda \mathbb{E}[X_1^j]. \tag{3.17}$$

This shows that the mean and variance of the compound Poisson random variable S are

$$\mathbb{E}[S] = \kappa_{S,1} = \lambda \mathbb{E}[X_1] \quad \text{and} \quad \text{Var}[S] = \kappa_{S,2} = \lambda \mathbb{E}[X_1^2], \tag{3.18}$$

and these are the same as given by (3.1) and (3.2). The coefficient of skewness of the compound Poisson random variable S is

$$\frac{\mathbb{E}[(S - \mathbb{E}[S])^3]}{(\text{SD}[S])^3} = \frac{\kappa_{S,3}}{\kappa_{S,2}^{3/2}} = \frac{\lambda \mathbb{E}[X_1^3]}{(\lambda \mathbb{E}[X_1^2])^{3/2}} = \frac{\mathbb{E}[X_1^3]}{\sqrt{\lambda}(\mathbb{E}[X_1^2])^{3/2}}. \tag{3.19}$$

Example 3.13   (i) Suppose that X_1 is degenerate at 1, so that $\Pr(X_1 = 1) = 1$ and $M_X(r) = e^r$. Then (3.16) gives the moment generating function of S as

$$M_S(r) = \exp\left(\lambda(e^r - 1)\right),$$

which we recognise from (2.2) as the moment generating function of a Poisson distribution with mean λ. We can also see the same thing from the convolution series (3.5) as follows. The n-fold convolution of the distribution function F_X of X_1 is

$$F_X^{\star n}(x) = \Pr(X_1 + \cdots + X_n \le x) = 1_{[n,\infty)}(x),$$

so that the distribution function of S is

$$F_S(x) = \sum_{n=0}^{\infty} \frac{e^{-\lambda}\lambda^n}{n!} 1_{[n,\infty)}(x).$$

Hence F_S has a jump of size $e^{-\lambda}\lambda^n/n!$ at n, $n = 0, 1, 2, \ldots$, and we recognise this as the distribution function of a Poisson distribution with mean λ.

If X_1 is degenerate at a (> 0), so that $\Pr(X_1 = a) = 1$, then F_S has a jump of size $e^{-\lambda}\lambda^n/n!$ at na, $n = 0, 1, 2, \ldots$.

(ii) If X_1 has an exponential distribution with mean μ, then, for $n \geq 1$, $X_1 + \cdots + X_n$ has a gamma distribution with density

$$f_{X_1+\cdots+X_n}(x) = \frac{x^{n-1}e^{-x/\mu}}{(n-1)!\mu^n}, \quad x > 0$$

(see §2.2.4). From (3.7), F_S consists of an atom of size $\Pr(N = 0) = e^{-\lambda}$ at 0 and of a part on $(0, \infty)$ with density

$$f_S(x) = \sum_{n=1}^{\infty} \frac{e^{-x/\mu}x^{n-1}}{(n-1)!\mu^n} \frac{e^{-\lambda}\lambda^n}{n!}$$

$$= \frac{\lambda e^{-\lambda - (x/\mu)}}{\mu} \sum_{n=1}^{\infty} \frac{1}{(n-1)!n!} \left(\frac{\lambda x}{\mu}\right)^{n-1}.$$

This can be written in terms of modified Bessel functions defined by

$$I_\nu(x) = \sum_{n=0}^{\infty} \frac{(x/2)^{2n+\nu}}{n!\,\Gamma(\nu + n + 1)}, \quad x \in \mathbb{R}, \nu \in \mathbb{R}.$$

Then

$$f_S(x) = e^{-\lambda - (x/\mu)} \sqrt{\frac{\lambda}{\mu x}} I_1\left(2\sqrt{\frac{\lambda x}{\mu}}\right).$$

(iii) We can use **R** to simulate observations from compound distributions in general, and we illustrate this below for a compound Poisson distribution.

Suppose that we wish to simulate n observations from a compound Poisson distribution, where the Poisson parameter is λ and where the claims are exponentially distributed with mean μ, i.e. we wish to simulate from a $CP(\lambda, \text{Exp}(1/\mu))$ distribution as in (ii) above.

Assume that the **R** objects n, lambda and mu contain the numerical values of n, λ and μ. The following **R** code produces the required simulated values:

```
total_claims = rep(0, n)
numclaims = rpois(n, lambda)
for(i in 1:n)
+ total_claims[i] = sum(rexp(numclaims[i], 1/mu))
```

where + at the beginning of a line is the continuation prompt in **R**. The above code works as follows. The first line initialises total_claims to be a vector of n zeros. The second line simulates n observations from a Poi(λ) distribution. We interpret the ith entry in the vector numclaims as being the number of claims in the ith Poisson random sum. For each i-value in turn, the remaining lines simulate the relevant number of exponential random variables, which are then added up to give a simulated value of the ith random sum.

Figure 3.1 shows a histogram of the result when $n = 5000$, $\lambda = 10$ and $\mu = 10$. By (3.18) and (3.19), this compound Poisson distribution has true mean, variance and coefficient of skewness given, respectively, by

$$\lambda\mu = 100, \quad \lambda \times 2\mu^2 = 2000 \text{ and } \frac{6\mu^3}{\sqrt{\lambda}(2\mu^2)^{3/2}} = 2.12.$$

Note the observed skewness in the sample of 5000 observations.

Independent compound Poisson random variables have the useful property that the sum of a fixed (i.e. non-random) number of them is also a compound Poisson random variable, as is shown in the following theorem.

Theorem 3.14 *Let $S_1, \ldots, S_m$ be independent compound Poisson random variables with Poisson parameters $\lambda_1, \ldots, \lambda_m$ and step distribution functions $F_1, \ldots, F_m$, respectively. Then $T = S_1 + \cdots + S_m$ has a compound Poisson distribution with Poisson parameter*

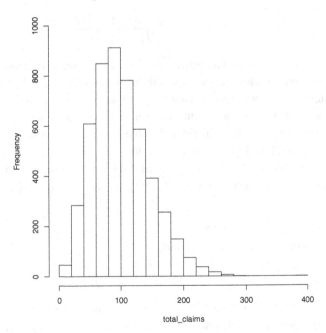

Figure 3.1. Histogram of a sample of 5000 observations simulated from a compound Poisson distribution (Poisson parameter $\lambda = 10$) and exponentially distributed claim sizes (mean $\mu = 10$).

$$\lambda = \lambda_1 + \cdots + \lambda_m$$

and step distribution function

$$F = \sum_{i=1}^{m} \frac{\lambda_i}{\lambda} F_i.$$

Proof Let $M_i(r)$ be the moment generating function belonging to F_i. From (3.16) S_i has the moment generating function $\mathbb{E}(e^{rS_i}) = \exp(\lambda_i(M_i(r) - 1))$. Then T has the moment generating function

$$M_T(r) = \mathbb{E}\left[\exp(r(S_1 + \cdots + S_m))\right]$$

$$= \prod_{i=1}^{m} \mathbb{E}[\exp(rS_i)]$$

$$= \prod_{i=1}^{m} \exp(\lambda_i(M_i(r) - 1))$$

$$= \exp\left(\left(\sum_{i=1}^{m} \lambda_i M_i(r)\right) - \lambda\right)$$

$$= \exp\left(\lambda\left(\left(\sum_{i=1}^{m} \frac{\lambda_i}{\lambda} M_i(r)\right) - 1\right)\right). \tag{3.20}$$

From the expression for the moment generating function of a finite mixture in (3.12), we observe that

$$\sum_{i=1}^{m} \frac{\lambda_i}{\lambda} M_i(r)$$

is the moment generating function belonging to the distribution function

$$F = \sum_{i=1}^{m} \frac{\lambda_i}{\lambda} F_i.$$

This is a finite mixture of the m distributions $F_1, \ldots, F_m$, with mixing proportions $(\lambda_1/\lambda), \ldots, (\lambda_m/\lambda)$. By comparison with (3.16), we recognise (3.20) as a compound Poisson moment generating function with Poisson parameter λ and step distribution function F. □

Example 3.15 Suppose that S_1 and S_2 are independent compound Poisson random variables, each with Poisson parameter 10, so that $\lambda_1 = \lambda_2 = 10$. For $i = 1, 2$, suppose that the step random variables for S_i are exponentially distributed with mean μ_i, where $\mu_1 = 1$ and $\mu_2 = 1/2$. Thus we have $F_1(x) = 1 - e^{-x}$ and $F_2(x) = 1 - e^{-2x}$. By Theorem 3.14, the random variable $S = S_1 + S_2$

has a compound Poisson distribution with Poisson parameter $\lambda = \lambda_1 + \lambda_2 = 20$ and with step distribution function

$$F(x) = \frac{\lambda_1}{\lambda}F_1(x) + \frac{\lambda_2}{\lambda}F_2(x) = \frac{1}{2}F_1(x) + \frac{1}{2}F_2(x)$$
$$= \frac{1}{2}(1 - e^{-x}) + \frac{1}{2}(1 - e^{-2x}) = 1 - \frac{1}{2}e^{-x} - \frac{1}{2}e^{-2x}.$$

The step distribution F is an equal mixture of an exponential distribution with mean 1 and an exponential distribution with mean $1/2$.

3.4.2 Compound mixed Poisson distributions

Here we consider the case where the counting random variable has a mixed Poisson distribution (as introduced in §2.3), in which case the resulting random sum is said to have a *compound mixed Poisson* distribution.

When fitting counting distributions to data, it is frequently observed that the data are not consistent with a distribution that has equal mean and variance, as is implied by the choice of a Poisson distribution; often the variance is larger than the mean (see the discussion in §2.1.2). In these cases, one way to find a more suitable distribution is to consider a *mixed Poisson* distribution as follows: given λ, suppose that N has a Poisson distribution with mean λ, i.e. $N \mid \lambda \sim \text{Poisson}(\lambda)$, and suppose that λ is itself a (positive) random variable.

Conditioning on λ, and using the fact that $N \mid \lambda$ is a Poisson distribution with mean λ, provides the technical tool for calculation of quantities for N. For example, using the conditional expectation and variance formulae (1.3) and (1.4), we have

$$\mathbb{E}[N] = \mathbb{E}[\mathbb{E}[N \mid \lambda]] = \mathbb{E}[\lambda]$$

and

$$\text{Var}[N] = \mathbb{E}[\text{Var}[N \mid \lambda]] + \text{Var}[\mathbb{E}[N \mid \lambda]] = \mathbb{E}[\lambda] + \text{Var}[\lambda].$$

Provided that $\text{Var}[\lambda] > 0$, this last expression shows that $\text{Var}[N] > \mathbb{E}[N]$. Writing F_λ for the distribution function of λ, the probabilities for N are given by

$$\Pr(N = n) = \int_{(0,\infty)} \Pr(N = n \mid \lambda)F_\lambda(d\lambda) = \int_{(0,\infty)} \frac{e^{-\lambda}\lambda^n}{n!}F_\lambda(d\lambda).$$

If M_λ is the moment generating function of λ, then we obtain (using the conditional expectation formula again)

$$G_N(z) = \mathbb{E}[z^N] = \mathbb{E}\left[\mathbb{E}[z^N \mid \lambda]\right] = \mathbb{E}[\exp(\lambda(z-1))] = M_\lambda(z-1), \quad (3.21)$$

so that the probability generating function $G_N(z)$ of the mixed Poisson is equal to the moment generating function M_λ of the mixing distribution, evaluated at $z - 1$. In the following example we consider a specific choice for the distribution of λ (see also §2.3).

Example 3.16   Suppose that $N \mid \lambda$ has a Poisson distribution with mean λ and that λ is exponentially distributed with mean $1/\nu$ ($\nu > 0$). Using (3.21), we obtain

$$G_N(z) = \frac{\nu}{\nu - (z - 1)} = \frac{\dfrac{\nu}{\nu + 1}}{1 - \dfrac{z}{\nu + 1}}.$$

From (2.5) we recognise this as the probability generating function of a geometric distribution with success probability $p = \nu/(\nu + 1)$, so that

$$\Pr(N = n) = \frac{\nu}{(\nu + 1)^{n+1}}, \qquad n = 0, 1, 2, \ldots.$$

The mean and variance of this geometric distribution are $(1 - p)/p = 1/\nu$ and $(1 - p)/(p^2) = (\nu + 1)/(\nu^2)$. We see that

$$\mathrm{Var}[N] = \frac{\nu + 1}{\nu^2} = \frac{\nu + 1}{\nu} \mathbb{E}[N] > \mathbb{E}[N],$$

and hence this mixed Poisson distribution has its variance greater than its mean.

When N is as in Example 3.16, then the mixed Poisson is a geometric distribution (a special case of the negative binomial distribution), and so the compound mixed Poisson distribution is, in this case, the same as a compound geometric distribution.

Formulae, such as those for the mean, variance, moment generating function, etc. for compound mixed Poisson distributions may be obtained by substituting the above formulae for N-quantities into the appropriate random sum formulae. In particular, from (3.1) and (3.2), we find that the mean and variance of the compound mixed Poisson random variable $S = X_1 + \cdots + X_N$ are

$$\mathbb{E}[S] = \mathbb{E}[N]\mathbb{E}[X_1] = \mathbb{E}[\lambda]\mathbb{E}[X_1]$$

and

$$\mathrm{Var}[S] = \mathbb{E}[N]\,\mathrm{Var}[X_1] + \mathrm{Var}[N]\left(\mathbb{E}[X_1^2]\right)^2$$
$$= \mathbb{E}[\lambda]\,\mathrm{Var}[X_1] + (\mathbb{E}[\lambda] + \mathrm{Var}[\lambda])\,(\mathbb{E}[X_1])^2.$$

3.4.3 Compound negative binomial distributions

Recall from §2.1.2 that the random variable N has a negative binomial distribution, $N \sim \text{nb}(k, p)$, if it has probability mass function

$$\Pr(N = n) = \frac{\Gamma(k + n)}{\Gamma(n + 1)\Gamma(k)} q^n p^k, \quad n = 0, 1, \ldots,$$

where $0 < p = 1 - q < 1$ and $k > 0$. The probability generating function of N is (see (2.3))

$$G_N(z) = \left(\frac{p}{1 - qz}\right)^k, \quad |z| < \frac{1}{q}.$$

A related random variable is defined by $M = N + k$, so that M takes values in $\{k, k + 1, \ldots\}$. Then M is a shifted version of N, with

$$\Pr(M = n) = \Pr(N = n - k) = \frac{\Gamma(n)}{\Gamma(n - k + 1)\Gamma(k)} q^{n-k} p^k, \quad n = k, k + 1, \ldots.$$
$$(3.22)$$

In the literature, this distribution is sometimes also called the negative binomial distribution (so care must be taken when looking up the negative binomial distributions in books to make sure that it is clear which of these two distributions is intended). We will use the notation $M \sim \widetilde{\text{nb}}(k, p)$ to denote this shifted negative binomial distribution. The probability generating function of M is

$$G_M(z) = \left(\frac{pz}{1 - qz}\right)^k, \quad |z| < \frac{1}{q}.$$

When $k = 1$, $N \sim \text{nb}(1, p)$ is a geometric $\text{geo}(p)$ distribution as in §2.1.3. When $k = 1$ in (3.22), the random variable M has a shifted geometric distribution, with probability mass function

$$\Pr(M = n) = q^{n-1} p, \quad n = 1, 2, \ldots.$$

This distribution is sometimes also called a geometric distribution in the literature, and we write $\widetilde{\text{geo}}(p)$ for this distribution.

Suppose that a random sum $S = X_1 + \cdots + X_N$ has counting random variable $N \sim \text{nb}(k, p)$. Then the random sum S is said to have a *compound negative binomial* distribution. If $k = 1$, then S is said to have a *compound geometric* distribution. The terms "compound negative binomial" and "compound geometric" are also sometimes used in the literature when the counting random variable has a $\widetilde{\text{nb}}(k, p)$ and $\widetilde{\text{geo}}(p)$ distribution, respectively. In this book, we will specify the counting distribution explicitly to avoid any possible confusion.

From (3.1) and (3.2), the mean and variance of the compound nb(k, p) random variable S are given by

$$\mathbb{E}[S] = \mathbb{E}[N]\mathbb{E}[X_1] = \frac{kq}{p}\mathbb{E}[X_1]$$

and

$$\begin{aligned}
\text{Var}[S] &= \mathbb{E}[N]\,\text{Var}[X_1] + \text{Var}[N]\,(\mathbb{E}[X_1])^2 \\
&= \frac{kq}{p}\,\text{Var}[X_1] + \frac{kq}{p^2}\,(\mathbb{E}[X_1])^2 \\
&= \frac{kq}{p}\mathbb{E}[X_1^2] + \frac{kq^2}{p^2}\,(\mathbb{E}[X_1])^2.
\end{aligned}$$

For the compound $\widetilde{\text{nb}}(k, p)$ case, the mean and variance of $\widetilde{S} = X_1 + \cdots + X_M$, where $M \sim \widetilde{\text{nb}}(k, p)$, are

$$\mathbb{E}[\widetilde{S}] = \frac{k}{p}\mathbb{E}[X_1]$$

and

$$\text{Var}[\widetilde{S}] = \frac{k}{p}\,\text{Var}[X_1] + \frac{kq}{p^2}\,(\mathbb{E}[X_1])^2 = \frac{k}{p}\mathbb{E}[X_1^2] + \frac{k(q-p)}{p^2}\,(\mathbb{E}[X_1])^2.$$

If the step distribution has moment generating function $M_X(r)$, then, from (3.8), the moment generating function of $S = X_1 + \cdots + X_N$ is $G_N(M_X(r))$, so that the compound nb(k, p) distribution has a moment generating function given by

$$\left(\frac{p}{1 - qM_X(r)}\right)^k. \tag{3.23}$$

Similarly, the moment generating function of a compound $\widetilde{\text{nb}}(k, p)$ distribution is given by

$$\left(\frac{pM_X(r)}{1 - qM_X(r)}\right)^k. \tag{3.24}$$

The simplest negative binomial is the geometric, when $k = 1$. For certain step distributions, there are easy explicit expressions for the resulting compound geometric distributions. Example 3.17 derives such an expression from the convolution series formula for F_S given in (3.7), whereas Example 3.18 derives an expression from the moment generating function formula in (3.8).

Example 3.17 Suppose that the counting random variable has a $\widetilde{\text{geo}}(p)$ distribution with $\text{Pr}(N = n) = q^{n-1}p$, $n = 1, 2, \ldots$, where $q = 1 - p$, so that $N \sim \widetilde{\text{nb}}(1, p)$, and suppose that X_1 has an exponential distribution with mean

μ. Since $\Pr(N = 0) = 0$ and $\Pr(X_1 = 0) = 0$, the compound distribution does not have an atom at zero, and consists solely of a distribution with a density f_S. From (3.7), and using the fact that the n-fold convolution of an exponential distribution with mean μ is a gamma distribution with parameters n and $1/\mu$ (as discussed in §2.2.4), we have

$$
\begin{aligned}
f_S(x) &= \sum_{n=1}^{\infty} \frac{x^{n-1} e^{-x/\mu}}{\mu^n (n-1)!} q^{n-1} p \\
&= \frac{p e^{-x/\mu}}{\mu} \sum_{n=1}^{\infty} \frac{1}{(n-1)!} \left(\frac{qx}{\mu} \right)^{n-1} \\
&= \frac{p}{\mu} e^{-px/\mu}.
\end{aligned}
$$

Thus, in this case, the random sum S has an exponential distribution with mean μ/p.

Example 3.18 Suppose that X_1 has an exponential distribution with mean μ, as in the previous example, so that its moment generating function is $M_X(r) = (1 - \mu r)^{-1}$. Suppose that N has a geometric distribution, where

$$
\Pr(N = n) = q^n p, \quad n = 0, 1, 2, \ldots,
$$

where $q = 1 - p$, so that $N \sim \text{nb}(1, p)$. We aim to find the distribution of S by substituting for G_N and M_X in (3.8) and then manipulating the resulting expression to obtain a moment generating function that we recognise. From (3.8), the compound distribution has a moment generating function

$$
M_S(r) = G_N[M_X(r)] = \frac{p}{1 - q \dfrac{1}{1 - \mu r}}.
$$

This expression may be rearranged as follows.

$$
\frac{p}{1 - q \dfrac{1}{1 - \mu r}} = \frac{p(1 - \mu r)}{p - \mu r} = p + q \frac{p}{p - \mu r} = p + q \frac{1}{1 - \dfrac{\mu}{p} r}.
$$

Using (3.12), we see that this is the moment generating function of a distribution that is a mixture of a distribution with moment generating function 1 and a distribution with moment generating function $[1 - r(\mu/p)]^{-1}$, with mixing proportions p and q, respectively. We recognise $[1 - r(\mu/p)]^{-1}$ as the moment generating function of an exponential random variable, W say, with mean μ/p and distribution function $F_W(x) = 1 - \exp[-x(p/\mu)]$, $x > 0$. For the first component of the mixture, we observe from (3.14) that 1 is the moment generating function of a random variable, Y say, that takes the value 0 with probability 1, and $F_Y(x) = 1_{[0,\infty)}(x)$. Hence the random sum S has distribution function

$$F_S(x) = pF_Y(x) + qF_W(x). \qquad (3.25)$$

From (3.25), we find

$$F_S(x) = \begin{cases} 0 & \text{if } x < 0 \\ 1 - q\exp(-\frac{p}{\mu}x) & \text{if } x \geq 0. \end{cases} \qquad (3.26)$$

The result in Example 3.18 can also be found via the infinite convolution power series in (3.7), and the result in Example 3.17 can be obtained via moment generating functions. It is a good exercise to check that you can do these for yourself (see Exercises 3.11 and 3.12).

These methods can be extended relatively easily to deal with compound negative binomials with $k = 2$ or with gamma step distributions with shape parameter $\alpha = 2$ (or both). The next example shows one way to approach the $k = 2$ case.

Example 3.19 Suppose that $N \sim \text{nb}(2, p)$, $0 < p = 1 - q < 1$, and that X_1 has an exponential distribution with mean μ. Then, from (3.8), the random sum S has moment generating function

$$M_S(r) = \left(\frac{p}{1 - q\dfrac{1}{1 - \mu r}} \right)^2.$$

From Example 3.18, the expression inside the brackets can be written

$$p + q\frac{1}{1 - \dfrac{\mu}{p}r},$$

so that

$$M_S(r) = \left(p + q\frac{1}{1 - \dfrac{\mu}{p}r} \right)^2$$

$$= p^2 + 2pq\frac{1}{1 - \dfrac{\mu}{p}r} + q^2\left(\frac{1}{1 - \dfrac{\mu}{p}r} \right)^2. \qquad (3.27)$$

We recognise $[1 - r(\mu/p)]^{-2}$ as the moment generating function of a gamma$(2, p/\mu)$ distribution with density $(p/\mu)^2 x e^{-px/\mu}$. Hence (3.27) is the moment generating function of a three-component mixture consisting of a unit mass at zero (with mixing proportion p^2), an exponential distribution with mean μ/p (with mixing proportion $2pq$) and a gamma$(2, p/\mu)$ distribution (with mixing proportion q^2). Then we have, for $x > 0$,

$$\Pr(S > x) = 2pqe^{-px/\mu} + q^2 \int_x^\infty \frac{p^2 t e^{-pt/\mu}}{\mu^2}\, dt$$

$$= \left(q^2(px + \mu) + 2pq\mu\right) \frac{e^{-px/\mu}}{\mu}. \tag{3.28}$$

3.4.4 Compound binomial distributions

Consider a group life insurance policy covering m lives, where, in a year, there is at most one claim on each life. If we assume that the probability of a claim is the same value, p say, for each life, and that the m lives are independent, then the number N of claims on the whole policy in one year has a binomial bi(m, p) distribution. If the sizes of the claims are iid random variables, independent of N, then the total amount S claimed on this policy in one year has a *compound binomial* distribution.

From (3.1) the mean of a compound binomial random variable S, with counting random variable $N \sim \text{bi}(m, p)$ and step random variables $X_1, X_2, \ldots$, is

$$\mathbb{E}[S] = \mathbb{E}[N]\mathbb{E}[X_1] = mp\mathbb{E}[X_1],$$

and, from (3.2), the variance of S is (with $q = 1 - p$)

$$\begin{aligned}
\text{Var}[S] &= \mathbb{E}[N]\,\text{Var}[X_1] + \text{Var}[N]\,(\mathbb{E}[X_1])^2 \\
&= mp\,\text{Var}[X_1] + mpq\,(\mathbb{E}[X_1])^2 \\
&= mp\mathbb{E}[X_1^2] - mp^2\,(\mathbb{E}[X_1])^2.
\end{aligned}$$

Using (3.8) and the binomial probability generating function in (2.7), we find that the moment generating function of a compound binomial distribution is

$$M_S(r) = G_N(M_X(r)) = (q + pM_X(r))^m,$$

where $M_X(r)$ is the moment generating function of the step random variables $X_1, X_2, \ldots$.

The next example shows how a particular compound binomial distribution is the same as a particular compound negative binomial distribution, and illustrates (again) the value of developing the facility of working with generating functions.

Example 3.20 Suppose that N has a binomial distribution $N \sim \text{bi}(m, p)$, with $0 < p = 1 - q < 1$, and that X_1 has an exponential distribution with mean μ. Let G_N be the probability generating function of N and let M_X be the moment generating function of X_1. Then the resulting compound binomial distribution has a moment generating function

$$M_S(r) = G_N(M_X(r)) = \left(q + p\frac{1}{1 - \mu r}\right)^m.$$

Reversing the argument in Example 3.18, we find

$$\begin{aligned}
q + p\frac{1}{1 - \mu r} &= \frac{q(1 - \mu r) + p}{1 - \mu r} \\
&= \frac{1 - q\mu r}{1 - \mu r} \\
&= q\left(\frac{q(1 - \mu r)}{1 - q\mu r}\right)^{-1} \\
&= q\left(1 - \frac{p}{1 - q\mu r}\right)^{-1}.
\end{aligned}$$

Hence the moment generating function of the random sum S is

$$M_S(r) = \left(\frac{q}{1 - p\dfrac{1}{1 - q\mu r}}\right)^m,$$

and, from (3.23), we recognise this as belonging to a compound negative binomial $\text{nb}(m, q)$ distribution with steps exponentially distributed with mean $q\mu$.

3.5 Numerical methods for compound distributions

In general, the distribution of the random sum $S = X_1 + \cdots + X_N$ is determined when we know the distributions of the counting random variable N and the step random variable X_1. In several of the examples in §3.4, it was possible to find simple explicit expressions for the compound distributions resulting from specific choices for the counting and step distributions. However, this is not usually feasible except for certain special cases, and in the general case we must resort to numerical methods or to approximations.

In this section, we consider the problem of numerically evaluating the distribution of S, given particular known distributions for N and X_1. We consider two numerical methods, the Panjer recursion algorithm and the fast Fourier transform algorithm, for calculating the compound distribution on a computer. These two methods are illustrated and compared in Example 3.22. In both cases, probability distributions are stored in one-dimensional arrays, and so we work with discrete distributions for N and X_1 (and so also for S).

3.5.1 Panjer recursion algorithm

This popular and much-used algorithm was proposed by Panjer (1981), and can be applied when the counting distribution satisfies a certain condition as follows. Let $p_n = \Pr(N = n)$, $n = 0, 1, 2, \ldots$, and suppose that the p_n satisfy *Panjer's recursion formula*,

$$p_n = \left(a + \frac{b}{n}\right) p_{n-1}, \qquad n = 1, 2, \ldots, \tag{3.29}$$

for some a and b. Many commonly used counting distributions satisfy this formula. For example, if N has a Poisson distribution with mean λ, then, for $n \geq 1$,

$$p_n = \frac{e^{-\lambda} \lambda^n}{n!} = \frac{\lambda}{n} p_{n-1},$$

so that (3.29) holds with $a = 0$ and $b = \lambda$. Check for yourself that (3.29) also holds for the binomial distribution $\text{bi}(m, p)$ (with $a = -p/(1 - p)$ and $b = (m + 1)p/(1 - p)$) and for the negative binomial distribution $\text{nb}(k, p)$ (with $a = 1 - p$ and $b = (k - 1)(1 - p)$). Not all distributions satisfy this recursion, for example the discrete uniform and the $\widetilde{\text{nb}}(k, p)$ negative binomial case, although sometimes the recursion algorithm can be generalised to cover these cases (see sect. 4.6 in Klugman *et al.* (1998)).

Assume that X_1 takes values in $\{1, 2, \ldots\}$ and write $f_k = \Pr(X_1 = k)$, $k = 1, 2, \ldots$, so that X_1 is a positive discrete random variable (see the discussion after the proof of the theorem below for how to proceed if $\Pr(X_1 = 0) > 0$ or if X_1 is not discrete).

We assume that the counting and step distributions are known, i.e. that a, b and $\{f_k\}_{k=1}^{\infty}$ are known. The resulting random sum is concentrated on $\{0, 1, 2, \ldots\}$ (because $N \geq 0$, $X_1 > 0$ and X_1 is concentrated on $\{1, 2, \ldots\}$), and we write $g_r = \Pr(S = r)$, $r = 0, 1, 2, \ldots$. The g_r are unknown, and the next result gives a recursion by which they may be calculated sequentially.

Theorem 3.21 *With notation as above, assume the p_n satisfy (3.29). Then the g_r satisfy*

(i) $g_0 = p_0$,

(ii) $g_r = \sum_{j=1}^{r} \left(a + \dfrac{bj}{r} \right) f_j g_{r-j}$, $r = 1, 2, \ldots$ (3.30)

Before proceeding with the proof, note that the right-hand side of (ii) in (3.30) involves $g_{r-1}, g_{r-2}, \ldots, g_0$, so that (ii) does indeed give g_r in terms of known quantities and of $g_0, g_1, \ldots, g_{r-1}$. Hence we can use (3.30) to find, first of all, g_0, then g_1 in terms of g_0, then g_2 in terms of g_1 and g_0, etc., i.e. we can use (3.30) to find each g_r in turn.

Proof   Since $\Pr(X_1 \leq 0) = 0$, the discussion below (3.6) yields $g_0 = \Pr(S = 0) = \Pr(N = 0) = p_0$, which gives (i).

For (ii), we first show that the probability generating function $G_N(z)$ of N satisfies

$$G_N'(z) = \frac{a+b}{1-az} G_N(z).$$ (3.31)

To see this, use (3.29) to obtain

$$G_N(z) - p_0 = \sum_{n=1}^{\infty} p_n z^n$$

$$= \sum_{n=1}^{\infty} z^n \left(a + \frac{b}{n} \right) p_{n-1}$$

$$= az G_N(z) + b \sum_{n=1}^{\infty} \frac{z^n}{n} p_{n-1},$$

which gives

$$(1 - az) G_N(z) = p_0 + b \sum_{n=1}^{\infty} \frac{z^n}{n} p_{n-1}.$$

Differentiate this with respect to z to obtain

$$-a G_N(z) + (1 - az) G_N'(z) = b G_N(z),$$

and this yields (3.31).

We now use (3.31) to find an expression for the probability generating function $G_S(z)$ of S. First note that from (3.9) we have $G_S(z) = G_N[G_X(z)]$, where $G_X(z)$ is the probability generating function of X_1. Differentiating this relationship with respect to z, and then substituting (3.31), we find

$$G_S'(z) = G_N'[G_X(z)] G_X'(z)$$

$$= \frac{a+b}{1 - a G_X(z)} G_N[G_X(z)] G_X'(z).$$

Hence

$$(1 - aG_X(z))\, G'_S(z) = (a+b)G_S(z)G'_X(z),$$

which, in terms of series, states that

$$\left(1 - a\sum_{k=1}^{\infty} f_k z^k\right)\left(\sum_{k=1}^{\infty} kg_k z^{k-1}\right) = (a+b)\left(\sum_{k=0}^{\infty} g_k z^k\right)\left(\sum_{k=1}^{\infty} kf_k z^{k-1}\right).$$

Equate coefficients of z^{r-1} on the left- and right-hand sides of this equation to obtain

$$rg_r - a\sum_{j=1}^{r-1} f_j(r-j)g_{r-j} = (a+b)\sum_{j=1}^{r} jf_j g_{r-j},$$

so that

$$\begin{aligned}
rg_r &= \sum_{j=1}^{r-1}(ar - aj)f_j g_{r-j} + \sum_{j=1}^{r}(aj + bj)f_j g_{r-j} \\
&= \sum_{j=1}^{r-1}(ar + bj)f_j g_{r-j} + (ar + br)f_r g_0 \\
&= \sum_{j=1}^{r}(ar + bj)f_j g_{r-j}.
\end{aligned}$$

Hence, on dividing by r we get

$$g_r = \sum_{j=1}^{r}\left(a + \frac{bj}{r}\right)f_j g_{r-j},$$

as required. $\square$

If X_1 has an atom at zero, so that $f_0 = \Pr(X_1 = 0) > 0$, then we obtain the corresponding recursion

$$\text{(i)} \quad g_0 = \sum_{n=0}^{\infty} p_n f_0^n,$$

$$\text{(ii)} \quad g_r = \frac{1}{1 - af_0}\sum_{j=1}^{r}\left(a + \frac{bj}{r}\right)f_j g_{r-j}, \quad r \geq 1 \tag{3.32}$$

(see Exercise 3.15). Note that this yields (3.30) when $f_0 = 0$.

If X_1 is not a discrete random variable, then we can still use the Panjer recursion algorithm if we first approximate the distribution of X_1 by a discrete distribution. One way to do this is to choose $h > 0$ and put

$$f_k = \Pr\big(X_1 \in ((k - 0.5)h, (k + 0.5)h]\big) \text{ for } k = 0, 1, 2, \ldots.$$

Here f_k is the mass given by the distribution of X_1 to the interval of width h centred on kh. This approximation is better if we choose a small value for h, so that the discretisation error is small. If we discretise in this way, then g_r given by the Panjer recursion algorithm will be the mass given to rh by the compound distribution belonging to N and the discrete distribution that is being used as an approximation to the true distribution of X_1. Thus g_r is an approximation to the probability that the random sum $S = X_1 + \cdots + X_N$ is in $((r-0.5)h, (r+0.5)h]$, so that $\{g_r\}_{r=0}^{\infty}$ is an approximation to the distribution of the true S. This method of discretisation is illustrated in Example 3.22 in Section 3.5.2.

3.5.2 The fast Fourier transform algorithm

Recall that the characteristic function of the random variable S is $\phi_S(\theta) = \mathbb{E}[e^{i\theta S}] = \int e^{i\theta x} F_S(dx)$. From (3.10), when $S = X_1 + \cdots + X_N$, its characteristic function $\phi_S(\theta)$ satisfies

$$\phi_S(\theta) = G_N[\phi_X(\theta)], \tag{3.33}$$

where ϕ_X is the characteristic function of X_1. This means that, given G_N and ϕ_X, we can find the Fourier transform of the distribution of S.

In terms of numerical approximations, working with Fourier transforms is relatively easy because of the wide availability of fast Fourier transform routines in software packages. In this subsection, we outline how to use the fast Fourier transform algorithm to calculate numerical approximations to F_S.

As we did for the Panjer recursion, we work with discrete random variables, discretising the distribution of X_1 if necessary, possibly as described in §3.5.1. The fast Fourier transform (FFT) algorithm takes as input a finite $1 \times m$-dimensional array $a = (a_0, \ldots, a_{m-1})$. The discrete Fourier transform of a is the function

$$\hat{a}(\theta) = \sum_{k=0}^{m-1} a_k e^{i\theta k}.$$

At first sight, it may seem necessary to store the function $\hat{a}(\theta)$ for all $\theta \in \mathbb{R}$ in order to keep complete information on a. However, it turns out that we can recover a completely from the m complex values $\hat{a}(\theta_0), \ldots, \hat{a}(\theta_{m-1})$, where $\theta_j = 2\pi j/m$, $j = 0, \ldots, m-1$, are called the *Fourier frequencies*. Hence

$$\hat{a}(\theta_j) = \sum_{k=0}^{m-1} a_k e^{i2\pi jk/m}. \tag{3.34}$$

The "recovery" of a from these m values of the transform at the Fourier frequencies is given by the inversion formula

$$a_k = \frac{1}{m} \sum_{j=0}^{m-1} \hat{a}(\theta_j) e^{-i2\pi kj/m}; \qquad (3.35)$$

see Exercise 3.16.

Hence, starting with the $1 \times m$-dimensional array $a = (a_0, \ldots, a_{m-1})$, the FFT returns the $1 \times m$-dimensional array $\hat{a} = (\hat{a}(\theta_0), \ldots, \hat{a}(\theta_{m-1}))$ (containing complex values). The inverse FFT starts with $\hat{a}$ and returns a. Comparison of (3.34) and (3.35) shows that the direct and inverse calculations are essentially of the same type, differing only in the presence or absence of an initial constant and the presence or absence of a minus sign in the exponent. The FFT algorithm exploits this, and makes use of various techniques to carry out the calculations efficiently. Naive calculation of (3.34) involves of the order of m^2 operations, whereas it turns out that if m is a power of 2, then the FFT algorithm uses approximately $m \log_2 m$ operations. We do not give the details of the algorithm here. Interested readers should see Brigham (1974) and Grübel (1989).

The FFT algorithm may be applied to probability distributions, since, if Y is a random variable whose distribution is concentrated on $\{0, 1, \ldots, m-1\}$ with $p_k = \Pr(Y = k)$, then $\hat{p}(\theta)$ is the characteristic function $\phi_Y(\theta)$ of Y.

We now describe how to apply the FFT algorithm to compound distributions. Our input ($1 \times m$-dimensional) array holds the distribution of X_1, which for now is assumed to be discrete and such that $\Pr(X_1 \geq m) = 0$. The algorithm returns an output ($1 \times m$-dimensional) array holding the distribution of S. Hence m must be big enough to hold (the bulk of) the distribution of S, and in addition m should be a power of 2 for efficient calculation. Let $f_k = \Pr(X_1 = k)$. Use the FFT algorithm to find $\phi_X(\theta_j)$, and then calculate $G_N(\phi_X(\theta_j)) = \phi_S(\theta_j)$, $j = 0, \ldots, m-1$. Finally, use the inverse FFT algorithm to find $g_k = \Pr(S = k)$, $k = 0, \ldots, m-1$.

As an example, we show how easily this can be carried out in a software package that contains a fast Fourier transform algorithm. We use the package **R**. Suppose we want to calculate the g_k for a compound Poisson distribution with known Poisson parameter λ and a particular claim-size distribution with known f_k. Suppose that the value of λ is stored in lambda, and that f is the $1 \times m$-dimensional array containing the values of $f_0, \ldots, f_{m-1}$. We aim to calculate the corresponding g_k, and we suppose that they will be stored in an array g. We use the transform relationship $\phi_S(\theta) = \exp(\lambda(\phi_X(\theta) - 1))$ to evaluate g very simply, in just one line in **R**:

```
g = Re(fft(exp(lambda*(fft(f) -1)),inverse=T)/m)
```

In **R**, the discrete Fourier transform of the array f is achieved using fft(f), and the inverse is achieved with the argument inverse=T. In **R**, this inverse transformation step does not divide by the array length m, which we require in (3.35), so we include this division by m explicitly in the **R** directive. Finally we take the real part of the inverse, because the g_k are real, and this makes for a neater output. If using a different package, care has to be taken because it may differ from **R** in the inclusion or otherwise of the constant in the relationships giving the direct and inverse transforms. It is important to get the constant in the right place, as we need the direct transform step in (3.34) to correspond to finding the characteristic function of the claim-size distribution.

In many applications, the distribution of X_1 is neither discrete nor concentrated on $\{0, \ldots, m - 1\}$ for any m. In these cases, before the FFT algorithm can be used, we must discretise and truncate the distribution of X_1. To do this, choose m (a power of 2) and h, and put $f_k = \Pr(X_1 \in ((k - 0.5)h, (k + 0.5)h])$ $k = 0, \ldots, m - 1$, and use this as an approximation to the distribution of X_1. Here, h needs to be small to make the discretisation error small and m needs to be large enough so that $[0, (m - 0.5)h]$ holds most of the distributions of X_1 and S. After applying the FFT algorithm, the resulting g_k is an approximation to $\Pr(S \in ((k - 0.5)h, (k + 0.5)h])$.

Example 3.22 We now illustrate the practical application of Panjer recursion and the FFT algorithm in finding the distribution of a particular random sum. We consider Example 3.18 again, where $\Pr(N = n) = q^n p, n = 0, 1, 2, \ldots$, with $0 < p = 1 - q < 1$, and where X_1 is exponentially distributed with mean μ. From Example 3.18, we already know that the distribution of the resulting random sum S is a mixture of a unit mass at zero and an exponential distribution with mean μ/p, with mixing proportions p and q, respectively, so in real applications we would not need to use numerical methods to find the distribution of this particular S. However, the fact that the true distribution of S is known means that this is a good example for assessing the performance of the two numerical procedures, as we do here.

For both calculation methods, we need to have discrete step distributions, so we first discretise the distribution of X_1. If we follow the method suggested after the proof of Theorem 3.21, then, for a given discretisation parameter h, we set

$$f_0 = \Pr(0 < X_1 \le h/2) = 1 - \exp\left(-\frac{h}{2\mu}\right)$$

and, for $k = 1, 2, \ldots,$

$$f_k = \Pr((k - 0.5)h < X_1 \leq (k + 0.5)h)$$
$$= \exp(-(k - 0.5)h/\mu)(1 - \exp(-h/\mu)).$$

Check that $\sum_{k=0}^{\infty} f_k = 1$. Then $(f_0, f_1, f_2, \ldots)$ is a discrete approximation to the distribution of X_1.

Panjer recursion algorithm

The distribution of N satisfies the Panjer recursion formula (3.29) with $a = q$ and $b = 0$, so we can use the Panjer recursion algorithm. Since $f_0 > 0$ we use the form of the recursion given in (3.32) and code this on a computer. We assume that the user specifies the values of p and μ, together with the values of the discretisation parameter h and of the number m specifying the length of the calculated $1 \times m$-dimensional array $(g_0, g_1, \ldots, g_{m-1})$, where g_r is our numerical approximation to $\Pr((r - 0.5)h < S \leq (r + 0.5)h)$. We see that

$$g_0 = \sum_{n=0}^{\infty} q^n p f_0^n = \frac{p}{1 - q f_0}.$$

We combine this with (3.32) to obtain a procedure for finding the $1 \times m$-dimensional array g containing the values g_0, ..., g_(m-1) of $g_0, \ldots, g_{m-1}$. Let f be the $1 \times m$-dimensional array containing the values f_0, ..., f_(m-1) of $f_0, \ldots, f_{m-1}$, and let p and q contain the values of p and q, respectively, all assumed to be known. In order to apply the Panjer recursion formula, we first find the value of g_0 and then find successive g_k recursively. *R* directives to carry this out are shown below:

```
g = rep(0,m)
g[1] = p/(1-q*f[1])
i = 1
while (i<=m-1){
g[i+1] = q*(f[2:(i+1)]%*%g[i:1])/(1-q*f[1])
i = i+1
}
```

Here f[j] is the *j*th entry in the array f, f[i:j] is the array $(f_{i-1}, \ldots, f_{j-1})$, and the operation %*% is a scalar product.

Note that rather than predetermining the number m of g_r to be calculated and finding f before the loop, we could instead evaluate each f_k as needed inside the loop and include a condition for stopping.

The numerical results are contained in Table 3.2 for the case where $\mu = 1$ (without loss of generality, we can work on a monetary scale where the unit is

Table 3.2. *Discretised approximations for the geometric/exponential example*

x	True	Panjer	FFT $m = 4096$	FFT $m = 8192$
0.00	0.091735162	0.091738925	0.091739889	0.091738926
0.02	0.001649890	0.001649904	0.001650866	0.001649904
0.04	0.001646893	0.001646907	0.001647867	0.001646907
0.06	0.001643902	0.001643915	0.001644874	0.001643916
0.08	0.001640915	0.001640929	0.001641886	0.001640929
⋮	⋮	⋮	⋮	⋮
9.98	0.0006671464	0.0006671444	0.0006675336	0.0006671446
10.00	0.0006659345	0.0006659325	0.0006663210	0.0006659327
10.02	0.0006647248	0.0006647228	0.0006651105	0.0006647230
⋮	⋮	⋮	⋮	⋮
64.76	4.586006e-06	4.585709e-06	4.588384e-06	4.585711e-06
64.78	4.577675e-06	4.577379e-06	4.580049e-06	4.577381e-06

equal to the mean size of a claim) and $p = 1/11$, so that $\mathbb{E}[N] = 10$, a small mean number of claims but useful for the purposes of this illustrative example. The discretisation parameter is $h = 0.02$. In the column labelled "true", we use Example 3.18 to calculate $\Pr(x - 0.5h < S \leq x + 0.5h)$, i.e. the discretised values of the true distribution of S, where $x = kh$ for some non-negative integer k. Results are shown in the table for a few small values of x, then for a few values of x in the region of $\mathbb{E}[S] = \mu_S = 10$, and finally for a few values of x in the region of $\mu_S + 5\sqrt{\text{Var}[S]} = 64.77$. In the column labelled "Panjer" the table also shows the corresponding values from the Panjer recursion algorithm.

The fast Fourier transform algorithm

We work with $(f_0, \ldots, f_{m-1})$ as for the Panjer recursion, with parameters h and m, making sure that m is a power of 2 for efficient use of the FFT algorithm. For the geometric/exponential example above, (3.33) becomes

$$\phi_S(\theta) = \frac{p}{1 - q\phi_X(\theta)}.$$

Assume that the values of p, μ, m and h are specified, and that the $1 \times m$-dimensional array f contains the values f_0, ..., f_(m-1) of $f_0, \ldots, f_{m-1}$. Then we use the FFT algorithm in $\boldsymbol{R}$ as follows to find the $1 \times m$-dimensional array g:

```
g = Re(fft(p*(1-q*fft(f))^{-1},inverse=T)/m)
```

Table 3.3. *Maximum errors for the geometric/exponential example*

	Panjer	FFT $m = 4096$	FFT $m = 8192$
Max absolute error	3.76×10^{-6}	4.73×10^{-6}	3.76×10^{-6}
Max percentage error	0.0084	0.059	0.0084

where $^$(-1) denotes the function that raises to the power -1. We run this for the same choices of p and μ as for the Panjer recursion algorithm, and with the same choice of discretisation parameter $h = 0.02$. For the choice of m, note that we need to make sure that the bulk of the distribution of S is contained on $[0, (m - 0.5)h)$. Preliminary calculations show that $\mathbb{E}[S] = 10$ and $\text{Var}[S] = 120$. If we choose $m = 4096 = 2^{12}$ then $(m - 0.5)h = 81.91$, which is more than six standard deviations above the mean of S. We first try the FFT algorithm with this choice of m, and repeat it with $m = 8192 = 2^{13}$ to see whether the resulting g-values are much affected. The results are in Table 3.2. We also note that the maximum absolute errors and the maximum percentage errors are as in Table 3.3.

In summary, for this example, with $m = 8192$ for the FFT algorithm, the results are similar for both the Panjer and the FFT algorithms, although when the FFT algorithm is used with $m = 4096$ it is not as accurate as the Panjer algorithm. It should be noted that, even with today's fast computers, the time taken for the Panjer algorithm is noticeably longer than that for the FFT algorithm, even when $m = 8192$. This is not important for the values given above, but we see that if, for example, we were carrying out a procedure where many thousands of such calculations are required, then these small differences in time become important. On the other hand, if interest is only in the values of g_r for not too large r, then the Panjer algorithm can be used just up to the required values, whereas the FFT algorithm would still need a large value of m to achieve accuracy in the low r-values. However, in many cases, it is the upper tail probabilities of S, and hence large r-values, that are of interest.

3.6 Approximations for compound distributions

There are various approximation formulae for the distribution of a random sum $S = \sum_{i=1}^{N} X_i$, and we discuss a few of them here. Although modern computing power means that numerical evaluation of compound distributions is much faster than it used to be, nevertheless quick and easy approximations may still be of some use, and, for example, asymptotic approximations can be helpful in showing how the tail of a compound distribution function depends on various parameters in the limit.

3.6.1 Approximations based on a few moments

Assume that N and X_1 have finite second moments, so that $S = X_1 + \cdots + X_N$ has finite mean μ_S and variance σ_S^2. A very simple approximation, based only on these two summary numbers, is the *normal approximation*, where we approximate the distribution of S by a $N(\mu_S, \sigma_S^2)$ distribution and $Pr(S \leq x)$ by $\Phi((x - \mu_S)/\sigma_S)$, where Φ is the standard normal distribution function. These probabilities are very easy to find, using, for example, standard normal tables, a calculator or a computer. However, the normal approximation is necessarily symmetric, and is likely to be rather crude as an approximation to the distribution of S, which typically is skew.

Motivated by the above drawback, but still aiming to have a simple approximation based on a few moments of S, we consider the *translated gamma approximation*, which is defined as follows. If Y has a gamma(α, ν) distribution and if k is a constant, then the random variable $k + Y$ is said to have a translated gamma distribution. We assume that S has a finite third moment and we let $\beta_S = \mathbb{E}[(S - \mu_S)^3]/\sigma_S^3$ be the coefficient of skewness of S. For the approximation, we choose k, α and ν so that the first three moments (or, equivalently, the first three cumulants) of $k + Y$ match those of S. Then the translated gamma approximation to the distribution of S has parameters

$$k = \mu_S - \frac{2\sigma_S}{|\beta_S|},$$

$$\alpha = \frac{4}{\beta_S^2},$$

$$\nu = \frac{2}{|\beta_S|\sigma_S}.$$

See Exercise 3.20, where you are asked to find these for yourself. Figure 3.2 compares the normal and translated gamma approximations to the true $Pr(S > x)$ when X_1 is exponentially distributed with mean 2, and $Pr(N = n) = (1 - p)^n p$, $n = 0, 1, 2, \ldots$, with $p = 0.1$. From the figure, we see that the translated gamma (which is very close to the true distribution function tail) is an improvement over the normal approximation. (See Exercise 3.21 to carry out the details of this approximation for this example.) It might be argued that we expect a gamma distribution function tail to be good for a compound

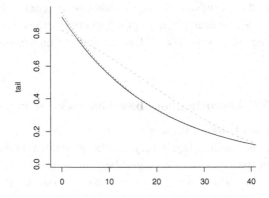

Figure 3.2. True (solid line), normal approximation (broken line) and translated gamma approximation (dotted line) for the tail of the compound geometric distribution function with $p = 0.1$ and X_1 exponentially distributed with mean 2.

geometric/exponential distribution, where the true tail of the distribution function of S has an exponential decay (see Example 3.18). You are asked to investigate the quality of the normal and translated gamma approximations for compound Poisson distributions in Exercise 3.22.

3.6.2 Asymptotic approximations

In this section, we illustrate asymptotic approximations by means of a particular case, that of the compound negative binomial distribution, where $N \sim$ nb(α, p). Let F_X be the distribution function of the step random variable X_1, and let M_X be its moment generating function. In order to express the asymptotic approximations, we write $g(x) \sim h(x)$ as $x \to \infty$ if $g(x)/h(x) \to 1$ as $x \to \infty$. The use of the symbol $\sim$ for asymptotic approximations is limited to the present section of the book. Elsewhere, the symbol $\sim$ is used for "is distributed as".

Theorem 3.23 *Suppose that*

$$\Pr(N = n) = \binom{\alpha + n - 1}{\alpha - 1} q^n p^\alpha, \quad n = 0, 1, \ldots,$$

where $q = 1 - p$, that $F_X(0^-) = \lim_{x \uparrow 0} F_X(x) = 0$ and that F_X is continuous. Suppose that there exists $\kappa > 0$ such that $M_X(\kappa) = 1/q$, and suppose that $v = qM_X'(\kappa) < \infty$. Then

$$\Pr(S > x) \sim \frac{p^\alpha x^{\alpha-1} e^{-\kappa x}}{v^\alpha \kappa \Gamma(\alpha)} \qquad as\ x \to \infty.$$

The proof of this result is beyond the scope of this book. The interested reader is referred to Embrechts *et al.* (1985b) (see also Embrechts *et al.* (1985a)). However, in the example below we work out the above asymptotic approximation for some special cases.

Example 3.24 When the step random variable is exponentially distributed with mean μ, we find that κ solves

$$\frac{1}{1 - \mu\kappa} = \frac{1}{q},$$

so that $\kappa = p/\mu$. In addition, ν is given by

$$\nu = qM'_X(\kappa) = \frac{q\mu}{(1 - \mu\kappa)^2} = \frac{\mu}{q}.$$

Then the asymptotic approximation is

$$\Pr(S > x) \sim \frac{p^{\alpha-1}q^\alpha x^{\alpha-1} e^{-px/\mu}}{\Gamma(\alpha)\mu^{\alpha-1}} \qquad \text{as } x \to \infty. \tag{3.36}$$

If $\alpha = 1$, so that N has a geometric distribution, this becomes

$$\Pr(S > x) \sim qe^{-px/\mu} \qquad \text{as } x \to \infty.$$

Looking back at the explicit expression for $F_S(x)$ in (3.26), we see that $\Pr(S > x) = qe^{-px/\mu}$, so that, in this case, the asymptotic approximation is equal to the exact tail probability. However, this is a special case. If, for example, we have $\alpha = 2$, then (3.36) becomes

$$\Pr(S > x) \sim \frac{pq^2 xe^{-px/\mu}}{\mu} \qquad \text{as } x \to \infty. \tag{3.37}$$

Equation (3.28), in Example 3.19, gives the true tail probability as

$$\Pr(S > x) = (q^2(px + \mu) + 2pq\mu)\frac{e^{-px/\mu}}{\mu},$$

so that the asymptotic approximation is not exact in this case. However, we can easily verify the result of Theorem 3.23 since

$$\frac{\Pr(S > x)}{\text{asymptotic approximation}} = \frac{q^2(px + \mu) + 2pq\mu}{pq^2 x}$$

$$= 1 + \frac{\mu(q + 2p)}{pqx},$$

and this final quantity converges to unity as x tends to infinity. Table 3.4 shows the true value of $\Pr(S > x)$, the asymptotic approximation, and the ratio of the

Table 3.4. Pr($S > x$), *the asymptotic approximation and the*
ratio (true/approximation) for the nb(2, 1/6) *case, with*
exponentially distributed claims with mean 1

x	Pr($S > x$)	Asymptotic approximation	Ratio
0	9.722222×10^{-1}	0.000000	–
20	1.172617×10^{-1}	8.257869×10^{-2}	1.420000
50	1.624720×10^{-3}	1.391027×10^{-3}	1.168000
100	7.248934×10^{-7}	6.687209×10^{-7}	1.084000
200	8.051953×10^{-14}	7.727402×10^{-14}	1.04200
400	5.267524×10^{-28}	5.159181×10^{-28}	1.021000

two, when $\alpha = 2$ and $p = 1/6$ (so that $\mathbb{E}[N] = 10$) and $\mu = 1$. For this example, we have $\mathbb{E}[S] = 10$ and $\text{Var}[S] = 70$. It is clear that the convergence to unity as $x \to \infty$ is not fast.

3.7 Statistics for compound distributions

In the preceding sections of this chapter, we have dealt with compound distributions arising from known distributions for the counting random variable N and the step random variable X_1. However, when we are applying the theory in practice, we do not have certain knowledge of the underlying distributions, but we have to use statistical methods to make inferences about them, and about the resulting quantities of interest, from the available data. For example, if we are interested in such quantities as the tail probabilities Pr($S > y$) for the random sum $S = X_1 + \cdots + X_N$, then our statements about such quantities will be subject to statistical uncertainty.

In this section, we take a (sometimes informal) look at statistical estimation for compound distributions. We do this through a particular case, where claims arrive in a Poisson process with rate λ (see §2.1.1 and §2.2.3 for properties of a Poisson process), and where claims are independent exponentially distributed random variables with mean μ. We suppose that our main aim is to make inferences about various characteristics of the distribution of the total amount S claimed in one time unit. This means that S is a random sum with the counting random variable N having a Poisson distribution with mean λ and with exponentially distributed step random variable X_1.

This is a fully parametric example, where we assume particular parametric families for the underlying distributions, and where these families are assumed known, although the parameter values are unknown and must be estimated from data. For this example, we assume that we observe n inter-claim arrival times $T_1, \ldots, T_n$ and n claim sizes $X_1, \ldots, X_n$, where the T_i are iid, and the X_i are iid and are independent of the T_i. We use these data to estimate λ and μ in a straightforward way, and then to make inferences about the distribution of S.

We use likelihood methods as described in §2.4 to estimate λ and μ. From properties of Poisson processes, we know that $T_1, \ldots, T_n$ are independent exponentially distributed with mean $1/\lambda$. The log-likelihood based on the n observations $T_1 = t_1, \ldots, T_n = t_n$ is

$$\ell_n(\lambda) = n \log \lambda - \lambda \sum_{i=1}^{n} t_i,$$

and

$$\frac{d\ell_n}{d\lambda} = \frac{\sum_{i=1}^{n} t_i}{\lambda} \left(\frac{n}{\sum_{i=1}^{n} t_i} - \lambda \right).$$

Hence the maximum likelihood estimator is $\hat{\lambda} = (\bar{T})^{-1}$, where $\bar{T} = \sum_{i=1}^{n} T_i / n$. Using results for maximum likelihood estimators (see, for example, sect. 4 of Morgan (2000), chap. 9 of Pawitan (2001) and sect. 5.5 of van der Vaart (1998)), we have (writing $\rightarrow_d$ for convergence in distribution)

$$\sqrt{n}(\hat{\lambda} - \lambda) \rightarrow_d N\left(0, (\mathbb{E}[-d^2\ell_1/d\lambda^2])^{-1}\right), \quad \text{as } n \rightarrow \infty,$$

so that, for large n, the distribution of $\hat{\lambda}$ is approximately normal with mean λ and variance λ^2/n. This means that an asymptotic $100(1 - \alpha)\%$ confidence interval for λ has end points $\hat{\lambda} \pm z_{\alpha/2}\lambda/\sqrt{n}$, where z_α is the upper $100\alpha\%$ standard normal percentage point. Since λ is unknown, this gives an approximate asymptotic $100(1 - \alpha)\%$ confidence interval for λ as

$$\left(\hat{\lambda}_L, \hat{\lambda}_U\right) = \left(\hat{\lambda} - \frac{z_{\alpha/2}\hat{\lambda}}{\sqrt{n}}, \hat{\lambda} + \frac{z_{\alpha/2}\hat{\lambda}}{\sqrt{n}}\right).$$

Repeating the process for μ, we find the maximum likelihood estimator for μ is $\hat{\mu} = \bar{X} = \sum_{i=1}^{n} X_i / n$, and that, for large n, $\hat{\mu}$ is approximately normally distributed with mean μ and variance μ^2/n. This gives rise to an approximate asymptotic $100(1 - \alpha)\%$ confidence interval for μ given by

$$(\hat{\mu}_L, \hat{\mu}_U) = \left(\hat{\mu} - \frac{z_{\alpha/2}\hat{\mu}}{\sqrt{n}}, \hat{\mu} + \frac{z_{\alpha/2}\hat{\mu}}{\sqrt{n}}\right).$$

So far, we have made inferences about the parameters of the input distributions, i.e. those of N and X_1. Now we use these to make inferences about the distribution of S. A first step is to estimate its mean, $\mathbb{E}[S] = \lambda\mu$, and a natural estimator is

$$\widehat{\mathbb{E}[S]} = \hat{\lambda}\hat{\mu}.$$

This estimate is a function of the original data, and so it is a random variable. We can demonstrate the variability of this estimate by means of a small simulation. Using simulation methods for the exponential distribution as explained in §2.2.3, we simulate $t_1, \ldots, t_n$ from an exponential distribution with mean $1/\lambda = 0.1$, and $x_1, \ldots, x_n$ from an exponential distribution with mean 1, and we evaluate $\widehat{\mathbb{E}[S]} = \hat{\lambda}\hat{\mu}$ for these data. If we repeat the whole simulation M times, so that we get M pairs of samples each of size n, then we can calculate M observations of $\hat{\lambda}\hat{\mu}$. Figure 3.3 shows the histogram of M such values, where $M = 1000$, for two sample sizes, $n = 30$ and $n = 300$. From the figure, we can see the effect of increasing the sample size: when $n = 300$, the histogram is more closely clustered about the true $\mathbb{E}[S]$ ($= \lambda\mu = 10$) than is the histogram for $n = 30$ (note the different scales on the two sets of axes). In addition, the histogram for $n = 300$ is more symmetric than that for $n = 30$.

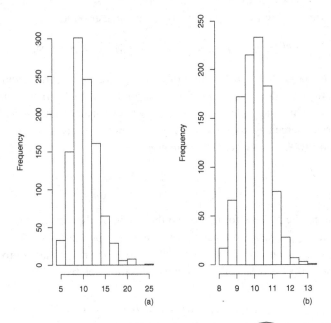

Figure 3.3. Histograms of 1000 simulated values of $\widehat{\mathbb{E}[S]}$ in the Poisson/exponential case where $\lambda = 10$ and $\mu = 1$, for sample sizes $n = 30$ (a) and $n = 300$ (b).

This example illustrates the variability in $\widehat{\mathbb{E}[S]} = \hat{\lambda}\hat{\mu}$ by simulating from known distributions. In practice, we only have one sample of T_i observations and one sample of X_i observations, and the true values of λ and μ are unknown. For this particular quantity, we are estimating a relatively simple function of λ and μ (i.e. their product). This means that we may obtain an approximate variance for $\hat{\lambda}\hat{\mu}$ directly, using the above approximate variances for $\hat{\lambda}$ and $\hat{\mu}$. Using the independence of $\hat{\lambda}$ and $\hat{\mu}$, we have

$$\text{Var}\left[\widehat{\mathbb{E}[S]}\right] = \text{Var}\left[\hat{\lambda}\hat{\mu}\right] = \mathbb{E}[\hat{\lambda}^2\hat{\mu}^2] - \left(\mathbb{E}[\hat{\lambda}\hat{\mu}]\right)^2$$
$$= \mathbb{E}[\hat{\lambda}^2]\mathbb{E}[\hat{\mu}^2] - \left(\mathbb{E}[\hat{\lambda}]\mathbb{E}[\hat{\mu}]\right)^2.$$

Using the above asymptotic normality results for $\hat{\lambda}$ and $\hat{\mu}$, we obtain

$$\text{Var}\left[\hat{\lambda}\hat{\mu}\right] \approx \left(\frac{\lambda^2}{n} + \lambda^2\right)\left(\frac{\mu^2}{n} + \mu^2\right) - \lambda^2\mu^2 = \lambda^2\mu^2\left(\frac{2}{n} + \frac{1}{n^2}\right). \quad (3.38)$$

Another approach to assessing the variability of $\hat{\lambda}\hat{\mu}$ is to find an asymptotic confidence interval for $\mathbb{E}[S]$ using the delta method. In the discussion below, we will use the delta method without proof; for more details see, for example, sect. 5.7 in DeGroot and Schervish (2002), chap. 5 of Morgan (2000) and chap. 3 of van der Vaart (1998). The delta method also works for the estimation of more complicated functions of λ and μ than their product, including ones where the above direct approach to the variance is not feasible. Here we illustrate the use of the delta method as applied to the estimation of $\lambda\mu$. First, let $\beta = (\lambda, \mu)^{\text{T}}$ and $\hat{\beta} = (\hat{\lambda}, \hat{\mu})^{\text{T}}$, where ^{T} denotes the transpose. Let ℓ_n be the log-likelihood of β based on $T_1 = t_1, \ldots, T_n = t_n$ and $X_1 = x_1, \ldots, X_n = x_n$, so that

$$\ell_n(\lambda, \mu) = n \log \lambda - \lambda \sum_{i=1}^n t_i - n \log \mu - \frac{1}{\mu} \sum_{i=1}^n x_i.$$

From results for maximum likelihood estimators (see Morgan (2000), Pawitan (2001) and van der Vaart (1998)), we have

$$\sqrt{n}(\hat{\beta} - \beta) \rightarrow_d \text{N}\left(\begin{pmatrix} 0 \\ 0 \end{pmatrix}, \Sigma = \left(\mathbb{E}\left[-\frac{\partial^2 \ell_1}{\partial\beta\partial\beta^{\text{T}}}\right]\right)^{-1}\right),$$

where $\partial^2\ell_1/\partial\beta\partial\beta^{\text{T}}$ is the 2×2 matrix of second derivatives of ℓ_1. We have $\partial^2\ell_1/\partial\lambda\partial\mu = 0$, $\partial^2\ell_1/\partial\lambda^2 = -1/\lambda^2$ and $\partial^2\ell_1/\partial\mu^2 = 1/(\mu^2) - (2x_1)/(\mu^3)$. On noting that

$$\mathbb{E}\left[-\frac{\partial^2 \ell_1}{\partial\mu^2}\right] = \frac{2\mathbb{E}[X_1]}{\mu^3} - \frac{1}{\mu^2} = \frac{1}{\mu^2},$$

we find that

$$\sqrt{n}\left(\left(\begin{array}{c} \hat{\lambda} \\ \hat{\mu} \end{array}\right) - \left(\begin{array}{c} \lambda \\ \mu \end{array}\right)\right) \to_d N\left(\left(\begin{array}{c} 0 \\ 0 \end{array}\right), \Sigma = \left(\begin{array}{cc} \lambda^2 & 0 \\ 0 & \mu^2 \end{array}\right)\right).$$

Let $g(\lambda, \mu) = \lambda\mu$. The delta method tells us that

$$\sqrt{n}(\hat{\lambda}\hat{\mu} - \lambda\mu) \to_d N(0, \sigma^2(\lambda, \mu)),$$

where

$$\sigma^2(\lambda, \mu) = (\partial g/\partial\lambda, \partial g/\partial\mu)\Sigma\left(\begin{array}{c} \partial g/\partial\lambda \\ \partial g/\partial\mu \end{array}\right), \tag{3.39}$$

and we find that $\sigma^2(\lambda, \mu) = 2\lambda^2\mu^2$. This gives

$$\hat{\lambda}\hat{\mu} \pm \frac{z_{\alpha/2} \sqrt{2}\hat{\lambda}\hat{\mu}}{\sqrt{n}} \tag{3.40}$$

as the end points of an approximate asymptotic $100(1-\alpha)\%$ confidence interval for $\mathbb{E}[S]$.

Note that the delta method gives

$$\text{Var}[\hat{\lambda}\hat{\mu}] \approx \frac{2\lambda^2\mu^2}{n}, \tag{3.41}$$

which agrees with (3.38) to order n^{-1}. The resulting approximate standard errors of $\hat{\lambda}\hat{\mu}$ are given by the square roots of these two variances. When $n = 30$, these are $0.2603\lambda\mu$ (from the square root of (3.38)) and $0.2582\lambda\mu$ (from the square root of (3.41)). When $n = 300$, the corresponding standard errors are $0.08172\lambda\mu$ and $0.08165\lambda\mu$. For both (3.38) and (3.41), the standard error when $n = 30$ is about three times bigger than that when $n = 300$, and this matches what is observed in Figure 3.3 (note the different scales on the horizontal axes in the two panels).

The above approaches are asymptotic. In this particular example, we know that $n\hat{\mu}$ has a gamma distribution with parameters n and $1/\mu$, and that $\hat{\lambda}/n$ is the reciprocal of a gamma random variable with parameters n and λ (so $\hat{\lambda}/n$ has an inverse gamma distribution). This means that for this example we can find exact confidence intervals for λ and μ (see Exercise 3.25).

Tail probabilities of the total amount claimed are also quantities of potential interest in insurance. Suppose that we wish to estimate $s_y = \text{Pr}(S > y)$ for some fixed (usually large) value y. Since s_y is determined by λ and μ, we may write $s_y = s_y(\lambda, \mu)$ for some function $s_y(\cdot, \cdot)$. An estimate for s_y is given by the plug-in estimator $\hat{s}_y = s_y(\hat{\lambda}, \hat{\mu})$.

In contrast to the $\mathbb{E}[S]$ case, the quantity of interest s_y is not a simple explicit function of λ and μ. In fact, for $y \geq 0$, and recalling that F_X is the distribution function of X_1, we have

$$s_y = \sum_{k=0}^{\infty} \frac{e^{-\lambda}\lambda^k}{k!}(1 - F_X^{\star k}(y))$$

$$= \sum_{k=1}^{\infty} \frac{e^{-\lambda}\lambda^k}{k!} \int_y^{\infty} \frac{t^{k-1}e^{-t/\mu}}{\mu^k(k-1)!}\,dt,$$

where the integral is the tail probability for the relevant gamma distribution, and where $1 - F_X^{\star 0}(y) = 0$. The above expression is quite complicated, and so we calculate the estimator $\hat{s}_y$ numerically, using either the Panjer recursion algorithm or the fast Fourier transform algorithm.

We now turn to the practical problem of quantifying the variation in $\hat{s}_y$ given one sample of T_i observations and one sample of X_i observations when the true parameter values are unknown, so that we can obtain confidence intervals for s_y.

In view of the complicated dependence of s_y on λ and μ, a possible approach is to use a technique called the parametric bootstrap, which gives confidence intervals for s_y as follows. We simulate n observations $t_1^*, \ldots, t_n^*$ from an exponential distribution with parameter $\hat{\lambda} = n/\sum_{i=1}^n t_i$, and we simulate n observations $x_1^*, \ldots, x_n^*$ from an exponential distribution with mean $\hat{\mu} = \sum_{i=1}^n x_i/n$. These *-samples are our bootstrap samples, which give rise to estimates

$$\lambda^* = n/\sum_{i=1}^{n} t_i^* \text{ and } \mu^* = \sum_{i=1}^{m} x_i^*/n,$$

and we obtain $s_y^* = s_y(\lambda^*, \mu^*)$. Repeating the sampling and subsequent calculations M times in total, we can find M values $\sqrt{n}(s_{y,1}^* - \hat{s}_y), \ldots, \sqrt{n}(s_{y,M}^* - \hat{s}_y)$. Next, we find the lower $\alpha/2$-quantile $q_{\alpha/2}^*$ and the upper $\alpha/2$-quantile $q_{1-(\alpha/2)}^*$ of these values, and we obtain an approximate $100(1-\alpha)\%$ confidence interval for s_y given by

$$\left(\hat{s}_y - \frac{q_{1-(\alpha/2)}^*}{\sqrt{n}}, \hat{s}_y - \frac{q_{\alpha/2}^*}{\sqrt{n}}\right).$$

There are other methods for obtaining bootstrap confidence intervals, and the interested reader is referred to Efron and Tibshirani (1993).

The above is a sketch of some possible ways to estimate various quantities relating to the distribution of a random sum given data on the counting and step random variables. There are many other statistical techniques and methods that could be used. For example, if we adopt a Bayesian approach, then prior distributions for the parameters are updated in the light of the available data into posterior distributions for the quantites of interest. See §4.4 and Klugman (1991) for Bayesian methods in insurance (mostly in the context of credibility theory).

Various non-parametric methods have also been proposed in the literature. Csörgő and Teugels (1990) consider, among other things, a non-parametric estimator for the asymptotic approximation to $\Pr(S > x)$ given in Theorem 3.23.

The above methods apply when we have data on the input distributions, i.e. on the distribution of N and/or on the distribution of X_1. An alternative scenario arises if we have iid data $S_1, \ldots S_n$ on the compound distribution itself, and then we can use standard statistical methods to estimate features of the distribution of S directly from the S-data; this is done in Case Study 7.1.

3.8 The individual risk model

So far in this chapter we have considered the collective risk model for a portfolio of short term insurance contracts. Now we turn our attention to an alternative model, the *individual risk model*, for the total amount claimed in a fixed time period on a portfolio of policies. This model was briefly introduced at the beginning of this chapter. In this section, we define it formally and study its properties and behaviour.

In the *individual risk model* there is a fixed number, n say, of policies, and, for $i = 1, \ldots, n$, the amount claimed on policy i in the fixed time period is a random variable Y_i. We assume that the Y_i are independent non-negative random variables, but they do not necessarily all have the same distribution. The total amount claimed on the whole portfolio in the fixed time period is

$$T = Y_1 + \cdots + Y_n.$$

For the individual risk model that we consider throughout the rest of this section, we specialise to the case where, for $i = 1, \ldots, n$, with probability q_i there is a claim on policy i during the time period and with probability $1 - q_i$ there is no claim on policy i. If there is a claim on policy i, then the amount claimed is a random variable Z_i. Thus in the individual risk model we assume that

$$Y_i = \begin{cases} 0 & \text{with probability } 1 - q_i \\ Z_i & \text{with probability } q_i, \end{cases} \tag{3.42}$$

where Z_i is a positive random variable. This means that the distribution of Y_i is a finite mixture (see §3.3) of an atom at zero (with mixing proportion $1 - q_i$)

and the distribution F_{Z_i} of Z_i (with mixing proportion q_i). With this form of the individual risk model, we are interested in quantities such as $\mathbb{E}[T]$, $\mathrm{Var}[T]$, $\mathrm{Pr}(T > x)$, etc.

When $Z_1, \ldots, Z_n$ all have the same distribution, this is called the *homogeneous individual risk model*; otherwise it is known as the *heterogeneous individual risk model*.

We observe that in the individual risk model we keep track of how much is claimed on each individual policy. This is in contrast to the collective risk model, where claims on the whole portfolio are modelled as iid random variables, irrespective of which policy gives rise to which claim. In other words, in the individual risk model we work at the level of individual policies, whereas in the collective risk model we work at the level of the whole portfolio.

Example 3.25 We give a simple example of the individual risk model that motivates the choice of the specific special form for the Y_i in (3.42). Consider a portfolio of n policies over a one-year time period, where for each policy there is at most one claim during the year (for example, this could be a portfolio of life insurance policies). For policy i, the probability of exactly one claim is q_i (independently of other policies), so that there are no claims on this policy during the year with probability $1 - q_i$. If a claim is made on policy i then it will be for a fixed known amount b_i (> 0). In this case Y_i is a discrete random variable with

$$Y_i = \begin{cases} 0 & \text{with probability } 1 - q_i \\ b_i & \text{with probability } q_i. \end{cases}$$

This could be a model for a portfolio of life insurance policies for n individuals, where the ith individual has mortality q_i and with benefit b_i to be paid when there is a claim on the life of individual i.

For this example, $T = Y_1 + \cdots + Y_n$ is also a discrete random variable. We have

$$\mathrm{Pr}(T = 0) = \prod_{i=1}^{n} (1 - q_i).$$

To find the possible values of the random variable T, note that if $n = 2$ then the distribution of T is concentrated on the set $\{0, b_1, b_2, b_1 + b_2\}$, although these four values need not be distinct. In general, the distribution of T is concentrated on the set

$$\{0\} \bigcup \left(\bigcup_{j=1}^{n} \{b_{i_1} + \cdots + b_{i_j} : i_1 < \cdots < i_j, i_1, \ldots, i_j \in \{1, \ldots, n\}\} \right);$$

these 2^n values need not all be distinct. In principle it is not difficult to calculate all possible values of T together with the associated probabilities. For an easy example, suppose that $n = 2$ and that b_1, b_2 and $b_1 + b_2$ are distinct. Then

$$T = \begin{cases} 0 & \text{with probability } (1 - q_1)(1 - q_2) \\ b_1 & \text{with probability } q_1(1 - q_2) \\ b_2 & \text{with probability } q_2(1 - q_1) \\ b_1 + b_2 & \text{with probability } q_1 q_2. \end{cases}$$

However, it can be tricky to do this by hand for large n.

3.8.1 The mean and variance for the individual risk model

In this section we find the mean and variance of T in the individual risk model. First let $\mathbb{E}[Z_i] = \mu_i$ and $\mathrm{Var}[Z_i] = \sigma_i^2$ (assumed finite). Then from (3.42) we have

$$\mathbb{E}[Y_i] = (1 - q_i) \times 0 + q_i \mathbb{E}[Z_i] = q_i \mu_i,$$

and so

$$\mathbb{E}(T) = \sum_{i=1}^{n} \mathbb{E}(Y_i) = \sum_{i=1}^{n} q_i \mu_i. \tag{3.43}$$

For the variance, it is easy to see that

$$\begin{aligned} \mathrm{Var}[Y_i] &= \mathbb{E}[Y_i^2] - \left(\mathbb{E}[Y_i]\right)^2 \\ &= (1 - q_i) \times 0^2 + q_i \mathbb{E}[Z_i^2] - q_i^2 \mu_i^2 \\ &= q_i(\sigma_i^2 + \mu_i^2) - q_i^2 \mu_i^2 \\ &= q_i \sigma_i^2 + q_i(1 - q_i)\mu_i^2. \end{aligned}$$

Using the independence of the policies, the variance of the total amount claimed is

$$\mathrm{Var}[T] = \sum_{i=1}^{n} \mathrm{Var}[Y_i] = \sum_{i=1}^{n} (q_i \sigma_i^2 + q_i(1 - q_i)\mu_i^2). \tag{3.44}$$

Example 3.26 In Example 3.25, we have an individual risk model where the Z_i in (3.42) are deterministic, i.e. $\Pr(Z_i = b_i) = 1$, where the b_i are known fixed positive values. In this case $\mu_i = \mathbb{E}[Z_i] = b_i$ and $\sigma_i^2 = \mathrm{Var}[Z_i] = 0$, so that $\mathbb{E}[Y_i] = q_i b_i$ and $\mathrm{Var}[Y_i] = q_i(1 - q_i)b_i^2$. From (3.43) and (3.44), the expectation and variance of the total amount claimed on the whole portfolio in this special case are given by

$$\mathbb{E}[T] = \sum_{i=1}^{n} q_i b_i \text{ and } \mathrm{Var}[T] = \sum_{i=1}^{n} q_i(1 - q_i)b_i^2.$$

3.8.2 The distribution function and moment generating function for the individual risk model

From (3.42), and using Definition 3.11 for the distribution function of a finite mixture, the distribution function of the amount Y_i claimed by individual i is

$$F_{Y_i}(x) = (1 - q_i)1_{[0,\infty)}(x) + q_i F_{Z_i}(x) \tag{3.45}$$

(recall from Definition 3.6 that $1_{[0,\infty)}(x)$ is the distribution function of a random variable that takes the value zero with probability 1). We aim to express the distribution of T in terms of a convolution product of the distributions of the Y_i, so we first explain what is meant by the convolution of two distribution functions. Let U and V be independent non-negative random variables with distribution functions F and G, respectively. The convolution product $F \star G$ is the distribution function of $U + V$, and is given by

$$F \star G(x) = \Pr(U + V \leq x) = \int_{[0,x]} F(x - y)G(dy), \quad x \geq 0.$$

When G has density g, this is

$$F \star G(x) = \int_0^x F(x - y)g(y)dy.$$

When $G = F$, we find that $F \star G = F^{\star 2}$, as in Definition 3.6. Now we turn our attention back to the individual risk model, where the random variables $Y_1, \ldots, Y_n$ are independent, so the total amount T claimed on the whole portfolio has the distribution function

$$F_T(x) = (F_{Y_1} \star \cdots \star F_{Y_n})(x). \tag{3.46}$$

Now consider moment generating functions. Let $M_{Z_i}(r)$ be the moment generating function of Z_i. Then, using the expression for the moment generating function of a finite mixture given in (3.12), the moment generating function of Y_i in (3.42) is

$$M_{Y_i}(r) = \mathbb{E}[e^{Y_i r}] = 1 - q_i + q_i \mathbb{E}[e^{Z_i r}] = 1 - q_i + q_i M_{Z_i}(r). \tag{3.47}$$

Using (1.8), we find that the moment generating function of T is

$$M_T(r) = \prod_{i=1}^n M_{Y_i}(r) = \prod_{i=1}^n (1 - q_i + q_i M_{Z_i}(r)). \tag{3.48}$$

Example 3.27 We continue with the special case in Examples 3.25 and 3.26, where $Z_i = b_i$ with probability 1. In this case, we have $F_{Z_i}(x) = 1_{[b_i,\infty)}(x)$, so that

$$F_{Y_i}(x) = (1 - q_i)1_{[0,\infty)}(x) + q_i 1_{[b_i,\infty)}(x),$$

which may be plugged into the right-hand side of (3.46) to give an expression for $F_T(x)$.

For the moment generating function, we first note that $M_{Z_i}(r) = e^{b_i r}$, and so

$$M_T(r) = \prod_{i=1}^{n}(1 - q_i + q_i e^{b_i r}),$$

completing this example.

In the general case given by (3.45), the random variable Y_i can be thought of in various ways. We have already mentioned that the distribution of Y_i is a finite mixture of an atom at zero (with probability $1 - q_i$) and the distribution F_{Z_i} (with probability q_i).

For an alternative representation, first consider the definition of a compound binomial random variable $S = X_1 + \cdots + X_N$, where $N \sim \text{bi}(m, p)$ and $X_1, X_2, \ldots$ are iid with distribution function F_X and moment generating function M_X (see §3.4.4). The moment generating function of S is

$$M_S(r) = (1 - p + pM_X(r))^m. \tag{3.49}$$

A special case of this is when $m = 1$, when N has a Bernoulli distribution with success probability p, i.e. $N \sim \text{bi}(1, p)$, in which case the moment generating function of S is

$$M_S(r) = 1 - p + pM_X(r).$$

Comparing the preceding expression with (3.47), it is clear that Y_i has a compound Bernoulli distribution with $p = q_i$ and $M_X = M_{Z_i}$, and (3.46) shows that the distribution of T is given by the convolution product of n compound Bernoulli distributions. This representation provides a nice link between the individual risk model and the compound distributions studied in earlier sections of this chapter. The next example shows how a compound binomial arises in a certain special case.

Example 3.28 Consider the homogeneous individual risk model, so that $F_{Z_1} = \cdots = F_{Z_n} = F_Z$, and let M_Z be the moment generating function associated with F_Z. Suppose further that $q_1 = \cdots = q_n = q$. Then from (3.48) we have

$$M_T(r) = (1 - q + qM_Z(r))^n.$$

By referring to (3.49), this shows that in this special case T has a compound binomial distribution with counting random variable having a $\text{bi}(n, q)$ distribution and steps having distribution function F_Z.

3.8.3 Approximations for the individual risk model

At its core, the individual risk model is conceptually simple, being the sum of n independent random variables, and as such is easy to deal with in principle. However, for large n, even the simplest individual risk model such as that in Examples 3.25, 3.26 and 3.27 can lead to tricky calculations. This motivates the use of approximations for the distribution of T for an individual risk model with known $q_1, \ldots, q_n$ and known distributions for $Z_1, \ldots, Z_n$ (and hence known μ_i and σ_i^2) in (3.42).

The most natural and straightforward approximation is the *normal approximation*, justified by the Central Limit Theorem for large n. Here, the distribution of T is approximated by a normal distribution with mean and variance the same as $\mathbb{E}[T]$ and $\mathrm{Var}[T]$, respectively, so, by (3.43) and (3.44), we obtain the approximation

$$N\left(\sum_{i=1}^{n} q_i\mu_i, \sum_{i=1}^{n}(q_i\sigma_i^2 + q_i(1-q_i)\mu_i^2) \right)$$

for the distribution of T. We then obtain approximations for quantities such as $\mathrm{Pr}(T > x)$ by calculating the corresponding quantities for the approximating normal distribution.

One of the main advantages of the normal approximation is that it is quick and easy to apply; another is that we only need to know the μ_i and the σ_i^2 for each i (instead of the whole distribution of Z_i). Disadvantages include the fact that the approximating normal distribution gives positive probability to the negative half-line, whereas the true distribution of T is concentrated on $[0, \infty)$. In addition, the approximating normal is a continuous distribution with no atoms, whereas the true distribution of T has an atom at zero of size $\prod_{i=1}^{n}(1 - q_i)$. In applications, q_i is often small, meaning $1 - q_i$ is not small, so that this atom at zero might not be negligible for small n.

Other approximations have been studied for the individual risk model. With the introduction of the popular Panjer recursion algorithm for numerical evaluation of various types of compound distributions, it became much easier to calculate the total claim amount for the collective risk model than for the individual risk model. This led to the use and study of *collective risk model approximations* for the individual risk model. One of the most common of these is the compound Poisson approximation, where we aim to use a $CP(\lambda, F)$ distribution to approximate the distribution of T for an individual risk model with given n, q_i and F_{Z_i} as in (3.42).

In order to specify this approximation completely, we need to choose appropriate λ and F for the compound Poisson. One way to do this is described in the following. Recall from §3.8.2 that Y_i has a compound Bernoulli distribution where the counting random variable has a bi$(1, q_i)$ distribution and the step random variable has distribution function F_{Z_i}. We replace the counting distribution bi$(1, q_i)$ by a Poi(λ_i) distribution, where $\lambda_i = q_i$, so that the distribution of Y_i is replaced by a CP(q_i, F_{Z_i}) distribution, which has the same mean, $q_i \mu_i$, as Y_i. Extending this, we replace the distributions of $Y_1, \ldots, Y_n$ by independent compound Poisson distributions CP$(q_1, F_{Z_1}), \ldots,$ CP(q_n, F_{Z_n}), respectively. Then we replace the true distribution of $T = Y_1 + \cdots + Y_n$ by the distribution of the sum of the n independent compound Poisson random variables. By Theorem 3.14, this sum has a compound Poisson distribution CP(q_+, F), where $q_+ = q_1 + \cdots + q_n$ and $F = \sum_{i=1}^{n} (q_i/q_+) F_{Z_i}$. We use this final CP$(q_+, F)$ as an approximation to the distribution of T:

$$F_T \approx \mathrm{CP}(q_+, F).$$

The step distribution of the approximating compound Poisson distribution is a finite mixture of the individual claim-size distributions (with distribution functions $F_{Z_1}, \ldots, F_{Z_n}$) with corresponding mixing proportions $(q_1/q_+), \ldots, (q_n/q_+)$. This mixture may be interpreted as follows. Suppose we know that we have a claim from the portfolio in question, but we do not know which policy the claim comes from. Then intuitively it comes from the ith policy with probability q_i/q_+. Given that the claim comes from policy i, we know that the claim size has distribution function F_{Z_i}. This is exactly what the mixture distribution is saying. The mean and variance of the approximating compound Poisson distribution are calculated in Exercise 3.31.

There are other ways to choose an approximating compound Poisson distribution, and there are other compound distribution approximations. These have been much studied, together with bounds on the distance between the true and approximating distributions (see, for example, sect. 4.6 in Rolski *et al.* (1999)). In addition to the above approximating models for the individual risk model, there is also the consideration of *numerical evaluation* of the distribution of the total amount claimed on the whole portfolio. We mention briefly two approaches here. Since the distribution of T is the convolution of n distribution functions, we may apply the fast Fourier transform algorithm to obtain numerical approximations to the distribution of T. Other approaches include recursion formulae, such as the De Pril recursion (see De Pril (1986), (1989)).

Exercises

The following exercises refer to a random sum $S = X_1 + \cdots + X_N$, where $X_1, X_2, \ldots$ are iid, independent of N. Recall that the random variable N is called the counting random variable, and that the random variable X_1 is called the step random variable.

3.1 If X_1 is non-negative and if $\Pr(X_1 = 0) = a > 0$, find $\Pr(S = 0)$.

3.2 Show by induction on k that if F has density f, then, for $k \geq 1$, $F^{\star k}$ has density f^{*k} (where $*$ and $\star$ are as in §3.2.1).

3.3 When X_1 has distribution function F_X, $F_X(0) = 0$ and $\Pr(N = 0) > 0$, show that the distribution function F_S of S is a finite mixture of a unit mass at zero and a proper distribution with distribution function

$$\widetilde{F}_S(x) = \sum_{n=1}^{\infty} F_X^{\star n}(x) \frac{\Pr(N = n)}{\Pr(N \geq 1)};$$

find the mixing proportions. Show that $\widetilde{F}_S$ is itself a compound distribution function.

3.4 Let X and Y be independent random variables with cumulant generating functions $K_X(r) = \log \mathbb{E}[e^{rX}]$ and $K_Y(r) = \log \mathbb{E}[e^{rY}]$. Let $\kappa_{X,j} = K_X^{(j)}(0)$ be the jth cumulant of X, and define $\kappa_{Y,j}$ similarly. Show that the cumulant generating function of $X + Y$ is $K_{X+Y}(r) = K_X(r) + K_Y(r)$, and hence show that cumulants of independent random variables are additive, i.e the jth cumulant of $X + Y$ is $\kappa_{X+Y,j} = \kappa_{X,j} + \kappa_{Y,j}$.

3.5 Let S be a random sum and suppose that the counting random variable N and step random variable X_1 have cumulant generating functions K_N and K_X, respectively. Use the relationship $K_S(r) = K_N(K_X(r))$ to find the first three cumulants of S in terms of the cumulants of X_1 and N.

3.6 Let S be the total claim amount in a year arising from a portfolio with N claims in a year and with iid positive claims $X_1, X_2, \ldots$.

(a) If N has a Poisson distribution with mean λ, show that the cumulant generating function of S is $K_S(r) = \lambda(M_X(r) - 1)$, where $M_X(r)$ is the moment generating function of X_1. Hence find expressions for the cumulants of S in terms of λ and the moments of X_1. What can you say about the skewness of the total claim amount?

(b) Suppose that N has a negative binomial distribution with parameters k (a positive integer) and p, so that $N \sim \text{nb}(k, p)$ and

$$\Pr(N = n) = \binom{k + n - 1}{n} p^k (1 - p)^n, \quad n = 0, 1, 2, \ldots.$$

Find expressions for the first three cumulants of S in terms of k, p, $q = 1 - p$ and the moments of X_1. Comment on the skewness of the total claim amount in this case.

3.7 Find the covariance between a random sum S and its counting random variable N.

3.8 (a) Consider a portfolio of n (n constant) independent motor insurance policies, where the aggregate claims S_i for policy i in one year has a compound Poisson distribution with Poisson parameter λ_i and claim-size distribution function F, and assume that F is known. Suppose that $\{\lambda_i\}$ are iid random variables, so that different people may have different accident rates. Find the expectation and variance of the total aggregate claims S in one year for the whole portfolio.

(b) Now consider a portfolio of n building insurance policies for buildings in a particular area where the risk of flooding varies from year to year. The total amount S_i claimed for policy i in one year has a compound Poisson distribution with Poisson parameter λ and claim-size distribution function F; assume that F is known. Suppose that λ is a random variable that varies from year to year, and that, given λ, the S_i are independent random variables. Find the expectation and variance of the total amount S claimed for the whole portfolio in one year, and compare your answers with those you obtained in (a).

3.9 Let N be the number of claims in a year on a portfolio of fire insurance policies. Suppose that N has a mixed Poisson distribution, so that the conditional distribution of N, given λ, is Poi(λ), and suppose that λ has a gamma distribution with mean μ and variance $\mu^2/2$. Find the distribution of N.

3.10 In a particular region, the number of severe weather events in a year is a random variable N with $\Pr(N = n) = (1 - p)^n p$, $n = 0, 1, 2, \ldots$, $0 < p < 1$. The ith severe weather event gives rise to M_i insurance claims, where the M_i are iid random variables, independent of N, with $\Pr(M_1 = k) = (1 - \tilde{p})^k \tilde{p}$, $k = 0, 1, 2, \ldots$, $0 < \tilde{p} < 1$. Find the distribution of the total number T of insurance claims arising from severe weather events in one year in this region.

3.11 For Example 3.17, use the moment generating formula (3.8) to obtain the distribution of the random sum resulting from an $\widetilde{\text{nb}}(1, p)$ counting distribution and an exponential step distribution with mean μ.

3.12 For Example 3.18, use the convolution series formula (3.7) to obtain the distribution of the random sum resulting from an nb$(1, p)$ counting distribution and an exponential step distribution with mean μ.

3.13 Suppose that $\Pr(N = n) = q^n p$, $n = 0, 1, 2, \ldots$, $0 < p < 1$, $q = 1 - p$, and that $X_1, X_2, \ldots$ are iid random variables, independent of N, with density $f(x) = \lambda^2 x e^{-\lambda x}$, $x > 0$. Show that the resulting compound distribution function has tail

$$\Pr(S > x) = \frac{\sqrt{q}e^{-\lambda x}}{2}\left((1 + \sqrt{q})e^{\lambda\sqrt{q}x} - (1 - \sqrt{q})e^{-\lambda\sqrt{q}x}\right).$$

Hint: Find the moment generating function of S as in Example 3.18. Then use partial fractions to write the second component as the sum of two recognisable moment generating functions.

3.14 Find the distribution function of S if $N \sim \text{nb}(2, p)$, with $\Pr(N = n) = (n + 1)(1 - p)^n p^2$, $n = 0, 1, 2, \ldots$, $0 < p < 1$ and X_1 has density $f(x) = \lambda^2 x e^{-\lambda x}$, $x > 0$.

3.15 Assume that $p_n = \Pr(N = n)$ and that the p_n satisfy (3.29), i.e. that

$$p_n = (a + b/n)\,p_{n-1} \text{ for } n = 1, 2, \ldots.$$

Suppose that $f_k = \Pr(X_1 = k)$ for $k = 0, 1, 2, \ldots$, with $f_0 > 0$ and $\sum_{k=0}^{\infty} f_k = 1$. Note that this is different from the assumptions for Theorem 3.21, where the distribution of X_1 is concentrated on $\{1, 2, \ldots\}$. Let $g_k = \Pr(S = k)$. Find g_0, and derive the recursion formula (3.32) for g_r, $r \geq 1$, in terms of a, b, the f_j and $g_0, \ldots, g_{r-1}$.

3.16 Prove that the discrete Fourier transform inversion formula (3.35) works, i.e. show that if you plug the right-hand side of (3.34) into the right-hand side of (3.35), then you do indeed obtain a_k.

3.17 Using a computer, carry out the Panjer recursion algorithm for the compound geometric distribution with exponentially distributed step random variable, with parameter values as for Table 3.2. Check that you can duplicate the values in the Panjer algorithm column of Table 3.2.

Using a package with a fast Fourier transform algorithm, experiment with its use and with how the package carries out the inverse, so that you can reproduce the numbers in the last two columns of Table 3.2.

Experiment with changing the discretisation parameter h. For example, you could try h large, for example 0.1, and h much smaller.

For the FFT algorithm, look at the effect of changing m. In particular, look for the wrap-around errors that occur when m and h are such that the interval $[0, mh)$ is not large enough to hold most of the distribution of S.

3.18 Using a computer, explore the use of the Panjer recursion algorithm and the FFT algorithm for compound Poisson distributions.

3.19 Suppose that N has a Poisson distribution with mean λ, and that X_1 is non-negative with moment generating function $M_X(r) < \infty$ for all r, and with finite moments $\mu_j = \mathbb{E}[X_1^j]$, $j = 1, 2, \ldots$. Define the standardised random variable Z by

$$Z = \frac{S - \lambda\mu_1}{\sqrt{\lambda\mu_2}}.$$

Show that the moment generating function of Z is

$$M_Z(r) = \exp\left[\lambda\left(\sum_{k=1}^{\infty} \frac{\mu_k r^k}{k!(\sqrt{\lambda\mu_2})^k} - \frac{\mu_1 r}{\sqrt{\lambda\mu_2}}\right)\right].$$

Show that as $\lambda \to \infty$, $M_Z(r)$ converges to $\exp\left(r^2/2\right)$ (the moment generating function of a standard normal distribution). This provides justification for the normal approximation to a compound Poisson distribution for large λ.

3.20 Let $X_1, X_2, \ldots$ be iid claim sizes, and let N be the number of claims in a time period, independent of the X_i. Let S be the resulting total amount claimed in one time period. For the translated gamma approximation, we approximate the distribution of S by that of $V = k + Y$, where k is a constant, Y has a gamma(α, ν) distribution, and k, α and ν are chosen so that V has the same mean, variance and coefficient of skewness as S.

Find the mean, variance and coefficient of skewness of V, and hence verify the expressions given in §3.6.1 for k, α and ν in terms of the mean, variance and coefficient of skewness of S.

3.21 Figure 3.2 shows the normal and translated gamma approximations for $\Pr(S > x)$, together with the true values, in the case where $\Pr(N = n) = q^n p$, $0 < p < 1$, $q = 1 - p$, and X_1 is exponentially distributed with mean μ, when $p = 0.1$ and $\mu = 2$. Work out the details of these approximations, and reproduce Figure 3.2 for yourself.

3.22 Investigate the quality of the normal and translated gamma approximations for compound Poisson distributions with various distributions for the step random variable. You will have to make use of numerical approximations for the "true" distribution.

3.23 For the set-up in Exercise 3.13, find the asymptotic approximation given by Theorem 3.23, and verify that

$$\Pr(S > x)/\text{asymptotic approximation} \to 1 \text{ as } x \to \infty.$$

3.24 *This exercise illustrates the effect of uncertainty in parameter values.* Consider a portfolio of n independent policies, whose premium each year is £c per policy. The total amount claimed from a single policy has a compound Poisson distribution with Poisson parameter λ and individual claims have a gamma(α, δ) distribution. The expense in settling a claim is a random variable uniformly distributed between £a and £b ($0 < a < b < \infty$), and is independent of the associated claim size. Let S be the total amount claimed together with the total expenses for the whole portfolio.

(a) Show that S has a compound Poisson distribution with Poisson parameter λn, and steps distributed as the sum of two independent random variables X and Y, where X has a gamma distribution and Y has a uniform distribution.

(b) Suppose that $c = 80$, $\lambda = 0.4$, $\alpha = 1$, $\delta = 0.01$, $a = 50$ and $b = 100$. Assume that the distribution of S may be approximated by a normal distribution. Find the number of policies the company must sell in a year to be at least 99% sure that the premium income will cover the cost of the claims and expenses.

(c) Suppose that α and δ are not known exactly, but we know $0.95 \leq \alpha \leq 1.05$ and $0.009 \leq \delta \leq 0.011$. Assume that all other parameter values are as in (b) and that the distribution of S may be approximated by a normal distribution. In the worst possible case (i.e. taking those parameter values leading to the largest values of E[S] and Var[S]), find the number of policies that the company must sell in one year to be at least 99% sure that the premium income will cover the cost of the claims and expenses.

3.25 Assume we have inter-claim arrival times $T_1, \ldots, T_n$, iid exponential random variables with mean $1/\lambda$, and claim sizes $X_1, \ldots, X_n$, iid exponential random variables with mean μ, as in §3.7. We found that the maximum likelihood estimators of λ and μ are $\hat{\lambda} = 1/\bar{T}$, where $\bar{T} = \sum_{i=1}^{n} T_i/n$, and $\hat{\mu} = \bar{X}$, where $\bar{X} = \sum_{i=1}^{n} X_i/n$, respectively.

(a) Find $\hat{\mu}_L$ and $\hat{\mu}_U$ in terms of percentage points of an appropriate χ^2 distribution and of the data, such that $\Pr\left((\hat{\mu}_L, \hat{\mu}_U) \ni \mu\right) = 1 - \alpha$ exactly for a given α, $0 < \alpha < 1$.

Hint: use the fact that $n\hat{\mu}$ has a gamma distribution, and recall Exercise 2.7(b).

(b) Find $\hat{\lambda}_L$ and $\hat{\lambda}_U$ such that $\Pr\left((\hat{\lambda}_L, \hat{\lambda}_U) \ni \lambda\right) = 1 - \alpha$.

3.26 *This and the following exercise use the same simulated samples, and you may wish to do them together.*

Consider the compound Poisson example in §3.7. Simulate samples $t_1, \ldots, t_n$ and $x_1, \ldots, x_n$ from exponential distributions with means 0.1 and 1, respectively, for $n = 100$. Using your samples, calculate $\widehat{\mathbb{E}[S]} = \hat{\lambda}\hat{\mu}$, where ˆ denotes the maximum likelihood estimator, and find the 95% confidence interval for $\mathbb{E}[S]$ given in (3.40). Does your confidence interval cover the true $\mathbb{E}[S] = 10$? Repeat the entire procedure a large number of times, and record the proportion of the confidence intervals that cover the true $\mathbb{E}[S]$. Repeat for $n = 30$ and $n = 300$.

3.27 Use the samples you simulated in Exercise 3.26 to find $\hat{s}_y = \Pr(S > y)$ (you will need to calculate $\hat{s}_y$ numerically from $\hat{\lambda}$ and $\hat{\mu}$) for various choices of y, such as $y = 10, 20, 30$. Calculate parametric bootstrap confidence intervals for $\hat{s}_y$, as in §3.7.

3.28 Consider an individual risk model (3.42), where Z_i has an exponential distribution with mean b_i. Calculate the mean and variance of the total amount T claimed on the whole portfolio. Compare your values with those obtained in Example 3.26.

3.29 In the individual risk model with Y_i as in (3.42), let $\mathbb{E}[Z_i] = \mu_i\ (> 0)$, $\text{Var}[Z_i] = \sigma_i^2$ and $\mathbb{E}[(Z_i - \mu_i)^3] = \beta_i$. Show that $T = Y_1 + \cdots Y_n$ has skewness

$$\mathbb{E}[(T - \mathbb{E}[T])^3] = \sum_{i=1}^{n} (q_i\beta_i + 3q_i(1 - q_i)\mu_i\sigma_i^2 + (1 - q_i)(1 - 2q_i)q_i\mu_i^3).$$

Show that it is possible to have T with negative skewness, even if Z_i has positive skewness for every i.

3.30 In a portfolio of n independent policies, suppose that the amount claimed on policy i in a particular year has a compound Poisson distribution $CP(\lambda_i, F_i)$.

(a) Show that the total amount claimed on the whole portfolio in a year has a compound Poisson distribution.

(b) By showing that Y_i can be written in the form of (3.42), show that this can be regarded as an individual risk model. Identify q_i and show that the distribution of Z_i is itself a compound distribution.

3.31 Let T be the total amount claimed on the whole portfolio in the individual risk model in (3.42). Suppose that the compound Poisson approximation is made as in §3.8.3. Let $\tilde{T}$ be a $CP(q_+, F)$ random variable, where q_+ and F are as in §3.8.3. Find $\mathbb{E}[\tilde{T}]$ and $\text{Var}[\tilde{T}]$, and compare your answers to the true values $\mathbb{E}[T]$ and $\text{Var}[T]$.

4

Model based pricing – setting premiums

An insurer is in business to provide insurance cover against specified risks. The insurer offers to provide a *policy* with certain benefits under particular conditions, and contracts to do this at a stated price, the *premium*. The customer may or may not accept the offer – if the contract is accepted and the premium is paid, the customer becomes the *policyholder*.

In this chapter we consider various pricing principles on which a general (non-life) insurer may base the premiums to be charged. We consider the premium based on the profile of the risks involved alone, not inflated or otherwise adjusted for the insurer's running costs or profit margins or other external considerations (such as competition in the marketplace).

In §4.1 we consider six principles based on the summary properties of the distribution of the random variable representing the risk. Two of these cases involve the insurer's view of the *utility* of the risk – an introduction to utility theory is given in Appendix A.

In §4.2 we consider the maximum and minimum premiums which are consistent with utility principles.

In §4.3 to §4.7 we consider an important branch of actuarial science called *credibility theory*, which consists of applications of Bayesian statistics in a general insurance context. The methods are concerned with setting a premium for a risk, taking into account the recent claims experience of the risk (and usually that of other comparable risks) – this methodology provides a major illustration of *experience rating*.

Let $S_1, S_2, \ldots, S_n$ be a sequence of iid random variables representing the aggregate claims from a risk (that is, the total of the sizes of all claims arising) in years $1, 2, \ldots, n$. Let S represent any one of the S_i. We will often use the simpler term "risk" for the "aggregate claims from a risk". The expected (mean) risk is $\mathbb{E}[S]$, and it is referred to as the *pure premium* for the risk.

Suppose the premium charged each year by the insurer to insure the risk is P. Consider three cases:

(a) $P > \mathbb{E}[S]$;
(b) $P = \mathbb{E}[S]$;
(c) $P < \mathbb{E}[S]$.

In case (a) the insurer is wisely charging more than the pure premium, but, even so, the cumulative surplus on this business may, or may not, at some point become negative. In case (b) the insurer is charging only the pure premium and making no provision for statistical variation in the risk, in particular for outcomes in which the risk exceeds its expected value. In case (c) the insurer is unwisely charging less than the expected risk, and the cumulative surplus on the business will eventually become negative (with probability 1), regardless of the initial surplus. This fate awaits the insurer in case (b) also.

The insurer recognises therefore that the premium must be set at a level above that of the expected risk (the mean aggregate claims). In other words, some premium "loading" is necessary to ensure $P - \mathbb{E}[S] > 0$. The quantity $P - \mathbb{E}[S]$ is the *risk loading* in the premium.

4.1 Premium calculation principles

A *premium calculation principle* is a rule for setting the premium P to be charged to cover a risk, S, say. The use of the different premium calculation principles which follow requires that we know various properties of the distribution of S. Let the mean and standard deviation of S be denoted $\mathbb{E}[S]$ and SD[S], respectively. In practice, of course, we will not know the distribution of S or its moments exactly, and we may have to estimate the moments or other properties of the distribution in the face of underlying parameter uncertainty.

4.1.1 The expected value principle (EVP)

This is the simplest premium calculation principle and sets P as

$$P = \mathbb{E}[S] + \alpha\,\mathbb{E}[S] = (1 + \alpha)\,\mathbb{E}[S] \tag{4.1}$$

for some $\alpha > 0$. The pure premium is increased by a percentage of the mean of the risk (that is, by a percentage of itself). To use this principle in practice, we need to know $\mathbb{E}[S]$. The value α is called the *relative security loading* on the pure premium $\mathbb{E}[S]$.

4.1.2 The standard deviation principle (SDP)

This principle sets P as

$$P = \mathbb{E}[S] + \alpha\, \mathrm{SD}[S] \qquad (4.2)$$

for some $\alpha > 0$. The pure premium is increased by a percentage of the standard deviation of the risk. To use this principle in practice, we need to know $\mathbb{E}[S]$ and $\mathrm{SD}[S]$ (or $\mathrm{Var}[S]$). Within the context of *SDP*, we may again refer to α as the *relative security loading*.

4.1.3 The variance principle (VP)

This principle sets P as

$$P = \mathbb{E}[S] + \alpha\mathrm{Var}[S] \qquad (4.3)$$

for some $\alpha > 0$. The pure premium is increased by a percentage of the variance of the risk. As in §4.1.2, to use this principle in practice we need to know $\mathbb{E}[S]$ and $\mathrm{Var}[S]$ (or $\mathrm{SD}[S]$). We may again refer to α as the *relative security loading*.

4.1.4 The quantile principle (QP)

This principle sets P to be a particular quantile of the distribution of S. For example, if we use the 95th percentile, P satisfies $\Pr(S \leq P) = 0.95$. The premium is set at a point which ensures a given size for the upper tail beyond P of the distribution of the risk. To use this principle in practice, we need to have a model for the distribution of S, or at least know the values of certain higher percentiles of the distribution.

Example 4.1 Suppose an insurer wants to have probability at least 0.95 of being able to cover a risk S. For illustrative purposes only, suppose the risk is modelled as having a normal distribution with known mean $\mathbb{E}[S]$ and standard deviation $\mathrm{SD}[S]$. The insurer requires $\Pr(S \leq P) = 0.95$, and the 95th percentile of the standard normal distribution is 1.6449, so we require P to satisfy

$$\frac{P - \mathbb{E}[S]}{\mathrm{SD}[S]} = 1.6449,$$

giving $P = \mathbb{E}[S] + (1.6449 \times \mathrm{SD}[S])$. In this case, the insurer's approach of using the quantile principle is equivalent to using the standard deviation principle (4.2) with relative security loading given by $\alpha = 1.6449$.

4.1.5 The zero utility principle (ZUP)

This principle sets P to satisfy

$$u(W) = \mathbb{E}[u(W + P - S)], \qquad (4.4)$$

where $u(\cdot)$ and W are the insurer's utility function and initial wealth, respectively. See Appendix A for an introduction to utility functions. From Appendix A, we note here that $u(W)$ represents the value the insurer places on having wealth W, and that $u(\cdot)$ satisfies $u'(x) \geq 0$ and $u''(x) \leq 0$ for all $x > 0$ (so that $u(\cdot)$ is concave).

In (4.4), the insurer sets the premium using a benchmark of indifference in utility terms: P is such that the insurer has "zero gain" in expected utility by insuring the risk. With no insurance in place, the insurer's utility is simply $u(W)$; with insurance in place, the insurer's wealth increases by the receipt of the premium P but decreases by the aggregate claims paid (the risk S), and so the insurer's utility is now a random variable (a function of the random variable S), with expected value $\mathbb{E}[u(W + P - S)]$.

To use this principle in practice, we need to know, in general, the insurer's utility function, initial wealth (but see §4.1.6) and certain expectation properties of the random variable S.

4.1.6 The exponential premium principle (EPP)

This is a special case of the zero utility principle (4.4), for the case in which the insurer's utility function is of exponential form, namely $u(x) = -e^{-\beta x}$, for some $\beta > 0$. In this case P satisfies

$$-e^{-\beta W} = \mathbb{E}[-e^{-\beta(W+P-S)}],$$

and so, noting that W and P are constants, we have $1 = e^{-\beta P}\mathbb{E}[e^{\beta S}]$, and hence P is given by

$$P = \frac{1}{\beta} \log \mathbb{E}[e^{\beta S}] = \frac{1}{\beta} \log M_S(\beta) \qquad (4.5)$$

when $M_S(\beta) < \infty$. Note that in this case the premium is independent of the insurer's initial wealth. To use this principle in practice, we need to know the parameter of the insurer's exponential utility function and the moment generating function of S. Using Jensen's inequality (see Appendix A), we can show, reassuringly, that $P \geq \mathbb{E}[S]$ (see also Exercise 4.1).

The function (4.5) giving the premium can be shown to be an increasing function of β, a property which has a direct interpretation in terms of the *risk aversion* of the insurer setting the premium. This property features in Example 4.2 and in Exercises 4.1 and 4.3.

Example 4.2 (i) An insurer sets the premiums for two risks, S_1 and S_2, using the exponential premium principle, with utility function $u(x) = -e^{-3x}$. Suppose $S_1 \sim N(8, 2)$ and $S_2 \sim N(6, 4)$. We use (4.5) with $\beta = 3$.

By the formula (2.10) for the moment generating function of a normal distribution, we have

$$M_{S_1}(3) = \exp\left(3 \times 8 + \frac{1}{2} \times 3^2 \times 2\right) = \exp(33),$$

$$M_{S_2}(3) = \exp\left(3 \times 6 + \frac{1}{2} \times 3^2 \times 4\right) = \exp(36).$$

So, the *EPP* premium for risk 1 is $33/3 = 11$ and that for risk 2 is $36/3 = 12$. The premium is higher for risk S_2, despite the fact that the expected aggregate claims for S_2 is lower. This reflects the greater uncertainty in the outcome for S_2, as summarised in the higher variance of the distribution.

(ii) An insurer sets the premium for a risk S using the exponential premium principle, with utility function $u(x) = -e^{-\beta x}$. Suppose $S \sim N(\mu, \sigma^2)$. Then

$$\log M_S(\beta) = \beta\mu + \frac{1}{2}\beta^2\sigma^2,$$

giving

$$P = \frac{1}{\beta}\left(\beta\mu + \frac{1}{2}\beta^2\sigma^2\right) = \mu + \frac{1}{2}\beta\sigma^2.$$

We note that in this case *EPP* produces a premium that is also consistent with the variance principle.

We also note that the premium is an increasing function of μ (higher expected claims leads to higher premium) and also of σ (higher uncertainty of claims leads to higher premium). It is also an increasing function of β, exhibiting the role of the utility function parameter as a measure of risk aversion – higher values of β correspond to higher risk aversion and the consequent setting of higher premiums. For more on risk aversion, see Appendix A.

Suppose instead that $S \sim \text{Exp}(1/\mu)$. By (2.15), $M_S(\beta) = (1 - \beta\mu)^{-1}$ for $\beta\mu < 1$. So

$$P = -\frac{1}{\beta}\log(1 - \beta\mu).$$

We note again that the premium is an increasing function of μ and also of β (to see this, look at a Maclaurin expansion for $\log(1 - \beta\mu)$).

4.1.7 Some desirable properties of premium calculation principles

There are various properties that are desirable for principles used in calculating premiums, and we give some of these below.

(1) Simplicity versus use of information

The characteristics of the distribution of the risk we need to know for *EVP*, *SDP*, *VP* and *QP* are as follows:

EVP: $\mathbb{E}[S]$;

SDP, *VP*: $\mathbb{E}[S]$, $\mathrm{SD}[S]$;

QP: one or more of the higher percentiles of the distribution.

So *EVP* is the simplest principle, followed by *SDP* and *VP*; *QP* in general requires a different kind of statistical information.

The use of *ZUP* requires a knowledge of the insurer's utility function (and with non-exponential-form utility functions also requires knowledge of the insurer's initial wealth); it also requires knowledge of certain expectation properties of the random variable S. The use of *EPP* requires knowledge of the moment generating function of S.

The other side of the "simplicity" coin is the "use of information". Being relatively simple principles, *EVP*, *SDP* and *VP* use only limited summary information about the distribution of S, specifically the first and second moments; *QP* uses information on quantiles, a different type of summary information. In general, *ZUP* and *EPP* require more information about the distribution of S. Since knowledge of the moment generating function means not just that we know all the moments of the distribution, but is equivalent to knowing the distribution function itself, using *EPP* effectively assumes a full knowledge of the distribution of the risk.

(2) Additivity

When the insurer takes a *portfolio* view of several risks, one can argue that the premium income for the portfolio should be equal to the sum of the premiums for the individual risks comprising the portfolio. To be precise, let P_1 and P_2 be the premiums for independent risks S_1 and S_2, respectively. The additivity property states that

$$P_{S_1+S_2} = P_{S_1} + P_{S_2}.$$

It is easy to see that this property holds for *EVP* and for *VP*, but that it does not hold for *SDP* (the variance of the sum of independent variables is the sum of the individual variances – standard deviations, however, do not in general "add up"). This property also holds for *EPP* (see Exercise 4.2).

A generalisation of the additivity property comes from noting that adding extra cover should not decrease the premium. So it makes sense to demand

$$P_{S_1+S_2} \geq P_{S_1}; \tag{4.6}$$

here S_2 is a non-negative random variable that is independent of S_1. The property (4.6) holds in general for *EVP, VP, SDP* and *EPP*.

(3) Scale invariance
This property states that if P is the premium for risk S then the premium for the risk cS should be cP, where c is a positive constant. This is obviously sensible – for example, if we double the risk, we should double the premium. This scaling property is well-illustrated when the insurer's business is re-expressed in another currency: suppose £1 = \$$c$, then the sterling risk S, when expressed in dollars, becomes the random variable cS, and the premium changes from £P to \$$cP$

The property holds for *EVP* and *SDP*, but does not hold for *VP* (see Exercise 4.2) or *EPP*.

The above properties are desirable from a theoretical standpoint. In practice, however, they may not be held to rigidly – an insurer will take into account the relevant "initial wealth" (through the capital/reserves supporting the business), the expected profits and the market factors when deciding premiums.

Example 4.3 The use of simulated data can help us to appreciate the use and effects of some of the premium calculation principles. Here we consider the information contained in a simulated sample of 10 000 observations of a risk, in this case an aggregate claims variable S. The variable has a compound Poisson $CP(\lambda, F_X)$ distribution, where X is exponentially distributed with mean μ.

In the simulation, λ was taken to be 0.1 and μ was taken to be 100. For this model, $Pr(S = 0) = \exp(-0.1) = 0.9048$, $\mathbb{E}[S] = 10$ and $Var[S] = 2000$. The simulated data set obtained consisted of 9013 values "$S = 0$" and 987 other values ranging from 0.13323 to 1183.1 – the data set has a major spike at value 0 (reflecting the fact that the claim frequency used was low) and has a long tail to the right (reflecting the tail of the exponential distribution).

Summary statistics of the simulated data are as follows:

sample mean 10.263; standard deviation 46.155; variance 2130.3; 95th percentile 68.158.

Using these sample statistics as estimates of the corresponding underlying population characteristics, we can see what premiums are produced using various principles.

(i) *EVP* with relative security loadings 0.5 (50%) and then 1.5 (150%):

$$P = 10.263 + 0.5 * 10.263 = 15.4 \text{ and } P = 10.263 + 1.5 * 10.263 = 25.7.$$

(ii) *SDP* with relative security loadings 0.2 and then 1.645:

$$P = 10.263 + 0.2*46.155 = 19.5 \text{ and } P = 10.263 + 1.645*46.155 = 86.2.$$

(iii) *VP* with relative security loading 0.01:

$$P = 10.263 + 0.01 * 2130.3 = 31.6.$$

(iv) *QP* using the 95th percentile:

$$P = 68.2.$$

The simulation was performed in **R**. A data set, here called simsamp, was created by issuing the command simsamp=simcomp(10000,0.1,100), which calls up and executes a function simcomp previously stored as a text file as follows:

```
simcomp=function(n, lambda, mu){
s=(1: n)*0
numclaims = rpois(n, lambda)
for(i in 1:n){
s[i]=sum(rexp(numclaims[i], 1/mu))}
s}
```

The method of simulation follows that in §3.4.1, but in the **R** code above we have illustrated how to build this into an **R** function.

We will return to the properties of premiums in Case Study 1 in Chapter 7.

4.1.8 Other premium calculation principles

There are several other premium calculation principles available, in particular one or two based on an adjustment to the distribution function of the risk which produces a weighted version of the original density function – the weight increases as x increases, producing a density with fatter tails than it had originally. Further information can be found in Dickson (2005).

4.2 Maximum and minimum premiums

First consider things from the point of view of an individual exposed to a possible loss and seeking insurance cover against that loss. Suppose the individual adopts a utility function $u(x)$ and has wealth W. Suppose also that the loss faced by the individual is represented by the random variable S and that insurance is available for a premium P. We assume that the individual behaves rationally, according to the expected utility criterion.

The individual's expected utility with no insurance cover is $\mathbb{E}[u(W - S)]$, and the individual's expected utility with insurance cover is

$$\mathbb{E}[u(W - P)] = u(W - P).$$

So the individual will buy the insurance (or be indifferent) if

$$u(W - P) \geq \mathbb{E}[u(W - S)].$$

As the premium P increases, the value of $u(W - P)$ decreases. It follows that the maximum premium the individual will be prepared to pay, say $P_{\max}$, satisfies

$$u(W - P_{\max}) = \mathbb{E}[u(W - S)]. \tag{4.7}$$

So, by Jensen's inequality, $P_{\max}$ satisfies

$$u(W - P_{\max}) \leq u(\mathbb{E}[W - S]) = u(W - \mathbb{E}[S]),$$

which implies that $P_{\max} \geq \mathbb{E}[S]$. The individual exposed to the loss is prepared to pay more than the expected loss – prepared, in fact, to pay the difference between the premium demanded and the expected loss to protect against a major (possibly catastrophic) loss.

Now consider things from the point of view of the insurer. Suppose the insurer adopts a utility function $v(x)$ and has wealth V (neither of which need be the same as for the individual). The insurer's expected utility if no cover is purchased is $v(V)$, and the insurer's expected utility if cover is purchased is $\mathbb{E}[v(V + P - S)]$. So the insurer will provide the cover (or be indifferent) if

$$\mathbb{E}[v(V + P - S)] \geq v(V).$$

As the premium P decreases, the value of $\mathbb{E}[v(V + P - S)]$ decreases. It follows that the minimum premium the insurer will accept, say $P_{\min}$, satisfies

$$\mathbb{E}[v(V + P_{\min} - S)] = v(V). \tag{4.8}$$

So, by Jensen's inequality, $P_{\min}$ satisfies

$$v(V) \leq v(\mathbb{E}[V + P_{\min} - S]) = v(V + P_{\min} - \mathbb{E}[S]),$$

which implies that $P_{min} \geq \mathbb{E}[S]$. The insurer will not provide cover for a premium less than $\mathbb{E}[S]$.

We conclude from this simple argument that an insurance contract between the individual wanting cover and the insurer is only possible if

$$P_{max} \geq P_{min} \geq \mathbb{E}[S].$$

Example 4.4 Suppose a risk S represents aggregate claims and has a compound Poisson distribution with rate (Poisson) parameter 0.1, with individual claim amounts having an exponential distribution with mean 100. (This is the same structure as we adopted for the variable S in Example 4.3 in which we used simulated data.) The party to be insured has initial wealth W and uses a utility function $u(x) = -e^{-0.005x}$.

The maximum premium the party to be insured will pay, P_{max}, satisfies (4.7), which leads to

$$e^{0.005 P_{max}} = \mathbb{E}[e^{0.005 S}].$$

Now $\mathbb{E}[e^{tS}]$ is the moment generating function of S, which, having a compound Poisson distribution with rate parameter 0.1, is given by (3.16) as

$$M_S(t) = \mathbb{E}[e^{tS}] = e^{0.1(M_X(t)-1)},$$

where the moment generating function of X, $M_X(t)$, is given by (2.15) as $M_X(t) = (1 - 100t)^{-1}$. So $M_X(0.005) = 2$ and $M_S(0.005) = e^{0.1}$. Hence P_{max} satisfies $e^{0.005 P_{max}} = e^{0.1}$, and so $0.005 P_{max} = 0.1$, giving $P_{max} = 20$. We note that $\mathbb{E}[S] = 0.1 * 100 = 10$ and that $P_{max} > \mathbb{E}[S]$.

4.3 Introduction to credibility theory

Credibility theory provides a framework for setting premiums in general insurance. It is an important illustration of *experience rating*, in which the premium is based on information coming from the recent past history of the risk and possibly that of other, comparable, risks with relevant information to add to the premium setting process. As more information becomes available (in the form of the latest year's claims experiences), the premium is updated. The method is easy to implement, and has found a range of useful areas of application.

The premium for a risk calculated on this basis is called a *credibility premium*, and, as we will see, is a weighted average of two quantities. One of these quantities summarises past information from the risk itself; the weight attached to this quantity is called the *credibility factor*. The other quantity summarises other relevant information, and is based mostly or entirely on information external to the risk itself.

In §4.5 we will consider the theory and methods of formal *Bayesian credibility theory*, which is rooted in Bayesian statistics, and then, in §4.6 and §4.7, we move on to consider the more widely applied *empirical* extensions of this methodology. There are two well-known models of *empirical Bayesian credibility theory* (EBCT). Model 1 (the Bühlmann model) provides an introduction to the basic method. The more useful model 2 (the Bühlmann–Straub model) provides a more practical extension of the basic method, incorporating as it does measures of *risk volumes* (measures of the volumes of business the claims experiences relate to).

We begin with a brief review of Bayesian estimation in §4.4, and consider three important models, all of which share some relevant characteristics, two of which find a natural place in the credibility framework.

4.4 Bayesian estimation

The methodology and practice of *Bayesian statistics* date from the second half of the eighteenth century. The methods are named after the Rev. Thomas Bayes (1702–1761), an English Presbyterian minister and mathematician who published work on probability (and other topics). His paper *Essay Towards Solving a Problem in the Doctrine of Chances* was read posthumously to the Royal Society of London in 1763 and was published in the Society's *Philosophical Transactions*.

Students of elementary probability are familar with *Bayes' rule* or *Bayes' theorem for events* for calculating probabilities conditional on the occurrence of some event E (with $\Pr(E) > 0$),

$$\Pr(A_i \mid E), \quad i = 1, 2, \ldots, m,$$

in the situation that the events $\{A_i, i = 1, 2, \ldots, m\}$ constitute a partition of the sample space and the probabilities $\Pr(A_i)$ (> 0) and $\Pr(E \mid A_i)$, $i = 1, 2, \ldots, m$, are known. The rule/theorem is usually stated as follows:

$$\Pr(A_i \mid E) = \frac{\Pr(A_i)\Pr(E \mid A_i)}{\sum\limits_{i=1}^{m} \Pr(A_i)\Pr(E \mid A_i)}, \quad i = 1, 2, \ldots, m. \tag{4.9}$$

The essential feature of Bayesian statistics is the use of *prior probabilities* or *probability distributions* to summarise the knowledge or beliefs we have concerning the values of the parameters in our model – separately from considering the information in the data. Bayesian statistics was out of favour in the nineteenth century as scientists did not really know how to handle prior probabilities. In the first half of the twentieth century, there were major developments in the theory and practice of frequentist statistics ("classical" methods, based on likelihood arguments). A modern Bayesian movement grew during the second half of the twentieth century, given rein by the availability of powerful computing facilities, which enabled new statistical methods to be developed and implemented. Proponents of Bayesian statistics argue that the Bayesian approach is philosophically consistent and has pragmatic advantages over other approaches, enabling it to be used to tackle problems in complex statistical situations and to produce sensible answers to questions. In the late twentieth century and in more recent years, there has been much increased interest, and extensive research, in Bayesian methods, with ever-widening applications being explored.

4.4.1 The posterior distribution

Suppose we have data $\underline{x} = (x_1, x_2, \ldots, x_n)$, where x_i is an observation of the random variable X_i, and where $X_1, \ldots, X_n$ are iid, distributed as a random variable X whose distribution depends on an unknown parameter θ. We want to estimate the value of θ (note that θ may be a vector of several individual parameters).

We have a model for the data, i.e. a probability distribution for X, and we write $f(\underline{x} \mid \theta)$ for the resulting joint probability density function of $\underline{X} = (X_1, \ldots, X_n)$, explicitly recognising the dependence on the value of the parameter θ. The function $f(\underline{x} \mid \theta)$ represents the information in the data relevant to estimating θ. Once the data values have been observed and so are known numbers, the function $f(\underline{x} \mid \theta)$ can be regarded as a function of θ, and in this context it is called the *likelihood function*. Here and in the following we use a generic notation $f(\cdot)$ for probability density functions; the argument of the function indicates which density function is intended.

We have some prior information about θ – from our own experience and knowledge and often that of others – and our current knowledge and beliefs about θ are summarised and expressed by adopting a probability distribution for the parameter, called the *prior distribution*, written as $f(\theta)$. The actual value of θ is then a value selected from this distribution.

The prior information and the likelihood are combined to produce an updated distribution called the *posterior distribution*, and we write $f(\theta \mid \underline{x})$ for the density of the posterior distribution. This distribution summarises our combined knowledge about θ (from the prior beliefs and the data as expressed in the likelihood), and Bayesian inference is based on it.

The main result, which follows, is usually called *Bayes' Theorem*, and gives the mechanism by which the posterior is calculated. It is a fundamental result on conditional probability distributions – expressed in our current notation it is as follows:

$$f(\theta \mid \underline{x}) = \frac{f(\theta)f(\underline{x} \mid \theta)}{f(\underline{x})} = \frac{f(\theta)f(\underline{x} \mid \theta)}{\int f(\theta)f(\underline{x} \mid \theta)d\theta}. \tag{4.10}$$

We often write this informally as

$$\text{posterior} = \frac{\text{prior} \times \text{likelihood}}{\text{marginal}},$$

where *marginal* is short for "marginal density of the data after averaging over the prior distribution of θ". As the denominator is a *number* (not a function of θ), the result is often expressed in proportional form:

$$f(\theta \mid \underline{x}) \propto f(\theta)f(\underline{x} \mid \theta), \tag{4.11}$$

and, once the right-hand side is established, the required scaling is sometimes obvious or easily calculated.

In a case in which the posterior distribution is in the same "family" of distributions as the prior, the prior is said to be a *conjugate prior* for the likelihood concerned. This will occur when the likelihood itself has the same functional form in θ as the prior.

The posterior distribution encompasses all our current information about the parameter θ, and we now regard the actual value of θ as being a value selected from this posterior distribution, rather than as coming from the prior.

Note that (4.10) is written in terms of probability density functions, but the ideas and methods may also be applied to discrete distributions; see §4.4.3. We will see how to use the posterior distribution after a short diversion to put Bayesian methodology into a wider context.

4.4.2 The wider context of decision theory

Bayesian methodology lies naturally within the general framework of *decision theory*, in which the consequences of alternative actions (decisions) are considered in the context of criteria based on the concept of a *loss function*. The loss function, which depends on the parameter (say θ) and the action chosen (say a), is denoted $L(\theta, a)$, and represents the loss, or penalty, we suffer if we take action a when the value of the parameter is θ. The possible values of the parameter θ are often referred to as the *states of nature*.

A well-known decision criterion is the *minimax* criterion, in which we choose the action for which the maximum possible loss that we can suffer as a result is minimised (that is, we choose the action which gives the "least worst" potential loss).

Another criterion is based on choosing the action which minimises the expected loss – where we take the expectation with respect to a probability distribution over the possible values of the parameter. The distribution we use is the prior distribution of the parameter, and the loss is now a random variable whose expected value is called the *Bayes loss*, $B(a)$, which is a function of the action concerned. In symbols: $B(a) = \mathbb{E}_\theta[L(\theta, a)]$, where the suffix notation $\mathbb{E}_\theta$ indicates that the expectation is with respect to θ. Using the *Bayes criterion*, we choose the action for which the Bayes loss is minimised.

With relevant data available – a random sample x from the distribution whose parameter is to be estimated – we can construct a *decision rule*, $d(x)$, which specifies which action to take for each possible value of an appropriate statistic computed from the sample data. This in turn gives us a *risk function*, $R(\theta, d)$, which is the expected value of the loss function averaged over the distribution of the data – it is a function only of the decision rule and the value of the parameter. In symbols: $R(\theta, d) = \mathbb{E}_X[L(\theta, d(X))] = \mathbb{E}_X[L(\theta, d)]$ for short. The data change our problem from one of a choice of action in the presence of a loss function to one of a choice of decision rule in the presence of a risk function.

It is not generally possible to find a decision rule which minimises the risk function $R(\theta, d)$ for all values of the parameter. So we employ decision criteria such as minimax or Bayes.

The Bayesian approach involves having a prior distribution for θ. The risk function is now a random variable. Using the Bayes criterion now, we choose the decision rule which minimises the *Bayes risk function*, $B(d)$, which is the expected value of the risk function with respect to the prior distribution of θ. In symbols: $B(d) = \mathbb{E}_\theta[R(\theta, d)]$

Appealing to a theorem which involves changing the order of the two averaging operations involved, it transpires that the rule which minimises the Bayes risk function $B(d)$ is the rule which minimises the expected loss with respect to the posterior distribution of the parameter. So the Bayesian decision rule is the rule which *minimises the posterior expected loss* (for more details, see Sect. 6.4 of DeGroot and Schervish (2002)). In the context of estimation, simple loss functions which are used include the following:

quadratic (squared error) loss	$L(\theta, d) = (\theta - d)^2,$
absolute loss	$L(\theta, d) = |\theta - d|,$
all or nothing (0–1) loss	$L(\theta, d) = 0$ for $d = \theta$ and $L(\theta, d) = 1$ for $d \neq \theta,$

Table 4.1. *Bayesian estimators for various loss functions*

Loss function	Bayesian estimator
Quadratic loss	the mean of the posterior distribution
Absolute loss	the median of the posterior distribution
All or nothing loss	the mode of the posterior distribution

and a decision rule d is an estimator of θ.

For each of the three loss functions given above, the Bayesian estimator can be identified easily with a summary characteristic of the level of the posterior distribution of the parameter, as indicated in Table 4.1.

We will assume in the following that we are estimating under a quadratic loss function, and thus that the Bayesian estimator will be the mean of the posterior distribution.

4.4.3 The binomial/beta model

The binomial/beta model provides a good introduction to Bayesian estimation, and illustrates some (but not all) of the general features of credibility premiums we will meet later.

Suppose we have a single observation x of a random variable X such that $X \mid p \sim \text{bi}(n, p)$, so that, given p, X is the total number of successes in n independent trials, each with success probability p. The likelihood is given by

$$f(x \mid p) = \Pr(X = x \mid p) = \binom{n}{x} p^x (1 - p)^{n-x} \propto p^x (1 - p)^{n-x}.$$

The data-based estimator of p (the maximum likelihood estimator, see §2.4) is given by

$$\hat{p} = \frac{X}{n}. \tag{4.12}$$

We adopt a beta distribution as the prior for p – this is the conjugate prior for the binomial likelihood. So, we take as the prior distribution beta(a, b) with both parameters known and with density function

$$f(p) = \frac{1}{B(a, b)} p^{a-1} (1 - p)^{b-1}, \quad 0 < p < 1, \quad a > 0, \quad b > 0,$$

where the beta function $B(a, b) = \Gamma(a)\Gamma(b)/\Gamma(a + b)$. Then the estimate of p using the prior mean, which we will denote $\tilde{p}$, is given by

$$\tilde{p} = \frac{a}{a + b}. \tag{4.13}$$

The posterior distribution of p is found using (4.11), giving

$$f(p \mid x) \propto p^{a-1}(1-p)^{b-1}p^x(1-p)^{n-x} = p^{a+x-1}(1-p)^{b+n-x-1}, \ 0 < p < 1.$$

We recognise this as the functional part of the beta$(a+x, b+n-x)$ density, so the posterior distribution is given by

$$p \mid x \sim \text{beta}(a+x, b+n-x).$$

The Bayesian estimator, which we will denote p^*, is the mean of the posterior distribution, so

$$p^* = \frac{X+a}{n+a+b}. \tag{4.14}$$

We note that the prior parameters have been transformed according to

$$(a, b) \rightarrow (a+x, b+n-x),$$

and that the data-based estimator X/n has been replaced by $(X+a)/(n+a+b)$. The value of the prior information can be thought of as being equivalent to having an additional $a+b$ trials, of which a were successes.

It is instructive – and important in the context of credibility theory – to note that the Bayesian estimator can be expressed as a weighted average of the data-based estimator and the prior mean as follows:

$$p^* = \frac{n}{n+a+b}\frac{X}{n} + \frac{a+b}{n+a+b}\frac{a}{a+b},$$

which in turn can be expressed as

$$p^* = Z\frac{X}{n} + (1-Z)\frac{a}{a+b}, \quad \text{where } Z = \frac{n}{n+a+b}. \tag{4.15}$$

We note that as n increases ($\Leftrightarrow$ more data), the weight attaching to the data-based estimator (X/n) increases and the weight attaching to the prior mean correspondingly decreases.

Example 4.5   Suppose we have $n = 20$ policies, on each of which there can be at most one claim in a year, and we observe a total of $x = 7$ claims in a year. Assume that, given $p\ (= \Pr(\text{claim}))$, the total number of claims $X \mid p \sim \text{bi}(20, p)$. By (4.12), the data-based estimate of $p = \Pr(\text{claim})$ is $\hat{p} = 7/20 = 0.35$.

(i)   Suppose we adopt a beta$(4, 6)$ prior for p. The prior estimate $\tilde{p}$ is the prior mean, so, by (4.13), $\tilde{p} = 4/10 = 0.4$. The posterior distribution of p is beta$(11, 19)$, so the Bayesian estimate is, by (4.14), $p^* = 11/30 = 0.367$. From (4.15), we can express p^* as

$$p^* = \frac{20}{30}\frac{7}{20} + \frac{10}{30}\frac{4}{10},$$

and we note that the weight attached to the data-based estimate is equal to $Z = 20/30 = 0.667$.

(ii) Suppose instead we adopt a beta$(2, 3)$ prior for p. The prior estimate $\tilde{p}$ is the prior mean, so $\tilde{p} = 2/5 = 0.4$. The posterior distribution of p is beta$(9, 16)$, so the Bayesian estimate is $p^* = 9/25 = 0.36$. We can express p^* as

$$p^* = \frac{20}{25}\frac{7}{20} + \frac{5}{25}\frac{2}{5},$$

and we note that the weight attached to the data-based estimate is equal to $Z = 20/25 = 0.8$.

The Bayesian estimates are different. The second (0.36) is lower and closer to the data-based estimate (0.35). The explanation comes from the fact that, while the priors have the same mean (0.4), they have different variances. The first prior has variance 0.0218 whereas the second has variance 0.04. This greater uncertainty about the value of p inherent in the second prior leads to the prior mean getting lower weight. The data-based estimate gets correspondingly higher weight ($Z = 0.8$ compared to $Z = 0.667$), pushing the estimate closer to the data-based estimate.

4.4.4 The Poisson/gamma model

The Poisson/gamma model provides another good introduction to Bayesian estimation, and is relevant to the use of credibility theory for estimating expected numbers of claims (rather than setting premiums).

Our data consist of n observations $x_1, \ldots, x_n$ of random variables $X_1, \ldots, X_n$, where, given λ, the X_i are iid Poisson random variables, i.e. $X_i \mid \lambda \sim \text{Poi}(\lambda)$, for $i = 1, \ldots, n$. The likelihood is given by

$$f(\underline{x} \mid \lambda) = \prod_{i=1}^{n} e^{-\lambda}\frac{\lambda^{x_i}}{x_i!} \propto e^{-n\lambda}\lambda^{\sum x_i}.$$

The data-based estimator of λ (the maximum likelihood estimator) is

$$\hat{\lambda} = \frac{\sum X_i}{n} = \overline{X}. \tag{4.16}$$

We adopt a gamma distribution as the prior for λ – this is the conjugate prior for the Poisson likelihood. So, taking as the prior distribution gamma(α, β), with both parameters known and with density function (see (2.19))

$$f(\lambda) = \frac{\beta^{\alpha}}{\Gamma(\alpha)}\lambda^{\alpha-1}e^{-\beta\lambda}, \quad \lambda > 0, \quad \alpha > 0, \quad \beta > 0,$$

we find that the estimate of λ using the prior mean is

$$\tilde{\lambda} = \frac{\alpha}{\beta}. \tag{4.17}$$

The posterior distribution of λ is found using (4.11), giving

$$f(\lambda \mid \underline{x}) \propto \lambda^{\alpha-1} e^{-\beta\lambda} e^{-n\lambda} \lambda^{\sum x_i} = \lambda^{\alpha+\sum x_i - 1} e^{-(\beta+n)\lambda}, \quad \lambda > 0.$$

We recognise this as the functional part of a gamma density: the posterior distribution is

$$\lambda \mid \underline{x} \sim \text{gamma}\left(\alpha + \sum x_i, \beta + n\right).$$

The Bayesian estimator is therefore given by

$$\lambda^* = \frac{\sum X_i + \alpha}{n + \beta}. \tag{4.18}$$

We note that the prior parameters have been transformed according to

$$(\alpha, \beta) \rightarrow \left(\alpha + \sum x_i, \beta + n\right),$$

and the data-based estimator $\sum X_i / n$ has been replaced by $(\sum X_i + \alpha)/(n + \beta)$. The value of the prior information can be thought of as being equivalent to having an additional β observations which sum to α.

As with the previous model considered in this section, the Bayesian estimator can be expressed as a weighted average of the data-based estimator and the prior mean, in this case

$$\lambda^* = \frac{n}{n+\beta} \frac{\sum X_i}{n} + \frac{\beta}{n+\beta} \frac{\alpha}{\beta},$$

which in turn can be expressed as follows:

$$\lambda^* = Z\overline{X} + (1 - Z)\frac{\alpha}{\beta}, \quad \text{where } Z = \frac{n}{n+\beta}. \tag{4.19}$$

We note that as n increases ($\Leftrightarrow$ more data), the weight attaching to the data-based estimator ($\overline{X}$) increases, and the weight attaching to the prior mean correspondingly decreases.

Example 4.6 Suppose we observe a total of $\sum x_i = 13$ claims on a group of $n = 50$ private motor policies. Assume that, given λ, the number of claims X_i on policy i are iid with $X_i \mid \lambda \sim \text{Poi}(\lambda)$, $i = 1, \ldots, n$. The data-based estimate of λ is, by (4.16), $\hat{\lambda} = 13/50 = 0.26$.

(i) Suppose we adopt a gamma(6, 30) prior for λ. The prior estimate $\tilde{\lambda}$ is the prior mean, so, by (4.17), $\tilde{\lambda} = 6/30 = 0.2$. The posterior distribution of λ is gamma(19, 80), so the Bayesian estimate is, by (4.18),

$$\lambda^* = 19/80 = 0.2375.$$

From (4.19), we can express λ^* as

$$\lambda^* = \frac{50}{80}\frac{13}{50} + \frac{30}{80}\frac{6}{30},$$

and we note that the weight attached to the data-based estimate is equal to $Z = 50/80 = 0.625$.

(ii) Suppose instead we adopt a gamma(2, 10) prior for λ. The prior estimate $\tilde{\lambda}$ is the prior mean, so $\tilde{\lambda} = 2/10 = 0.2$. The posterior distribution of λ is gamma(15, 60), so the Bayesian estimate is $\lambda^* = 15/60 = 0.25$. We can express λ^* as

$$\lambda^* = \frac{50}{60}\frac{13}{50} + \frac{10}{60}\frac{2}{10},$$

and we note that the weight attached to the data-based estimate is equal to $Z = 50/60 = 0.833$.

The Bayesian estimates are different. The second (0.25) is higher and closer to the data-based estimate (0.26). The explanation comes from the fact that, while the priors have the same mean (0.2), they have different variances. The first prior has variance 0.0067 whereas the second has variance 0.02. This greater uncertainty about the value of λ inherent in the second prior leads to the prior mean getting lower weight. The data-based estimate gets correspondingly higher weight ($Z = 0.833$ compared to $Z = 0.625$), pushing the Bayesian estimate closer to the data-based estimate.

4.4.5 The normal/normal model

The normal/normal model is a "standard" part of the theory of Bayesian estimation, and it provides the clearest motivation for the methodology of credibility theory for estimating premiums. The model illustrates the general features of credibility premiums that we will meet later.

Our data consist of n observations $x_1, \ldots, x_n$ of random variables $X_1, \ldots, X_n$, where, given μ, we assume that the X_i are iid normally distributed random variables, $X_i \mid \mu \sim N(\mu, \sigma^2)$, $i = 1, \ldots, n$, where σ is known. In this set-up, we are concerned with estimating μ, and, to keep our analysis as simple as possible, we assume that the other parameter σ is known. The likelihood is given by

$$f(\underline{x} \mid \mu) = \prod_{i=1}^{n} \frac{1}{\sigma \sqrt{2\pi}} \exp\left[-\frac{1}{2\sigma^2}(x_i - \mu)^2\right] \propto \exp\left[-\frac{1}{2\sigma^2}\sum(x_i - \mu)^2\right].$$

The data-based estimator of μ (the maximum likelihood estimator) is given by

$$\hat{\mu} = \frac{\sum X_i}{n} = \overline{X}. \tag{4.20}$$

We adopt a normal distribution as the prior for μ – this is the conjugate prior for the normal likelihood. So, taking as the prior distribution $N(\mu_0, \sigma_0^2)$, with both parameters known and with density function

$$f(\mu) \propto \exp\left[-\frac{1}{2\sigma_0^2}(\mu - \mu_0)^2\right],$$

the estimate of μ using the prior mean is given by

$$\tilde{\mu} = \mu_0. \tag{4.21}$$

The posterior distribution of μ is found using (4.11), giving

$$f(\mu \mid \underline{x}) \propto \exp\left[-\frac{1}{2\sigma_0^2}(\mu - \mu_0)^2\right]\exp\left[-\frac{1}{2\sigma^2}\sum(x_i - \mu)^2\right] = \exp[-A/2],$$

where, as a function of μ,

$$A = \frac{1}{\sigma_0^2 \sigma^2}\left[\sigma^2(\mu - \mu_0)^2 + \sigma_0^2\sum(x_i - \mu)^2\right]$$

$$\propto \frac{1}{\sigma_0^2 \sigma^2}[(\sigma^2 + n\sigma_0^2)\mu^2 - 2(\mu_0\sigma^2 + n\overline{x}\sigma_0^2)\mu]$$

$$\propto \frac{\sigma^2 + n\sigma_0^2}{\sigma_0^2 \sigma^2}\left(\mu - \frac{\mu_0\sigma^2 + n\overline{x}\sigma_0^2}{\sigma^2 + n\sigma_0^2}\right)^2.$$

We recognise this as the functional part of another normal density, whose mean and variance we can identify from this expression. Thus the posterior distribution of $\mu \mid \underline{x}$ is normal with mean

$$\mu^* = \frac{\dfrac{n\overline{x}}{\sigma^2} + \dfrac{\mu_0}{\sigma_0^2}}{\dfrac{n}{\sigma^2} + \dfrac{1}{\sigma_0^2}}$$

and variance

$$\frac{1}{\dfrac{n}{\sigma^2} + \dfrac{1}{\sigma_0^2}}. \tag{4.22}$$

The Bayesian estimator of μ is therefore given by

$$\mu^* = \frac{\dfrac{n\overline{X}}{\sigma^2} + \dfrac{\mu_0}{\sigma_0^2}}{\dfrac{n}{\sigma^2} + \dfrac{1}{\sigma_0^2}}. \tag{4.23}$$

It is again instructive – and important in the context of credibility theory – to note that the Bayesian estimator can be expressed as a weighted average of the data-based estimator and the prior mean as follows:

$$\mu^* = \frac{\dfrac{n}{\sigma^2}}{\dfrac{n}{\sigma^2} + \dfrac{1}{\sigma_0^2}}\overline{X} + \frac{\dfrac{1}{\sigma_0^2}}{\dfrac{n}{\sigma^2} + \dfrac{1}{\sigma_0^2}}\mu_0,$$

which in turn can be expressed as

$$\mu^* = Z\overline{X} + (1-Z)\mu_0, \quad Z = \frac{\dfrac{n}{\sigma^2}}{\dfrac{n}{\sigma^2} + \dfrac{1}{\sigma_0^2}}. \tag{4.24}$$

As with the previous models, we note that as n increases ($\Leftrightarrow$ more data), the weight attaching to the data-based estimator ($\overline{X}$) increases, and the weight attaching to the prior mean correspondingly decreases.

We note also that, for fixed n and σ, the weight Z is an increasing function of σ_0. Large prior variance equates to somewhat uninformative prior information (greater uncertainty about the value of μ), so less weight is given to the prior mean and more weight is given to the sample mean (which incorporates the information in the data).

For fixed n and σ_0, the weight Z is a decreasing function of σ^2. Large variance for the data equates to somewhat unreliable information from that source, so less weight is given to the sample mean and more weight is given to the prior mean.

The posterior variance in (4.22) has the interesting property that its reciprocal is the sum of the reciprocals of the variance of the data-based estimate $\overline{X}$ (which has variance σ^2/n) and the variance of the prior distribution (σ_0^2):

$$(\text{posterior variance})^{-1} = (\text{variance}(\overline{X}))^{-1} + (\text{prior variance})^{-1}.$$

It is worth examining the marginal (unconditional) distribution of X in the model $X \mid \mu \sim N(\mu, \sigma^2)$ with prior $\mu \sim N(\mu_0, \sigma_0^2)$. We can easily obtain the mean and variance using the conditional expectation and variance formulae in (1.3) and (1.4):

$$\mathbb{E}[X] = \mathbb{E}[\mathbb{E}[X \mid \mu]] = \mathbb{E}[\mu] = \mu_0$$

and

$$\text{Var}[X] = \mathbb{E}[\text{Var}[X \mid \mu]] + \text{Var}[\mathbb{E}[X \mid \mu]] = \mathbb{E}[\sigma^2] + \text{Var}[\mu] = \sigma^2 + \sigma_0{}^2.$$

We note that, by averaging over the distribution of μ, the variance of X is increased (from σ^2 to $\sigma^2 + \sigma_0{}^2$). We also note (see Exercise 2.18) that the marginal distribution of X (the mixture distribution) is also normal, so we can summarise the findings as follows:

$$X \mid \mu \sim \text{N}(\mu, \sigma^2) \text{ and } \mu \sim \text{N}(\mu_0, \sigma_0{}^2) \Rightarrow X \sim \text{N}(\mu_0, \sigma^2 + \sigma_0{}^2).$$

These results are used in connection with averaging the distributions of claims over different risk parameters in a heterogeneous portfolio.

Example 4.7 Suppose we observe total claims of $\sum x_i = £42\,600$ on a group of 20 general insurance claims. Using a normal model with mean μ and standard deviation $\sigma = £600$ (assumed known) for the conditional distribution of a single claim amount, given μ, then the data-based estimate of μ is, by (4.20), $\hat{\mu} = \bar{x} = £42\,600/20 = £2130$.

(i) Suppose we adopt a $\text{N}(2400, 150^2)$ prior for μ. The prior estimate $\tilde{\mu}$ is the prior mean, so, by (4.21), $\tilde{\mu} = £2400$. From the general results given above, the posterior distribution of μ is normal with mean £2250 and variance 100^2. The Bayesian estimate is, by (4.23), $\mu^* = £2250$. From (4.24), we can express μ^* as

$$\frac{5}{9} \times 2130 + \frac{4}{9} \times 2400,$$

and the weight attached to the data-based estimate is $Z = 5/9 = 0.556$.

(ii) Suppose instead we adopt a $\text{N}(2400, 250^2)$ prior for μ. The prior estimate $\tilde{\mu}$ is the prior mean, so $\tilde{\mu} = £2400$. The posterior distribution of μ is normal with mean £2190 and variance 118.2^2, so the Bayesian estimate is $\mu^* = £2190$. We can express μ^* as

$$\frac{125}{161} \times 2130 + \frac{36}{161} \times 2400,$$

and the weight attached to the data-based estimate is $Z = 125/161 = 0.776$.

The Bayesian estimates are different. The second (£2190) is lower and closer to the data-based estimate (£2130). The explanation comes from the fact that, while the priors have the same mean (£2400), they have different variances. The first prior has variance 150^2 whereas the second has variance 250^2. This greater uncertainty about the value of μ inherent in the second prior leads to the prior mean getting lower weight. The data-based estimate gets correspondingly higher weight ($Z = 0.776$ compared to $Z = 0.556$), moving the Bayesian estimate closer to the data-based estimate.

4.5 Bayesian credibility theory

We set the premium for next year for a risk covered by a general insurance contract by estimating a parameter in a probability model. The estimate is based on information from two sources:

(a) recent past data from the risk itself;
(b) relevant information from other sources.

We will have one quantity summarising the information from each source. We want the method to be easy to implement, and we insist that

- the premium is a weighted average of the two quantities from sources (a) and (b) above; and
- the premium lends itself to easy regular updating.

For source (a) we will use the statistic $\bar{x}$, the mean annual aggregate claims on the risk over the recent past. For source (b) we will use the best estimate of the premium based on the experience of other "similar" or "comparable" risks; call this estimate μ. The data from these other risks are referred to as the *collateral data*. The premium set in this way is called a *credibility premium* and is given by the formula

$$Z\bar{x} + (1 - Z)\mu. \tag{4.25}$$

In this formula, the statistic $\bar{x}$ comes from the *individual* experience, that is from the risk itself; the quantity μ comes from the *collective* experience, as represented by the collateral data. The quantity Z, $0 < Z < 1$, is the weight attached to the data from the risk itself and is called the *credibility factor*.

If the other risks (those providing the collateral data) have similar claims experience amongst themselves and to the risk we are considering, then the collateral data will be reliable and useful to us in setting the premium for the risk, and so Z will be (relatively) low. If, on the other hand, the other risks have rather different claims experience amongst themselves and from the risk we are considering, then the collateral data will not be so relevant or useful to us in setting the premium for the risk we are considering, and so Z will be (relatively) high.

In principle, Z should increase over time, as it will be based on more data relating directly to the risk itself.

There are two methods used for evaluating Z and the premium, namely

- Bayesian credibility theory;
- empirical Bayesian credibility theory.

Bayesian credibility theory is the subject of this section, and, as we will see, the method can be applied to estimating claim numbers as well as estimating premiums. In Bayesian credibility theory, the input from the collateral data is represented by a prior distribution for the parameter.

Empirical Bayesian credibility theory is the subject of §4.6 and §4.7. In this approach there is no prior distribution involved and the collateral data is provided by actual quantitative data from a group of comparable risks.

4.5.1 Bayesian credibility estimates under the Poisson/gamma model

In this case we are concerned with estimating only the *expected number of claims* (not with estimating premiums). Suppose that the number of claims, X, which arise under a risk in a year can be modelled as $X \mid \lambda \sim \text{Poi}(\lambda)$. Our problem in this case is to estimate $\mathbb{E}[X \mid \lambda]$, that is to estimate λ, given data on claim numbers for the past n years.

As we have indicated, the Bayesian approach is to regard λ as a random variable with some known prior distribution, with the actual value of λ being a value selected from this distribution. As in §4.4.4, we adopt a gamma prior for λ: $\lambda \sim \text{gamma}(\alpha, \beta)$.

Initially, for a "new" risk, we have no data from the risk itself, so the first problem is to decide how to estimate λ in this situation. Our solution is to use the *mean of the prior distribution*, giving our first estimate as $\tilde{\lambda} = \alpha/\beta$.

Suppose now that we do have data from the risk, in the form of $\underline{x} = (x_1, x_2, \ldots, x_n)$, the numbers of claims over the past n years – our model is $X_i \mid \lambda \sim \text{Poi}(\lambda)$. These observations change our information set – we can review our estimate of λ, incorporating the new information. We now view the value of λ as coming from the posterior distribution of λ given the data $\underline{x}$. From §4.4.4 we have

$$\lambda \mid \underline{x} \sim \text{gamma}\left(\alpha + \sum x_i, \beta + n\right).$$

Noting (4.18) and (4.19), the Bayesian estimate is given by

$$\lambda^* = \frac{\sum x_i + \alpha}{n + \beta} = \frac{n}{n + \beta} \frac{\sum x_i}{n} + \frac{\beta}{n + \beta} \frac{\alpha}{\beta} = Z\bar{x} + (1 - Z)\frac{\alpha}{\beta}, \qquad (4.26)$$

where the credibility factor $Z = n/(n + \beta)$. The value α/β is the estimate of λ if we have no data from the risk itself; $\bar{x}$ is the estimate of λ using only data from the risk itself. The Bayesian estimate is a weighted average of these two separate estimates and is in the form of a *credibility estimate*.

We note that $0 \leq Z \leq 1$, and with $n = 0$ the estimate reverts to the prior mean. The factor Z is an increasing function of n (more data from the risk leads to more weight being attached to the estimate from that source, $\bar{x}$).

Example 4.8 Suppose that the number of claims, X, that arise under a risk in a year can be modelled as $X \mid \lambda \sim \text{Poi}(\lambda)$. We have data on the observed number of claims for each of the past eight years, and we estimate λ at the *start of each year*. (The claims data used in this example were simulated from a Poisson distribution with mean $\lambda = 100$.)

Table 4.2 shows the credibility factors and estimates of λ under two different assumed gamma prior distributions (gamma(80, 1) and gamma(400, 5)), both of which have mean 80. In each case the initial estimate is the prior mean. For example, at the start of year 4 we have $n = 3$ years data available, with a total of 282 claims and a mean of 94 claims. At this point, using the first prior, $Z = 3/(3 + 1) = 0.75$, and the Bayesian estimate of λ is therefore

$$\lambda^* = 0.75 \times 94 + 0.25 \times 80 = 90.5,$$

which is, of course, the mean of the gamma(362, 4) posterior distribution.

We note that the estimates rise from the original $\lambda^* = 80$ and move towards the value of λ actually used to simulate the claim numbers data ($\lambda = 100$). This occurs faster, and more clearly, with the first prior. The main difference between the priors is that the first has variance $80/1^2 = 80$ whereas the second

Table 4.2. *Credibility factors and estimates for Example 4.8*

Year	Number of claims	$\alpha = 80, \beta = 1$		$\alpha = 400, \beta = 5$	
		Z	Estimate of λ	Z	Estimate of λ
1	97	0	80	0	80
2	101	0.500	88.5	0.167	82.8
3	84	0.667	92.7	0.286	85.4
4	111	0.750	90.5	0.375	85.2
5	95	0.800	94.6	0.444	88.1
6	105	0.833	94.7	0.5	88.8
7	110	0.857	96.1	0.545	90.3
8	101	0.875	97.9	0.583	91.9
9	–	0.889	98.2	0.615	92.6

has variance $400/5^2 = 16$. The second prior has much lower variance, which corresponds to the fact that we have stronger prior belief in the initial estimate of 80. This is reflected in the much lower values of the credibility factors (and the estimates) at all stages and the slower convergence to 100 – the developing sequence of estimates is more reluctant to move away from 80 in the direction of the overall sample mean (= 100.5 after eight years) and the true value of λ.

4.5.2 Bayesian credibility premiums under the normal/normal model

We now return to the problem of estimating premiums. Suppose that the aggregate claims, X, that arise under a risk in a year can be modelled as

$$X \mid \theta \sim N(\theta, \sigma_1^2),$$

where σ_1 is known. The pure premium for the risk is $\mathbb{E}[X \mid \theta] = \theta$, which is fixed but unknown and is called the *risk parameter* for the risk concerned. Our problem is to estimate $\mathbb{E}[X \mid \theta]$, that is to estimate θ, given relevant data $\underline{x}$.

As we know, the Bayesian approach is to regard θ as a random variable with some known prior distribution, with the actual value of θ being a value selected from this distribution. Here, as in §4.4.5, we adopt a normal prior (but with altered notation, to fit in with the usual conventions of credibility theory): $\theta \sim N(\mu, \sigma_2^2)$.

Before we have any data from the risk itself, our estimate of θ is the prior mean, so $\tilde{\theta} = \mu$. Suppose now we do have data from the risk, in the form of $\underline{x} = (x_1, x_2, \ldots, x_n)$, the aggregate claims over the past n years – our model is $X_i \mid \theta \sim N(\theta, \sigma_1^2)$. These observations change our information set – we can review our estimate of θ, incorporating the new information. We now view the value of θ as coming from the posterior distribution of θ given the data $\underline{x}$.

From §4.4.5 we have that $\theta \mid \underline{x}$ is normal with mean

$$\theta^* = \frac{\dfrac{n\bar{x}}{\sigma_1^2} + \dfrac{\mu}{\sigma_2^2}}{\dfrac{n}{\sigma_1^2} + \dfrac{1}{\sigma_2^2}}$$

and variance

$$\frac{1}{\dfrac{n}{\sigma_1^2} + \dfrac{1}{\sigma_2^2}}.$$

Noting (4.23) and (4.24), the Bayesian estimator of θ is therefore given by

$$\theta^* = \frac{\dfrac{n\overline{X}}{\sigma_1^2} + \dfrac{\mu}{\sigma_2^2}}{\dfrac{n}{\sigma_1^2} + \dfrac{1}{\sigma_2^2}} = Z\overline{X} + (1 - Z)\mu, \tag{4.27}$$

where

$$Z = \frac{\dfrac{n}{\sigma_1^2}}{\dfrac{n}{\sigma_1^2} + \dfrac{1}{\sigma_2^2}} = \frac{n}{n + \dfrac{\sigma_1^2}{\sigma_2^2}}.$$

The value μ is the estimate of θ if we have no data from the risk itself; $\overline{x}$ is the estimate of θ using only data from the risk itself. The Bayesian estimate is a weighted average of these two separate estimates, and is in the form of a *credibility estimate* – it is the *credibility premium*. The credibility factor is

$$Z = \frac{n}{n + \sigma_1^2/\sigma_2^2}.$$

We note that $0 \leq Z \leq 1$, and with $n = 0$ the estimate reverts to the prior mean; Z is an increasing function of n (more data from the risk leads to more weight being attached to the estimate from that source, $\overline{x}$).

There are two sources of variation in this model. The first, summarised by the risk variance σ_1^2, is the variation in claims from year to year for the risk concerned – it is *internal* to the risk and is called the *within risk variance*. The second, summarised by the prior variance σ_2^2, is the variation in the expected claims from risk to risk across a (real or imagined) collective of comparable risks – it is *external* to the risk and is called the *between risk variance*. With this interpretation we have the following very useful result, which informs the whole of our work on credibility theory:

$$\text{credibility factor} \quad Z = \frac{n}{n + \dfrac{\text{within risk variance}}{\text{between risk variance}}}. \tag{4.28}$$

Recalling points made in §4.4.5, we note that, for fixed n and σ_1, the weight Z is an increasing function of σ_2. With a larger prior variance, less weight is given to the prior mean and more weight is given to the sample mean.

For fixed n and σ_2, the weight Z is a decreasing function of σ_1. With a larger variance for the data from the risk, less weight is given to the sample mean and more weight is given to the prior mean.

Table 4.3. *Credibility factors and premiums for Example 4.9*

Year	Claims	$\mu = 300,\ \sigma_2^2 = 400$		$\mu = 300,\ \sigma_2^2 = 100$	
		Z	Premium	Z	Premium
1	310	0	300	0	300
2	343	0.889	309	0.667	307
3	332	0.941	325	0.800	321
4	348	0.960	327	0.857	324
5	315	0.970	332	0.889	330
6	–	0.976	329	0.909	327

Example 4.9 Suppose the aggregate claims, X, which arise under a risk in a year can be modelled as $X \mid \theta \sim N(\theta, 50)$. From experience with other business, we adopt as a prior $\theta \sim N(300, 400)$. We have the claims figures for the past five years for the risk (£, thousands: 310, 343, 332, 348, 315).

Table 4.3 shows the credibility factors and premiums for the risk under the prior specified above, and, for comparison, with a second prior, namely $\theta \sim N(300, 100)$. The premiums are set at the *start of each year*. For example, at the start of year 3 we have $n = 2$ years data available, with a total claim amount of 653 and a mean of 326.5. At this point, using the first prior, $Z = 2/[2 + (50/400)] = 16/17 = 0.941$, and the estimate of θ is therefore $\theta^* = 0.941 \times 326.5 + 0.059 \times 300 = 324.9$, which is, of course, the mean of the posterior distribution, which is $N(324.9, 23.53)$.

The first prior has higher variance than the second, and so is less informative about θ; the higher variance corresponds to greater risk-to-risk variation. We have less useful "external" information and must rely more on the data from the risk itself. The ratio of "between" to "within" risk variance using the first prior is 8:1, while using the second prior it is only 2:1. With the first prior more weight is given to the data from the risk itself – this is reflected in the higher values of the credibility factors (and the estimates) at all stages.

A question of independence Let $X_1, X_2, \ldots$ represent the aggregate claims for a risk in successive years. The risk has its own risk parameter θ, which is fixed, but unknown. Our main structural *assumption* is as follows: given θ, the random variables $X_1, X_2, \ldots$ are a sequence of independent and identically distributed random variables, and we also have $X_i \mid \theta \sim N(\theta, \sigma_1^2)$. The independence property can be expressed in other words by stating that the X_i are *conditionally independent* given θ.

Consider now the marginal (unconditional) distribution of X_i. As noted in §4.4.5 (and using our current notation), X_i has a normal distribution with $\mathbb{E}[X_i] = \mu$ and $\text{Var}[X_i] = \sigma_1{}^2 + \sigma_2{}^2$. So the X_i are *identically distributed*.

Now, using the conditional expectation formula (1.3), we find that

$$\mathbb{E}[X_i X_j] = \mathbb{E}[\mathbb{E}[X_i X_j \mid \theta]] = \mathbb{E}[\mathbb{E}[X_i \mid \theta]\mathbb{E}[X_j \mid \theta]],$$

by conditional independence. But $\mathbb{E}[X_i \mid \theta] = \theta$ and hence

$$\mathbb{E}[X_i X_j] = \mathbb{E}[\theta^2] = \text{Var}[\theta] + \{\mathbb{E}[\theta]\}^2 = \sigma_2^2 + \mu^2.$$

Now, since $\mathbb{E}[X_i]\mathbb{E}[X_j] = \mu^2$, we have $\mathbb{E}[X_i X_j] \neq \mathbb{E}[X_i]\mathbb{E}[X_j]$ if $\sigma_2^2 > 0$, and it follows that the unconditional X_i are *not independent*. This shows that the X_i are *conditionally independent* but are not (unconditionally) *independent*.

So, if we know the risk parameter for a risk, successive claim amount variables are independent. But if we do not know the risk parameter for the risk, successive claim amount variables are not independent (but have dependence through the risk parameter).

As a way to illustrate this, let $\sigma_1^2 = 1$ for simplicity, and suppose that you are told the value of X_1. You know that $X_1 \mid \theta \sim N(\theta, 1)$, and so knowing the value of X_1 gives you information about possible values of θ. For example, suppose $x_1 = 15$. Since this value comes from a $N(\theta, 1)$ distribution it is very likely that θ has a value somewhere between 12 and 18 (otherwise it is most unlikely that you would have observed $x_1 = 15$). So you know something about θ, the risk parameter for the risk involved, and this in turn tells you something about the possible values of X_2. Knowledge of X_1 gives you knowledge of X_2, and so X_1 and X_2 are not independent. However, if θ is known, then X_1 tells us nothing *further* about X_2.

Comments on the Bayesian approach With both the Poisson/gamma and the normal/normal models the Bayesian estimate of the quantity concerned is a weighted average of an estimate based on the data from the risk itself and an estimate based on "external" information. So it is in the form we require for it be considered a *credibility estimate*. The fact that it *is* in the form we require is not guaranteed and will not always be the case – with some distributions we cannot make the Bayesian estimate fit the form of a credibility estimate.

A general difficulty with credibility estimates is that we need to know the value of the credibility factor Z. In the two models above, we were able to solve the problem, provided we knew the values of various parameters, in particular parameters summarising our prior beliefs.

In the following section we consider an "empirical" approach to the calculation of credibility premiums, in which the use of a prior is abandoned and the information therein is replaced by quantitative collateral information.

4.6 Empirical Bayesian credibility theory: Model 1 – the Bühlmann model

This model was formulated by Hans Bühlmann; see Bühlmann (1967). We again want to estimate the premium for a risk, and we now set the risk in a collective of comparable risks for which we have relevant data over the past several years. The information from the *collateral data* enables us to proceed using a similar approach to that which we have previously used, but with the advantage that we can allow the collateral data to provide the *between risk* information, rather than relying on a parametric model for the distribution of the data and a fully specified prior distribution to summarise our information/beliefs about the unknown parameter. The method is referred to as an *empirical Bayesian* method, and in this context is called *empirical Bayesian credibility theory*, which we will call EBCT for short. In summary:

EBCT = Bayesian credibility theory

  – distributional assumptions

  + collateral data.

In this section we consider the first of two EBCT models, called EBCT Model 1 – this serves as an introduction to the more widely applicable and useful EBCT Model 2, which we consider in §4.7

Let $X_1, X_2, \ldots, X_n$ represent the aggregate amounts (or numbers) of claims that arise under a risk over n successive years. We want to estimate the pure premium for the risk, which is the expected value of X_{n+1}, given that we have observed the values of $X_1, X_2, \ldots, X_n$. The basic structure of our model is that the distribution of each X_i, $i = 1, 2, \ldots, n$, depends on the value of a parameter θ, which is fixed for that risk, but is unknown. The parameter θ is called the *risk parameter* for the risk. We regard θ as a random variable with unknown distribution.

We make the following structural *assumption* concerning a single risk with risk parameter θ.

Assumption Given θ, the X_i, $i = 1, 2, \ldots, n$, are iid.

Note that this assumption is a statement about *conditional* distributions. We are assuming that the aggregate claims for the risk are identically distributed from year to year. This is actually a stronger assumption than is required – all we need is stationarity in first and second moments (constant mean and variance from year to year). It follows from our assumption that the (unconditional) X_i are identically distributed – it does not follow from our assumption that they are independent.

Table 4.4. *Structure of Bayesian and empirical Bayesian models*

	Normal/normal	Poisson/gamma	EBCT
Prior	$\theta \sim N(\mu, \sigma_2^2)$	$\lambda \sim \text{gamma}(\alpha, \beta)$	none
Conditional mean of X_i	θ	λ	$m(\theta)$
Conditional variance of X_i	σ_1^2	λ	$s^2(\theta)$

Since, given θ, the distribution of X_i does not depend on i, we can introduce notation for the mean and variance of the conditional distribution using symbols which depend only on θ. We will adopt the symbols in general use, so we define

$$m(\theta) = \mathbb{E}[X_i \mid \theta] \quad \text{and} \quad s^2(\theta) = \text{Var}[X_i \mid \theta]. \tag{4.29}$$

The pure premium/estimator for the risk is $\mathbb{E}[X_i \mid \theta] = m(\theta)$, and so we can now state that our problem is to estimate $m(\theta)$, given data $\underline{x} = (x_1, x_2, \ldots, x_n)$.

In Table 4.4 we give the earlier normal/normal and Poisson/gamma Bayesian models alongside the new empirical structure. Using the table to compare the normal/normal model (as in §4.5.2) and EBCT, we find that

$\mathbb{E}[m(\theta)]$ in EBCT corresponds to $\mathbb{E}[\mathbb{E}[X_i \mid \theta]] = \mathbb{E}[\theta] = \mu$;

$\text{Var}[m(\theta)]$ in EBCT corresponds to $\text{Var}[\mathbb{E}[X_i \mid \theta]] = \text{Var}[\theta] = \sigma_2^2$, the between risk variance; and

$\mathbb{E}[s^2(\theta)]$ in EBCT corresponds to $\mathbb{E}[\text{Var}[X_i \mid \theta]] = \mathbb{E}[\sigma_1^2] = \sigma_1^2$, the within risk variance.

The credibility premium in the normal/normal model is $Z\bar{x} + (1 - Z)\mu$, where

$$Z = \frac{n}{n + \dfrac{\sigma_1^2}{\sigma_2^2}} = \frac{n}{n + \dfrac{\text{within risk variance}}{\text{between risk variance}}} .$$

A similar analogy between the Poisson/gamma model and EBCT can be demonstrated, noting that Z can be expressed as

$$Z = \frac{n}{n + \beta} = \frac{n}{n + \dfrac{\alpha/\beta}{\alpha/\beta^2}} .$$

The above comparisons suggest that we tentatively adopt the formula

$$Z\bar{X} + (1 - Z)\mathbb{E}[m(\theta)] \tag{4.30}$$

for use in EBCT for the credibility premium/estimator, where the credibility factor Z is given by

$$Z = \frac{n}{n + \dfrac{\text{within risk variance}}{\text{between risk variance}}} = \frac{n}{n + \dfrac{\mathbb{E}[s^2(\theta)]}{\text{Var}[m(\theta)]}}. \qquad (4.31)$$

Theorem 4.11 in the following shows that we can justify this adoption under the criterion of using the "best" linear estimator of $m(\theta)$, that is the estimator which is the linear function of the observations with minimum mean square error.

As a preliminary to the proof of the theorem, we first establish some expectations that we will require. In doing so, we use $\mathbb{E}[Xg(\theta) \mid \theta] = g(\theta)\mathbb{E}[X \mid \theta]$ (see, for example, Sect. 7.7 of Grimmett and Stirzaker (2001)).

Lemma 4.10   *With the set-up and notation of EBCT Model 1, we have*

(i) $\mathbb{E}[X_i] = \mathbb{E}[\overline{X}] = \mathbb{E}[m(\theta)]$;

(ii) $\mathbb{E}[X_i m(\theta)] = \mathbb{E}[\overline{X} m(\theta)] = \mathbb{E}[m^2(\theta)]$;

(iii) $\mathbb{E}[\overline{X}^2] = \dfrac{1}{n}\mathbb{E}[s^2(\theta)] + \mathbb{E}[m^2(\theta)]$.

Proof   For (i), note that the conditional expectation formula (1.3) implies that $\mathbb{E}[X_i] = \mathbb{E}[\mathbb{E}[X_i \mid \theta]] = \mathbb{E}[m(\theta)]$, and it follows that

$$\mathbb{E}[\overline{X}] = \frac{1}{n}\mathbb{E}\left[\sum_{i=1}^{n} X_i\right] = \frac{1}{n}n\mathbb{E}[m(\theta)] = \mathbb{E}[m(\theta)].$$

For (ii), using the conditional expectation formula (1.3) again, we have

$$\mathbb{E}[X_i m(\theta)] = \mathbb{E}[\mathbb{E}[X_i m(\theta) \mid \theta]] = \mathbb{E}[m(\theta)\mathbb{E}[X_i \mid \theta]] = \mathbb{E}[m^2(\theta)],$$

and hence $\mathbb{E}[\overline{X} m(\theta)] = \mathbb{E}[m^2(\theta)]$.

For (iii), by conditional independence, note that, for $i \neq j$,

$$\mathbb{E}[X_i X_j] = \mathbb{E}[\mathbb{E}[X_i X_j \mid \theta]] = \mathbb{E}[\mathbb{E}[X_i \mid \theta]\mathbb{E}[X_j \mid \theta]],$$

and so $\mathbb{E}[X_i X_j] = \mathbb{E}[m^2(\theta)]$. Further, we have

$$\begin{aligned}
\mathbb{E}[X_i^2] &= \mathbb{E}[\mathbb{E}[X_i^2 \mid \theta]] = \mathbb{E}[\text{Var}[X_i \mid \theta] + \{\mathbb{E}[X_i \mid \theta]\}^2] \\
&= \mathbb{E}[s^2(\theta)] + \mathbb{E}[m^2(\theta)].
\end{aligned}$$

It follows that

$$\mathbb{E}[\overline{X}^2] = \frac{1}{n^2}\mathbb{E}\left[\sum_{i=1}^{n}X_i\sum_{j=1}^{n}X_j\right]$$

$$= \frac{1}{n^2}\left\{n\mathbb{E}[X_i^2] + n(n-1)\mathbb{E}[X_iX_j]\right\} \text{ for } i \neq j$$

$$= \frac{1}{n^2}\left\{n\mathbb{E}[s^2(\theta)] + n\mathbb{E}[m^2(\theta)] + n(n-1)\mathbb{E}[m^2(\theta)]\right\}$$

$$= \frac{1}{n}\mathbb{E}[s^2(\theta)] + \mathbb{E}[m^2(\theta)],$$

as required. Alternatively, one can show $\text{Var}[\overline{X} \mid \theta] = \frac{1}{n}s^2(\theta)$, and use this to find $\text{Var}[\overline{X}]$ and hence $\mathbb{E}[\overline{X}^2]$.   □

We now *derive* the credibility premium/estimator; that is, we verify that the form of the optimum linear estimator $a_0 + a_1X_1 + a_2X_2 + \cdots + a_nX_n$ of $m(\theta)$ is indeed that of the credibility estimator tentatively adopted above.

Theorem 4.11 *Let $X_1, X_2, \ldots, X_n$ be a sequence of random variables, each of whose distribution depends on a parameter θ, and which, given θ, are iid, with $\mathbb{E}[X_i \mid \theta] = m(\theta)$ and $\text{Var}[X_i \mid \theta] = s^2(\theta)$. Let $a_0, a_1, \ldots, a_n$ be constants. Then the estimator $a_0 + \sum_{j=1}^{n} a_jX_j$ of $m(\theta)$ for which*

$$\mathbb{E}\left[\left\{m(\theta) - a_0 - \sum_{j=1}^{n}a_jX_j\right\}^2\right]$$

is minimised is given by

$$Z\overline{X} + (1-Z)\mathbb{E}[m(\theta)], \quad \text{where } \cdot Z = \frac{n}{n + \dfrac{\mathbb{E}[s^2(\theta)]}{\text{Var}[m(\theta)]}}. \tag{4.32}$$

Proof The problem is symmetric in the X_i, and so $a_1 = a_2 = \cdots = a_n = \tilde{a}$, say, so that

$$a_0 + \sum_{j=1}^{n}a_jX_j = a_o + \tilde{a}\sum_{i=1}^{n}X_j,$$

which means that the estimator is of the form $a + b\overline{X}$. The problem is therefore to find a and b such that

$$S = \mathbb{E}\left[\left\{m(\theta) - a - b\overline{X}\right\}^2\right]$$

is minimised. Taking the partial derivative of S with respect to a gives

$$\frac{\partial S}{\partial a} = 0 \Rightarrow \mathbb{E}[m(\theta) - a - b\overline{X}] = 0.$$

Using Lemma 4.10(i), we have $a + b\mathbb{E}[m(\theta)] = \mathbb{E}[m(\theta)]$, and hence

$$a = (1 - b)\mathbb{E}[m(\theta)]. \tag{4.33}$$

Taking the partial derivative of S with respect to b gives

$$\frac{\partial S}{\partial b} = 0 \Rightarrow \mathbb{E}[\overline{X}\{m(\theta) - a - b\overline{X}\}] = 0,$$

so, using Lemma 4.10(i), (ii) and (iii), we have

$$a\mathbb{E}[m(\theta)] + b\left(\frac{1}{n}\mathbb{E}[s^2(\theta)] + \mathbb{E}[m^2(\theta)]\right) = \mathbb{E}[m^2(\theta)]. \tag{4.34}$$

Solving (4.33) and (4.34) gives

$$b = \frac{n}{n + \dfrac{\mathbb{E}[s^2(\theta)]}{\mathrm{Var}[m(\theta)]}},$$

and so, denoting b by Z, we have that the best estimator is given by

$$Z\overline{X} + (1 - Z)\mathbb{E}[m(\theta)],$$

and the result is proved. $\square$

The estimator involves three quantities, $\mathbb{E}[m(\theta)]$, $\mathbb{E}[s^2(\theta)]$ and $\mathrm{Var}[m(\theta)]$, which we have to estimate from collateral data. These quantities are sometimes referred to as the three *structural parameters*.

We suppose now that the risk we are interested in is one of a collective of a fixed number N of comparable risks. Our data now consist of values x_{ij} of random variables X_{ij}, where X_{ij} represents the aggregate amount (or number) of claims for risk number i in year j, $i = 1, 2, \ldots, N$, $j = 1, 2, \ldots, n$, as in Table 4.5. For convenience, we are making the (perhaps rash) assumption that we have complete data – the same number of years data for each risk.

For each risk, say risk i, the distribution of each X_{ij}, $j = 1, 2, \ldots, n$, depends as before on the value of a *risk parameter* θ_i, which is fixed for that risk, but unknown. We regard θ_i as a random variable with an unknown distribution function. Each risk has its own risk parameter, which is fixed for that risk over the period of years we are considering. It is very important to recognise that the risks are heterogeneous – different risks have different risk parameters – and we will set appropriate premiums which reflect this.

For each risk, say risk i, we make the following structural assumption.

Table 4.5. *Collective of risks*

		Year			
		1	2	· ·	n
Risk	1	X_{11}	X_{12}	· ·	X_{1n}
	2	X_{21}	X_{22}	· ·	X_{2n}
	·	·	·	· · ·	·
	·	·	·	· · ·	·
	N	X_{N1}	X_{N2}	· ·	X_{Nn}

Assumption 1 Given θ_i, the X_{ij}, $j = 1, 2, \ldots, n$, are iid.

This is exactly the same assumption we made earlier for a single risk, and gives us the within risk structure we require.

We need to make an assumption to give us appropriate between risk structure, and it is conveniently expressed as follows.

Assumption 2 For different risks i, j ($i \neq j$), the pairs of variables (θ_i, X_{il}) and (θ_j, X_{jk}), $l, k = 1, 2, \ldots, n$, are iid.

It follows from assumption 2 that any two variables in different rows in Table 4.5 are iid. It also follows from assumption 2 that the risk parameters θ_i, $i = 1, 2, \ldots, N$, are iid. The "identicality" of the distributions is a formal statement which firms up what we mean by the references to a collective of "comparable risks". We can think of the values of the risk parameters for the different risks as coming from some common underlying distribution.

For risk i define $m(\theta_i) = \mathbb{E}[X_{ij} \mid \theta_i]$ and $s^2(\theta_i) = \text{Var}[X_{ij} \mid \theta_i]$; these do not depend on j (because of our assumptions). We identify $m(\theta_i)$ and $s^2(\theta_i)$ as the mean and variance of the amounts (or numbers) of claims for risk i (row i in Table 4.5).

Now, since $\theta_i, i = 1, 2, \ldots, N$, are identically distributed, it follows that none of $\mathbb{E}[m(\theta_i)]$, $\mathbb{E}[s^2(\theta_i)]$ or $\text{Var}[m(\theta_i)]$ depend on i. So we write them as $\mathbb{E}[m(\theta)]$, $\mathbb{E}[s^2(\theta)]$ and $\text{Var}[m(\theta)]$, respectively, bringing us back to the three structural parameters we have already met:

$\mathbb{E}[m(\theta)]$, the expected value (the average) of the risk means;
$\mathbb{E}[s^2(\theta)]$, the expected value (the average) of the risk variances – it is is the
　　　　average variance *within risks*;
$\text{Var}[m(\theta)]$, the variance of the risk means – it is the variance *between risks*.

We now seek estimators of the three structural parameters. The first two, the estimators of $\mathbb{E}[m(\theta)]$ and $\mathbb{E}[s^2(\theta)]$, are quite easy to identify, given the physical nature of the parameters. The third, the estimator of $\mathrm{Var}[m(\theta)]$, is less easy to justify, but we can show that all three "usual" estimators (that is, the estimators in everyday use) are unbiased for the parameters concerned.

We next define notation. Let

$$\overline{X}_i = \frac{1}{n} \sum_{j=1}^{n} X_{ij}$$

be the mean amount (or number) of claims for risk i over the n years for which we have data (the mean for row i in Table 4.5), and let

$$\overline{X} = \frac{1}{N} \sum_{i=1}^{N} \overline{X}_i = \frac{1}{Nn} \sum_{i=1}^{N} \sum_{j=1}^{n} X_{ij}$$

be the overall mean amount (or number) of claims for all years and all risks involved. It is important to note our *notation* here: the mean amount (or number) of claims for an individual risk (risk i) is now denoted $\overline{X}_i$, and $\overline{X}$ denotes the overall mean amount (or number) of claims for all risks involved.

The usual estimators of the structural parameters are given in Table 4.6. The estimator of $\mathbb{E}[m(\theta)]$ is the overall mean of the claims data for all the risks in the collective. The estimator of $\mathbb{E}[s^2(\theta)]$ is the mean of the individual risk sample variances. The estimator of $\mathrm{Var}[m(\theta)]$ is the sample variance of the risk means corrected for bias – the correction is a reduction given by the estimator of $\mathbb{E}[s^2(\theta)]$ divided by n, the number of years data for each risk we have available.

Table 4.6. *Usual estimators of the structural parameters in*
EBCT Model 1

Structural parameter	Estimator
$\mathbb{E}[m(\theta)]$	$\overline{X}$
$\mathbb{E}[s^2(\theta)]$	$\dfrac{1}{N} \sum_{i=1}^{N} \dfrac{1}{n-1} \sum_{j=1}^{n} (X_{ij} - \overline{X}_i)^2$
$\mathrm{Var}[m(\theta)]$	$\dfrac{1}{N-1} \sum_{i=1}^{N} (\overline{X}_i - \overline{X})^2 - \dfrac{1}{Nn} \sum_{i=1}^{N} \dfrac{1}{n-1} \sum_{j=1}^{n} (X_{ij} - \overline{X}_i)^2$

It is easy to verify that $\overline{X}$ is unbiased for $\mathbb{E}[m(\theta)]$:

$$\mathbb{E}[\overline{X}] = \frac{1}{Nn} \sum_{i=1}^{N} \sum_{j=1}^{n} \mathbb{E}[X_{ij}] = \frac{1}{Nn} \sum_{i=1}^{N} \sum_{j=1}^{n} \mathbb{E}[\mathbb{E}(X_{ij} \mid \theta_i)]$$

$$= \frac{1}{Nn} \sum_{i=1}^{N} \sum_{j=1}^{n} \mathbb{E}[m(\theta_i)] = \frac{1}{N} \sum_{i=1}^{N} \mathbb{E}[m(\theta_i)]$$

$$= \frac{1}{N} N\mathbb{E}[m(\theta)] = \mathbb{E}[m(\theta)].$$

The verification of the unbiasedness of the other two estimators is deferred to Exercise 4.13 (some hints are given).

Comments on the calculation of the credibility premium for a risk

(i) The credibility factor

$$Z = \frac{n}{n + \dfrac{\mathbb{E}[s^2(\theta)]}{\text{Var}[m(\theta)]}}$$

is the same for all risks in the collective – it only has to be calculated once. Its value is between 0 and 1, and it is an increasing function of n.

(ii) A large value of $\mathbb{E}[s^2(\theta)]$ implies large variability from year to year within risks. This implies low credibility for the data from the particular risk concerned, which implies a low value of the credibility factor Z.

(iii) A large value of $\text{Var}[m(\theta)]$ implies large variability between risks. This implies that data from other risks are not very relevant/informative/reliable, which implies high credibility for the data from the individual risk concerned, and hence we have a high value of the credibility factor Z.

(iv) $\mathbb{E}[s^2(\theta)]$ and $\text{Var}[m(\theta)]$ are positive parameters. While the estimator of the former parameter is always positive, that of the latter can be negative. If this occurs we take a pragmatic approach – we set $\text{Var}[m(\theta)] = 0$; then $Z = 0$ and the credibility estimate for risk i is just the overall mean $\overline{X}$.

To sum up, the credibility premium for risk i in the collective is given by

$$Z \times \{\text{mean for risk } i\} + (1 - Z) \times \{\text{estimate of } \mathbb{E}[m(\theta)]\},$$

that is

$$Z\overline{X}_i + (1 - Z)\overline{X}. \tag{4.35}$$

Example 4.12   Table 4.7 gives the aggregate claims in five successive years from comparable policies covering the estate (buildings, vehicles, stock) of

Table 4.7. *Aggregate claims for the four risks in Example 4.12*

Risk	Year				
	1	2	3	4	5
1	146	151	132	96	136
2	108	94	107	135	93
3	130	142	106	150	95
4	157	175	129	138	159

Table 4.8. *Sample means and variances for the four risks in Example 4.12*

Risk	Risk mean	Risk variance
1	132.2	467.2
2	107.4	287.3
3	124.6	549.8
4	151.6	331.8

four medium-sized businesses. The claims are inflation-adjusted and are in units of £1000.

We will calculate the credibility premium to be charged in the coming year (year 6) for each risk, giving full details for risk 1. We are assuming that the conditions which have held for the past five years justify our adoption of the structural assumptions that underpin EBCT Model 1, and that these conditions continue to hold in the coming year.

First we calculate the sample mean and variance for each risk. The values are given in Table 4.8.

The estimate of $\mathbb{E}[m(\theta)]$ is the mean of the four risk means (the overall mean), namely $\bar{x} = (132.2 + 107.4 + 124.6 + 151.6)/4 = 128.95$.

The estimate of $\mathbb{E}[s^2(\theta)]$ is the mean of the four risk sample variances, namely $(467.2 + 287.3 + 549.8 + 331.8)/4 = 409.025$.

The estimate of $\text{Var}[m(\theta)]$ is an adjusted version of the sample variance for the four risk means, which is 335.637; the estimate is

$$335.637 - 409.025/5 = 253.83.$$

This gives the credibility factor as

$$Z = \frac{5}{5 + 409.03/253.83} = 0.756.$$

Table 4.9. *Credibility premiums*
for the four risks in Example 4.12

Risk	Premium (£)
1	131 410
2	112 650
3	125 660
4	146 080

The credibility premium for risk i is therefore given by

$$0.756\bar{x}_i + 0.244 \times 128.95 = 0.756\bar{x}_i + 31.46.$$

The credibility premium for risk 1 is

$$0.756 \times 132.2 + 31.46 = 131.40 \, (= \pounds131\,400).$$

The credibility premiums for all four risks (calculated in *R* with greater accuracy throughout) are given in Table 4.9.

It is easy to check that the mean of the credibility premiums equals $\bar{x}$, the mean of the risk means (the overall mean of the claims data – the estimate of $\mathbb{E}[m(\theta)]$). This will always be the case (see Exercise 4.14), and reflects the fact that *overall* the insurer receives the appropriate total pure premium for the group of risks. As emphasised earlier, the risks are *heterogeneous* – they have different risk parameters – and the credibility premiums for individual risks vary, reflecting the claims experience of the risks. The higher the value of the mean claims in the available history of the risk ($\bar{x}_i$), the higher the credibility premium for that risk. But *over all the risks* things average out as they should.

4.7 Empirical Bayesian credibility theory: Model 2 – the Bühlmann–Straub model

This model was formulated by Bühlmann and Straub; see Bühlmann and Straub (1970).

The model we have discussed in §4.6 (EBCT Model 1) clearly shows similarities with a "pure" Bayesian approach and is a necessary and useful introduction to "empirical" credibility methods. However, it involves rather restrictive assumptions and is not very useful in practice.

EBCT Model 2 – the Bühlmann–Straub model – encompasses a major generalisation of Model 1 by allowing for changing levels of business (changing *risk volumes*). It is easy to see why this is such an important and practical extension and improvement. The risk during one year may relate to cover for a small business with four shops and three delivery vans on the road – the business may do well, expand, and next year have six shops and four vans on the road. The increased estate (buildings, vans, stock) and general activity is not taken account of by Model 1 but *is* taken account of by Model 2.

With this recognition of changing risk volumes, it is inappropriate now to assume that, given the risk parameter, the claims variables are identically distributed. The assumptions we do make for Model 2 are most conveniently expressed in a manner which makes them less restrictive than was the case for Model 1 – and these assumptions are made not about the claims variables themselves, but about the variables representing claims *per unit of risk volume*.

So, let $Y_1, Y_2, \ldots, Y_n$ represent the aggregate claims in n successive years for a risk, and let $P_1, P_2, \ldots, P_n$ be corresponding risk volumes. These risk volumes are known numbers (not random variables) and can be quantified in various ways – for example, numbers of policies in a changing portfolio, numbers of shops in a chain, numbers of vehicles in a fleet, etc. A sensible general measure which can be used – perhaps obvious once mentioned – is the annual premium income the insurer has charged to cover the risk over recent years (provided the premiums were set sensibly to reflect the risk).

We introduce X_i to represent the aggregate claims in year i scaled to take account of the volume of business, that is

$$X_i = Y_i/P_i, \quad i = 1, 2, \ldots, n, \tag{4.36}$$

so X_i is the aggregate claims per unit of risk volume in year i.

The basic structure of this model is that the distribution of each variable X_i, $i = 1, 2, \ldots, n$, depends on the value of a risk parameter θ, which is fixed for that risk but unknown, and is regarded as a random variable with unknown distribution function. It is not appropriate to assume that the X_i are identically distributed, either conditionally (given θ), or unconditionally.

Assumptions

(1) Given θ, the X_i, $i = 1, 2, \ldots, n$, are independent.
(2) $\mathbb{E}[X_i \mid \theta]$ does not depend on i.
(3) $P_i \text{Var}[X_i \mid \theta]$ does not depend on i.

Under these assumptions we define

$$m(\theta) = \mathbb{E}[X_i|\theta] \quad \text{and} \quad s^2(\theta) = P_i \text{Var}[X_i|\theta]. \tag{4.37}$$

To motivate assumption (3), consider a risk which consists of a portfolio of independent policies – suppose the number of policies in force in year i is P_i (a known number). Suppose also that, for each policy, the aggregate claims in any given year have mean $m(\theta)$ and variance $s^2(\theta)$, where θ is the risk parameter for all policies involved. The aggregate claims in year i is Y_i, and let $X_i = Y_i/P_i$, as in (4.36). It is clear that

$$\mathbb{E}[Y_i \mid \theta] = P_i m(\theta) \text{ and } \mathrm{Var}[Y_i \mid \theta] = P_i s^2(\theta).$$

Hence, $\mathrm{Var}[X_i \mid \theta] = s^2(\theta)/P_i$, and so $P_i \mathrm{Var}[X_i \mid \theta] = s^2(\theta)$, thus satisfying the stated assumption (3).

The pure premium per unit of risk is $m(\theta)$, and we want to estimate the expected value of the aggregate claims in the coming year, namely $\mathbb{E}[Y_{n+1} \mid \theta] = P_{n+1} m(\theta)$ (we assume we know P_{n+1} at the start of year $n + 1$). So our problem is again to estimate $m(\theta)$, given the data $(y_1, P_1), (y_2, P_2), \ldots, (y_n, P_n)$, and of course, derived from these, $x_1, x_2, \ldots, x_n$.

As in the case of Model 1 we want to find the estimator of $m(\theta)$ which is the linear function of the observations $X_1, X_2, \ldots, X_n$ with minimum mean square error; we want to choose $a_0, a_1, \ldots, a_n$ to optimise the estimator of $m(\theta)$ given by $a_0 + a_1 X_1 + a_2 X_2 + \cdots + a_n X_n$.

Theorem 4.14 below gives the derivation of the credibility premium. Again, as in §4.6, and as a preliminary to the proof, we give some expectations we will require (the verifications of the first three are the same as the verifications in the proof of §4.6; the fourth is different, and is deferred to Exercise 4.20).

Lemma 4.13 *With the set-up and notation of EBCT Model 2, we have*

(i) $\mathbb{E}[X_i] = \mathbb{E}[m(\theta)]$;
(ii) $\mathbb{E}[X_i m(\theta)] = \mathbb{E}[m^2(\theta)]$;
(iii) $\mathbb{E}[X_i X_j] = \mathbb{E}[m^2(\theta)]$ *for* $i \neq j$;
(iv) $\mathbb{E}[X_i^2] = \frac{1}{P_i}\mathbb{E}[s^2(\theta)] + \mathbb{E}[m^2(\theta)]$.

Unlike the situation in the previous analysis, the constants $a_1, a_2, \ldots, a_n$ in this case are not equal (since the X_i are not identically distributed).

Theorem 4.14 *Let $X_1, X_2, \ldots, X_n$ be a sequence of random variables, each of whose distribution depends on a parameter θ, and which, given θ, are independent, with $\mathbb{E}[X_i \mid \theta] = m(\theta)$ and $P_i \mathrm{Var}[X_i \mid \theta] = s^2(\theta)$, $i = 1, \ldots, n$.*
Then the estimator $a_0 + \sum_{j=1}^{n} a_j X_j$ of $m(\theta)$ for which

$$\mathbb{E}\left[\left\{m(\theta) - a_0 - \sum_{j=1}^{n} a_j X_j\right\}^2\right]$$

is minimised is given by

$$Z\overline{X} + (1 - Z)\mathbb{E}[m(\theta)],$$

where

$$\overline{X} = \frac{\sum_{i=1}^{n} P_i X_i}{\sum_{i=1}^{n} P_i} \quad \text{and} \quad Z = \frac{\sum_{i=1}^{n} P_i}{\sum_{i=1}^{n} P_i + \frac{\mathbb{E}[s^2(\theta)]}{\mathrm{Var}[m(\theta)]}}.$$

Proof Let $S = \mathbb{E}\left[\{m(\theta) - a_0 - a_1 X_1 - a_2 X_2 - \cdots - a_n X_n\}^2\right]$. Taking the partial derivative of S with respect to a_0 gives

$$\frac{\partial S}{\partial a_0} = 0 \Rightarrow \mathbb{E}\left[m(\theta) - a_0 - a_1 X_1 - a_2 X_2 - \cdots - a_n X_n\right] = 0.$$

Noting that $\mathbb{E}[X_i] = \mathbb{E}[m(\theta)]$ (by Lemma 4.13(i)), this gives

$$a_0 = (1 - a_1 - a_2 - \cdots - a_n)\mathbb{E}[m(\theta)].$$

For $j = 1, 2, \ldots, n$ we have

$$\frac{\partial S}{\partial a_j} = 0 \Rightarrow \mathbb{E}\left[X_j\{m(\theta) - a_0 - a_1 X_1 - a_2 X_2 - \cdots - a_n X_n\}\right] = 0.$$

This gives

$$\mathbb{E}[X_j m(\theta)] = a_0 \mathbb{E}[X_j] + \sum_{i \neq j} a_i \mathbb{E}[X_i X_j] + a_j \mathbb{E}[X_j^2],$$

from which, using Lemma 4.13, we have

$$\mathbb{E}[m^2(\theta)] = a_0 \mathbb{E}[m(\theta)] + \sum_{i=1}^{n} a_i \mathbb{E}[m^2(\theta)] + a_j \mathbb{E}[s^2(\theta)]/P_j.$$

Using the expression for a_0 above and Lemma 4.13, and after some algebra, we find

$$a_j = \left(1 - \sum_{i=1}^{n} a_i\right)\frac{\mathrm{Var}[m(\theta)]}{\mathbb{E}[s^2(\theta)]}P_j.$$

Hence, summing from $j = 1$ to n, we have

$$\sum_{j=1}^{n} a_j = \left(1 - \sum_{i=1}^{n} a_i\right)\frac{\mathrm{Var}[m(\theta)]}{\mathbb{E}[s^2(\theta)]}\sum_{j=1}^{n} P_j,$$

which gives

$$a_0 = \frac{\mathbb{E}[m(\theta)]\dfrac{\mathbb{E}[s^2(\theta)]}{\mathrm{Var}[m(\theta)]}}{\sum_{i=1}^{n} P_i + \frac{\mathbb{E}[s^2(\theta)]}{\mathrm{Var}[m(\theta)]}},$$

and

$$a_j = \frac{P_j}{\sum_{i=1}^{n} P_i + \frac{\mathbb{E}[s^2(\theta)]}{\mathrm{Var}[m(\theta)]}}, \quad j = 1, 2, \ldots, n.$$

Putting these expressions into $a_0 + a_1 X_1 + \cdots + a_n X_n$, we find that the "best" linear estimator is given by

$$\frac{\mathbb{E}[m(\theta)]\frac{\mathbb{E}[s^2(\theta)]}{\mathrm{Var}[m(\theta)]} + \sum_{i=1}^{n} P_i X_i}{\sum_{i=1}^{n} P_i + \frac{\mathbb{E}[s^2(\theta)]}{\mathrm{Var}[m(\theta)]}}.$$

Table 4.10. *Collective of risks for Model 2*

		Year			
		1	2	$\cdot \quad \cdot$	n
Risk	1	$Y_{11}; P_{11}$	$Y_{12}; P_{12}$	$\cdot \quad \cdot$	$Y_{1n}; P_{1n}$
	2	$Y_{21}; P_{21}$	$Y_{22}; P_{22}$	$\cdot \quad \cdot$	$Y_{2n}; P_{2n}$
	$\cdot$	$\cdot$	$\cdot$	$\cdot \quad \cdot \quad \cdot$	$\cdot$
	N	$Y_{N1}; P_{N1}$	$Y_{N2}; P_{N2}$	$\cdot \quad \cdot$	$Y_{Nn}; P_{Nn}$

This may be written as

$$Z\overline{X} + (1 - Z)\mathbb{E}[m(\theta)],$$

where

$$Z = \frac{\sum_{i=1}^{n} P_i}{\sum_{i=1}^{n} P_i + \frac{\mathbb{E}[s^2(\theta)]}{\text{Var}[m(\theta)]}} \quad \text{and} \quad \overline{X} = \frac{\sum_{i=1}^{n} P_i X_i}{\sum_{i=1}^{n} P_i},$$

and the result is proved. $\qquad \square$

Notes

(1) $\overline{X}$ can of course also be written as $\overline{X} = \dfrac{\sum_{i=1}^{n} Y_i}{\sum_{i=1}^{n} P_i}$.

(2) The coefficient a_j of X_j in the optimal estimator is proportional to P_j, the risk volume for that year – this makes good sense as the claims data for years with higher risk volumes should have greater influence on the value of the credibility premium.

(3) In the case that the risk volumes are all equal, Model 2 is, in practice, the same as Model 1; putting all the risk volumes P_i equal to 1 gives

$$Z = \frac{n}{n + \dfrac{\mathbb{E}[s^2(\theta)]}{\text{Var}[m(\theta)]}},$$

which is exactly the expression we had in the case of Model 1.

We now consider how to estimate the three structural parameters $\mathbb{E}[m(\theta)]$, $\mathbb{E}[s^2(\theta)]$ and $\text{Var}[m(\theta)]$ using data from a collective of N (fixed) comparable risks. Our data consist of values (y_{ij}, P_{ij}), where y_{ij} is an observation of a random variable Y_{ij} which represents the aggregate claims for risk i in year j, $i = 1, 2, \ldots, N$, $j = 1, 2, \ldots n$. We present the data in the cells in Table 4.10 in the form $Y_{ij}; P_{ij}$.

For each i and j define $X_{ij} = Y_{ij}/P_{ij}$. For each risk, say risk i, the distribution of each X_{ij}, $j = 1, 2, \ldots, n$, depends on a risk parameter θ_i, which is fixed for that risk, but unknown.

For each risk, say risk i, we make the following distributional assumptions.

Assumptions

(1) Given θ_i, the X_{ij}, $j = 1, 2, \ldots, n$, are independent.
(2) $\mathbb{E}[X_{ij} \mid \theta_i]$ does not depend on j.
(3) $P_{ij}\text{Var}[X_{ij} \mid \theta_i]$ does not depend on j.

Under these assumptions we define

$$m(\theta_i) = \mathbb{E}[X_{ij} \mid \theta_i] \text{ and } s^2(\theta_i) = P_{ij}\text{Var}[X_{ij} \mid \theta_i].$$

Each risk therefore has the same structure as that of the single risk we considered earlier – this gives us the within risk structure we require.

We now make assumptions to give us the appropriate between risk structure:

Assumptions (continued)

(4) For different risks $i \neq j$, the pairs of variables (θ_i, X_{il}) and (θ_j, X_{jk}), $l, k = 1, 2, \ldots, n$, are independent.
(5) The risk parameters θ_i, $i = 1, 2, \ldots, N$, are iid.

Since θ_i, $i = 1, 2, \ldots, N$, are identically distributed, it follows that none of $\mathbb{E}[m(\theta_i)]$, $\mathbb{E}[s^2(\theta_i)]$ or $\text{Var}[m(\theta_i)]$ depend on i, and so we write them as $\mathbb{E}[m(\theta)]$, $\mathbb{E}[s^2(\theta)]$ and $\text{Var}[m(\theta)]$, respectively.

We now seek estimators of these three structural parameters, and at this point it is helpful to introduce some new notation – we will adopt the statistical convention of using a clear point ($\bullet$) in place of a subscript to indicate that we have summed over that subscript (so, for example, the sum of $x_{31}, x_{32}, x_{33}, \ldots, x_{3n}$ is denoted $x_{3\bullet}$, the sum of $y_{15}, y_{25}, y_{35}, \ldots, y_{m5}$ is denoted $y_{\bullet 5}$, and the sum of z_{ij} over all values of i and j is denoted $z_{\bullet\bullet}$).

We now have

$$\overline{X}_i = \frac{\sum_{j=1}^n P_{ij}X_{ij}}{\sum_{j=1}^n P_{ij}} = \frac{\sum_{j=1}^n Y_{ij}}{\sum_{j=1}^n P_{ij}} = \frac{Y_{i\bullet}}{P_{i\bullet}}, \tag{4.38}$$

$$\overline{X} = \frac{\sum_{i=1}^N \sum_{j=1}^n P_{ij}X_{ij}}{\sum_{i=1}^N \sum_{j=1}^n P_{ij}} = \frac{\sum_{i=1}^N P_{i\bullet}\overline{X}_i}{\sum_{i=1}^N \sum_{j=1}^n P_{ij}} = \frac{\sum_{i=1}^N \sum_{j=1}^n Y_{ij}}{\sum_{i=1}^N \sum_{j=1}^n P_{ij}} = \frac{Y_{\bullet\bullet}}{P_{\bullet\bullet}} \tag{4.39}$$

and finally

$$P^* = \frac{1}{Nn - 1} \sum_{i=1}^N P_{i\bullet}\left(1 - \frac{P_{i\bullet}}{P_{\bullet\bullet}}\right). \tag{4.40}$$

Table 4.11. *Usual estimators of the structural parameters in EBCT Model 2*

Structural parameter	Estimator
$\mathbb{E}[m(\theta)]$	$\overline{X}$
$\mathbb{E}[s^2(\theta)]$	$\dfrac{1}{N}\displaystyle\sum_{i=1}^{N}\dfrac{1}{n-1}\sum_{j=1}^{n}P_{ij}(X_{ij}-\overline{X}_i)^2$
$\text{Var}[m(\theta)]$	$\dfrac{1}{P_*}\left\{\dfrac{1}{Nn-1}\displaystyle\sum_{i=1}^{N}\sum_{j=1}^{n}P_{ij}(X_{ij}-\overline{X})^2-\dfrac{1}{N}\sum_{i=1}^{N}\dfrac{1}{n-1}\sum_{j=1}^{n}P_{ij}(X_{ij}-\overline{X}_i)^2\right\}$

It is important to note our *notation* here: the mean of the claims (per unit of risk volume) for an individual risk (risk i) is now denoted $\overline{X}_i$ and is a *weighted mean*, the weights being the risk volumes (see definition of $\overline{X}_i$ above). The symbol $\overline{X}$ now denotes the overall (weighted) mean claims (per unit of risk volume) for all risks involved.

The credibility premium for risk i now appears as

$$Z_i\overline{X}_i + (1 - Z_i)\mathbb{E}[m(\theta)], \tag{4.41}$$

where $\overline{X}_i$ is given by (4.38) and

$$Z_i = \frac{P_{i\bullet}}{P_{i\bullet} + \dfrac{\mathbb{E}[s^2(\theta)]}{\text{Var}[m(\theta)]}}. \tag{4.42}$$

We give the usual estimators for the structural parameters in Table 4.11. These estimators are unbiased – it is easy to verify that $\overline{X}$ is unbiased for $\mathbb{E}[m(\theta)]$:

$$\mathbb{E}[X_{ij}] = \mathbb{E}[\mathbb{E}(X_{ij} \mid \theta_i)] = \mathbb{E}[m(\theta_i)] = \mathbb{E}[m(\theta)],$$

and it follows immediately that

$$\mathbb{E}[\overline{X}] = \mathbb{E}\left[\frac{\sum_{i=1}^{N}\sum_{j=1}^{n}P_{ij}X_{ij}}{\sum_{i=1}^{N}\sum_{j=1}^{n}P_{ij}}\right] = \mathbb{E}[m(\theta)].$$

We defer the unbiasedness of the estimator of $\mathbb{E}[s^2(\theta)]$ to Exercise 4.21.

Comments

(1) The estimators revert to those of Model 1 in the case that the risk volumes are the same for all risks and years. Setting $P_{ij} = 1$ for all i, j we have $\sum_{j=1}^{n}P_{ij} = n$ and $P^* = [n(N - 1)]/(Nn - 1)$ (see Exercise 4.22).

Table 4.12. *Aggregate claims/volumes of business for*
the four risks in Example 4.15

Risk	1	2	3	4	5
1	33 ; 4	26 ; 4	28 ; 5	41 ; 5	34 ; 5
2	22 ; 3	16 ; 2	19 ; 3	29 ; 4	33 ; 5
3	114 ; 16	117 ; 19	116 ; 18	171 ; 22	139 ; 22
4	77 ; 8	74 ; 8	59 ; 7	86 ; 10	98 ; 12

(2) While the estimates of $\mathbb{E}[m(\theta)]$, $\mathbb{E}[s^2(\theta)]$ and $\text{Var}[m(\theta)]$ only have to be calculated once for the collective (and hence the expression $\mathbb{E}[s^2(\theta)]/\text{Var}[m(\theta)]$ is the same for all risks), the credibility factors Z_i are different for different risks, since the expression for Z_i involves $P_{i\bullet}$, the total risk volume for that risk.

(3) Z_i is an increasing function of $P_{i\bullet}$ – high risk volume for a risk implies high credibility factor for that risk.

(4) As with Model 1, this model can be applied to estimating the *expected number of claims* in the coming year rather than finding the credibility premium – this is the case if the variables Y_{ij} represent the *number of claims* for the risk rather than the aggregate claims.

To sum up, the credibility premium for risk i in the collective is given by

$$Z_i \overline{X}_i + (1 - Z_i)\overline{X},$$

where

$$\overline{X}_i = \frac{\sum_{j=1}^n P_{ij} X_{ij}}{\sum_{j=1}^n P_{ij}}, \quad \text{and} \quad Z_i = \frac{\sum_{j=1}^n P_{ij}}{\sum_{j=1}^n P_{ij} + \frac{\mathbb{E}[s^2(\theta)]}{\text{Var}[m(\theta)]}},$$

with the structural parameters estimated as above.

Example 4.15　Table 4.12 gives the aggregate claims Y_{ij}, $i = 1, 2, 3, 4$, $j = 1, 2, 3, 4, 5$, in five successive years from comparable policies covering the estate (buildings, vehicles, stock) of four medium-sized companies. The level of activity for each company has been changing from year to year, and for each company and year loss adjusters have given a quantitative assessment (P_{ij}) of the relative volume of business (the exposure covered by the policy). The claims are inflation-adjusted and are in units of £1000. The cells in the table contain the data in the form Y_{ij}; P_{ij}.

We will calculate the credibility premium per unit of risk volume to be charged in the coming year for each risk. We are assuming that the conditions which have held for the past five years justify our adopting the structural

Table 4.13. *The X_{ij} for the four risks in Example 4.15*

Risk	1	2	3	4	5
1	8.2500	6.5000	5.6000	8.2000	6.8000
2	7.3333	8.0000	6.3333	7.2500	6.6000
3	7.1250	6.1579	6.4444	7.7727	6.3182
4	9.6250	9.2500	8.4286	8.6000	8.1667

Table 4.14. *Intermediate calculations for Example 4.15*

Risk	$Y_{i\bullet}$	$P_{i\bullet}$	$\overline{X}_i$	$\sum_{j=1}^{5} P_{ij}(X_{ij} - \overline{X}_i)^2$	$\sum_{j=1}^{5} P_{ij}(X_{ij} - \overline{X})^2$
1	162	23	7.0435	24.407	26.148
2	119	17	7.0000	4.7167	6.4431
3	657	97	6.7732	37.653	66.516
4	394	45	8.7556	13.155	106.06

assumptions which underpin EBCT Model 2 (for example, we are assuming that the claims per unit of risk volume from year to year for a particular company have constant mean).

The calculations for this example were done using **R** (see the computing recipes listed after this example), and the intermediate results are quoted to five significant figures.

First we calculate the X_{ij} (these are given in Table 4.13). Next we calculate the values of $Y_{i\bullet}, P_{i\bullet}, \overline{X}_i$ (using (4.38)) and $\overline{X}$ (using (4.39)), followed by the sums we require for the calculation of the estimates; these intermediate values are shown in Table 4.14.

For the calculations in the last column of Table 4.14 we require the value

$$\overline{X} = Y_{\bullet\bullet}/P_{\bullet\bullet} = 1332/182 = 7.3187.$$

We will also require P^*, which, by (4.40), is given by

$$\{23(1 - 23/182) + 17(1 - 17/182) + 97(1 - 97/182) + 45(1 - 45/182)\}/19$$
$$= 6.0359.$$

From Table 4.11, the estimate of $\mathbb{E}[m(\theta)]$ is 7.3187, and the estimate of $\mathbb{E}[s^2(\theta)]$ is given by

$$(24.407 + 4.7167 + 37.653 + 13.155)/(4 \times 4) = 4.9957.$$

Further, the estimate of $\text{Var}[m(\theta)]$ is given by

$$[(26.148 + 6.4431 + 66.516 + 106.06)/19 - 4.9957]/6.0359 = 0.96134,$$

Table 4.15. *Credibility factors and*
premiums for the four risks in Example 4.15

Risk	Z_i	Premium per unit of risk (£)
1	0.8157	7094
2	0.7659	7075
3	0.9492	6801
4	0.8965	8607

(note that the correct value from R is 0.96137). The ratio $\mathbb{E}[s^2(\theta)]/\mathrm{Var}[m(\theta)]$ from R is 5.1965. The credibility factor for risk 1 is $23/(23 + 5.1965) = 0.81570$ (by (4.42)), and the credibility premium per unit of risk volume is

$$0.81570 \times 7.0435 + 0.18430 \times 7.3187 = 7.094.$$

The credibility factors and premiums for all four companies are given in Table 4.15.

If the assessments of the risk volumes for companies 1, 2, 3 and 4 for the coming year are 5, 6, 24 and 11, respectively, then the insurer's credibility premiums will be £35 470, £42 450, £163 220 and £94 680.

Computing recipes in R for EBCT Model 2

Recipe 1: using a simple sequence of elementary step-by-step calculations for Example 4.15

Note that # is the comment symbol in R.

```
n=5 # number of years' data for each risk
N=4 # number of risks in the collective
y1=c(33,26,28,41,34) # claims for risk 1
p1=c(4,4,5,5,5) # risk volumes for risk 1
x1=y1/p1 # claims per unit risk volume for risk 1
c12 = sum(y1)    # total claims for risk 1
c13 = sum(p1)    # total risk volumes for risk 1
x1bar=c12/c13
```

So x1bar contains $\overline{X}_1$. Carry out similar commands for y2, p2, x2, c22, c23, x2bar (containing $\overline{X}_2$); y3, p3, x3, c32, c33, x3bar (containing $\overline{X}_3$); and y4, p4, x4, c42, c43, x4bar (containing $\overline{X}_4$).

```
c2=c(c12,c22,c32,c42)
c3=c(c13,c23,c33,c43)
```

```
c5=c6=1:N*0 # set up two N-element vectors
xbar=sum(c2)/sum(c3)
c5[1]=sum(p1*(x1-x1bar)^2)
c6[1]=sum(p1*(x1-xbar)^ 2)
```

So `xbar`, `c5[1]` and `c6[1]` contain

$$\overline{X}, \sum_{j=1}^{5} P_{1j}(X_{1j} - \overline{X}_1)^2 \text{ and } \sum_{j=1}^{5} P_{1j}(X_{1j} - \overline{X})^2,$$

respectively. Carrry out similar commands for `c5[2]`, `c5[3]`, `c5[4]`, `c6[2]`, `c6[3]` and `c6[4]`.

```
pstar=sum(c3*(1-c3/sum(c3)))/(N*n-1)
e1 = xbar
e2 = sum(c5)/(N*(n-1))
e3 = (sum(c6)/(N*n-1) - e2)/pstar
z1 = c13/(c13 + e2/e3)
prem1 = z1*x1bar + (1-z1)*xbar
```

So `e1`, `e2` and `e3` contain the estimates of $\mathbb{E}[m(\theta)]$, $\mathbb{E}[s^2(\theta)]$ and $\text{Var}[m(\theta)]$, respectively, and `z1` and `prem1` contain the credibility factor and credibility premium, respectively, for risk 1. Carry out similar commands for `z2`, `prem2`, `z3`, `prem3`, `z4` and `prem4`.

Recipe 2: via a function which uses matrices and a simple loop
First the claims and risk volumes are entered into matrices (here of dimension 4×5 and named `mex411y` and `mex411p`). The data are entered column by column.

```
mex411y=matrix(c(33,22,114,77,26,...,139,98),4,5)
mex411p=matrix(c(4,3,16,8,4,...,22,12),4,5)
```

where ... denotes other values to be entered. A vector containing the credibility premiums, here called `credpremiums`, is created issuing the command

```
credpremiums=ebctmodel2(4,5,mex411y,mex411p)
```

which calls up and executes a function called `ebctmodel2` previously stored as a text file as follows:

```
ebctmodel2=function(N,n,my,mp){
mx=my/mp
c2=apply(my,1,sum)
c3=apply(mp,1,sum)
```

```
c5=c6=1:N*0
xibar=c2/c3
xbar=sum(c2)/sum(c3)
for (i in 1:N){
c5[i]=sum(mp[i,]*(mx[i,]-xibar[i])^ 2)
c6[i]=sum(mp[i,]*(mx[i,]-xbar)^2)}
pstar=sum(c3*(1-c3/sum(c3)))/(N*n-1)
e1 = xbar
e2 = sum(c5)/(N*(n-1))
e3 = (sum(c6)/(N*n-1) - e2)/pstar
z=c3/(c3+e2/e3)
prem=z*xibar+(1-z)*xbar
prem}
```

Exercises

4.1   Let P be the premium for a risk S calculated using the exponential premium principle, with utility function parameter β (> 0).

    (a) By using Jensen's inequality (see Appendix A) on $\mathbb{E}[-e^{\beta S}]$ show that $P \geq \mathbb{E}[S]$.

    (b) By expanding $\mathbb{E}[e^{\beta S}]$ as far as the term in β^2, show that, for small β, the results of using the exponential premium principle can be approximated by using the variance principle.

4.2   Show that

    (a) the exponential premium calculation principle satisfies the additivity property for independent risks,

    (b) the expected value and the standard deviation premium calculation principles satisfy the scale invariance property, but the variance principle does not.

4.3   Let $u(x)$ be a utility function (with $u'(x) > 0$ and $u''(x) < 0$). The *risk aversion* of an individual or company using this utility function and with wealth x can be measured by the function

$$r(x) = \frac{-u''(x)}{u'(x)},$$

where higher values of $r(x)$ correspond to greater risk aversion (see Appendix A).

    Find the risk aversion $r(x)$ under

    (a) the log utility function $u(x) = \beta \log x$,

(b) the exponential utility function $u(x) = -e^{-\beta x}$,

(c) the fractional power utility function $u(x) = \beta x^{1/2}$,

where in all cases $x > 0$ and $\beta > 0$.

In each case state whether or not the risk aversion depends on

(1) current wealth,

(2) the parameter β.

If there is dependence, comment on the nature of that dependence.

4.4 An insurer sets the premium P for a risk as the 90th percentile of the distribution of the risk. Suppose the risk, in units of £1000, has a Pa(5, 16) distribution (and hence has mean £4000).

(a) Calculate the premium P.

(b) Calculate the relative security loadings required if the same premium were to be arrived at using (1) the expected value principle and (2) the standard deviation principle.

4.5 Consider an individual who adopts $u(x) = -e^{-0.004x}$ as the (exponential) utility function, and makes decisions according to the expected utility criterion. Suppose the individual buys insurance to cover a loss which has a normal distribution with mean μ = £5000 and standard deviation σ. Find the range of values of σ for which the premium the individual will be prepared to pay is £6000 or less.

4.6 Consider an individual who adopts $u(x) = -e^{-0.0001x}$ as the (exponential) utility function and makes decisions according to the expected utility criterion. Suppose the individual buys insurance to cover a loss which has a compound Poisson distribution with rate parameter 2 and an individual claim size which is constant at £5000. Find the maximum premium the individual will be prepared to pay.

4.7 Repeat Exercise 4.6, but in the case in which the individual claim size (£) is uniformly distributed on (0, 10 000).

4.8 Let S be a risk and suppose we model S as $S \sim N(\mu, \sigma^2)$. The insurer wants to fix a premium P for insuring the risk; let us call $\Pr(S \le P)$ the insurer's *security level*.

Show that

(a) setting P using the expected value principle with relative security loading α gives the insurer a security level of $\Phi(\alpha\mu/\sigma)$, and

(b) setting P using the variance principle with relative security loading α gives the insurer a security level of $\Phi(\alpha\sigma)$.

4.9 Suppose the number of claims which arise in a year on a group of policies is modelled as $X \sim \text{Poi}(\lambda)$ and that we observe a total of 14 claims over a six year period. Suppose also we adopt a gamma(6, 3) distribution as a prior distribution for λ.

(a) State the maximum likelihood estimate of λ and the prior mean.

(b) State the posterior distribution of λ, find the mode of this distribution, and hence state the Bayesian estimate of λ under all or nothing loss.

(c) Recall the result of Exercise 2.7(b).

    (1) Using tables, find the Bayesian estimate of λ under absolute error loss.

    (2) Using tables, find an equal-tailed 95% Bayesian interval estimate of λ, that is an interval (λ_L, λ_U), such that

$$\Pr(\lambda > \lambda_U \mid \underline{x}) = \Pr(\lambda < \lambda_L \mid \underline{x}) = 0.025.$$

(d) Find the credibility estimate (the Bayesian estimate under squared-error loss) of λ.

4.10 Suppose we model the annual claims which arise under a risk, X, in units of £1000, as $X \mid \theta \sim N(\theta, 0.5^2)$, and we adopt a $N(2, 0.2^2)$ prior for θ. You observe a total claim amount of £16 240 over a period of seven years.

(a) Find the marginal (unconditional) distribution of X.

(b) Find the credibility factor and the credibility premium for the risk.

(c) Find an equal-tailed 95% Bayesian interval estimate of θ.

4.11 A general insurer examines the records for a collective of five separate risks, each of which has been in existence for at least eight years. The mean and variance of the aggregate claims over the past eight years (adjusted for inflation) for each risk are given in Table 4.16. Calculate the credibility premiums for all five risks (using EBCT Model 1).

4.12 Table 4.17 gives the values of X_{ij}, the aggregate claims for each of three risks in a collective for each of the past five years.

(a) Show that the credibility premium for risk i is given by

$$0.9749\overline{X}_i + 1.792$$

and calculate the premium for each risk.

Table 4.16. *Means and variances of the aggregate claims in Exercise 4.11*

Risk	Within risk mean	Within risk variance
1	121	246
2	104	187
3	130	223
4	107	159
5	118	204

Table 4.17. *Aggregate claims for*
Exercise 4.12

Risk i	Year j				
	1	2	3	4	5
1	76	65	77	68	74
2	59	54	62	50	56
3	95	81	89	82	83

(b) Explain why the credibility premiums depend almost entirely on the means for the individual risks.

4.13 In EBCT Model 1 show that the second two estimators in Table 4.6 are unbiased for the parameters concerned by answering (a) and (b).

(a) Show that

$$\frac{1}{N} \sum_{i=1}^{N} \frac{1}{n-1} \sum_{j=1}^{n} (X_{ij} - \overline{X}_i)^2$$

is unbiased for $\mathbb{E}[s^2(\theta)]$.

Hint: For fixed i, and given θ_i, the variables X_{ij}, $j = 1, 2, \ldots, n$, are iid and hence constitute a random sample with mean $\overline{X}_i$ from a distribution with variance $\mathrm{Var}[X_{ij} \mid \theta_i] = s^2(\theta_i)$. Now use the standard result that the sample variance of a random sample of observations is an unbiased estimator of the population variance, that is

$$\mathbb{E}\left[\frac{1}{n-1} \sum_{j=1}^{n} \left(X_{ij} - \overline{X}_i \right)^2 \mid \theta_i \right] = s^2(\theta_i).$$

(b) Show that

$$\frac{1}{N-1} \sum_{i=1}^{N} (\overline{X}_i - \overline{X})^2 - \frac{1}{Nn} \sum_{i=1}^{N} \frac{1}{n-1} \sum_{j=1}^{n} (X_{ij} - \overline{X}_i)^2$$

is unbiased for $\mathrm{Var}[m(\theta)]$.

Hint: The $\overline{X}_i$ are iid, with mean $\overline{X}$. Use the

$$\mathbb{E}[\text{sample variance}] = \text{population variance}$$

result as used in part (a), together with the conditional variance formula (1.4).

4.14 Suppose the credibility premiums for a collective of risks are calculated using EBCT Model 1. Show that, in general, the mean of the credibility

Table 4.18. *Data for Exercise 4.15*

Risk	1	2	3	4	5	m(i)	ss(i)
	\multicolumn{5}{c}{Year}						
1	44	49	53	43	61	50	216
2	70	74	61	83	72	72	250
3	54	67	49	44	61	55	338

Table 4.19. *Numbers of claims for Exercise 4.16*

Risk	1	2	3	4	5	6
	\multicolumn{6}{c}{Year}					
1	127	156	166	141	123	151
2	150	134	123	141	147	127
3	138	179	176	150	154	193
4	176	158	181	129	182	110

premiums equals the mean of the observed risk means (the overall mean of the claims data).

4.15 Table 4.18 shows the values of x_{ij}, $i = 1, 2, 3$, $j = 1, 2, 3, 4, 5$, the annual aggregate claims for the past five years for three risks, together with some summary statistics, where $m(i)$ denotes the mean claims for risk i, that is $m(i) = \bar{x}_i$, and $ss(i)$ denotes $\sum_{j=1}^{5}(x_{ij} - m(i))^2$. Calculate credibility estimates of the pure premium for the coming year for risk 1 using:

(a) the normal/normal Bayesian model with $X \mid \theta \sim N(\theta, 100)$ and a $N(65, 200)$ prior distribution for θ, and

(b) EBCT Model 1.

4.16 Table 4.19 gives the annual numbers of claims for the past six years for four different risks in a collective. Table 4.20 gives some summary statistics.

(a) Using the Poisson/gamma model with prior parameters $\alpha = 150$ and $\beta = 1$, calculate a credibility estimate of the number of claims for the coming year for each risk.

(b) Using the Poisson/gamma model with prior parameters $\alpha = 450$ and $\beta = 3$, calculate a credibility estimate of the number of claims for the coming year for each risk.

Table 4.20. *Summary statistics for Exercise 4.16*

Risk	$\overline{X}_i$	$\sum\limits_{j=1}^{6}(X_{ij} - \overline{X}_i)^2$
1	144	1416
2	137	590
3	165	2176
4	156	4550

Table 4.21. *Aggregate claims/volumes of business for Exercise 4.17*

			Year			
Risk	1	2	3	4	5	6
1	32 ; 6	19 ; 5	22 ; 5	28 ; 6	19 ; 6	29 ; 7
2	56 ; 11	71 ; 10	77 ; 10	56 ; 12	91 ; 14	89 ; 14
3	23 ; 4	22 ; 5	32 ; 6	29 ; 7	37 ; 7	52 ; 8
4	27 ; 8	30 ; 8	23 ; 8	33 ; 8	29 ; 8	31 ; 8
5	79 ; 14	130 ; 15	115 ; 16	91 ; 18	141 ; 18	118 ; 18

(c) Using EBCT Model 1, estimate the number of claims for the coming year for each risk.

(d) Comment on the differences between the answers to parts (a), (b) and (c).

4.17 Table 4.21 gives the aggregate claims Y_{ij}, $i = 1, 2, 3, 4, 5$, $j = 1, 2, 3, 4, 5, 6$, in six successive years from five comparable portfolios. For each portfolio and year, we have a measure (P_{ij}) of the volume of business (the exposure covered by the policies in the portfolio). The claims are inflation-adjusted and are in units of £1000. The cells in the table contain the data in the form Y_{ij} ; P_{ij}.

(a) Using EBCT Model 2, calculate the credibility premiums per unit of risk volume for the coming year for all five risks.

(b) Supposing the risk volumes for the coming year for the five portfolios are assessed to be 8, 16, 10, 8 and 20, respectively, calculate the total value of the five premiums for next year.

4.18 The mean claims (£, to the nearest £10) over the past ten years for each risk in a collective of six comparable risks is given in Table 4.22.

Table 4.22. *Mean claims for Exercise 4.18*

Risk	1	2	3	4	5	6
Mean claims	5120	6230	7470	8230	9670	11 050

Table 4.23. *Volumes of business for Exercise 4.18*

Risks	1–2	3–8	9–12
Vol. of business per year in years 1–6	5	5	8
Vol. of business per year in years 7–10	8	10	10

(a) Pure premiums for the coming year's cover for all six risks are to be calculated using EBCT Model 1, using the experience of the group of six risks as the collateral information. The pure premium for next year for risk 1 has been calculated on this basis and is £5486.

   (1) Determine the value of the credibility factor.
   (2) Determine the value of the mean of the within risk variances of the claims over the past ten years for these six risks.
   (3) Calculate the pure premiums for the coming year for risks 2, 3 and 4.

(b) Suppose now that you are told that the six risks in part (a) are in fact part of a larger collective of 12 comparable risks (the additional risks being numbered 7–12). In addition you have information on the volume of business corresponding to these risks as given in Table 4.23. Pure premiums for the coming year's cover are now to be calculated using EBCT Model 2 using the experience of the group of all 12 risks as the collateral information. The pure premiums per unit of risk volume for next year for risks 1 and 2 have been calculated on this basis and are £876 and £1029, respectively. Calculate the pure premiums per unit of risk volume for the coming year for risks 3 and 4.

4.19 Table 4.24 shows the aggregate claims (£), denoted Y_{ij}, for each of three risks over five years, together with some summary statistics, where $m(i)$ denotes $\bar{y}_i$ and $ss(i)$ denotes $\sum_{j=1}^{5}(y_{ij} - m(i))^2$.

(a) Using EBCT Model 1, calculate the credibility premium for the coming year for each of these three risks.

(b) Now suppose you have available a risk volume, denoted P_{ij}, corresponding to the aggregate claims Y_{ij}. You are given the values of the

Table 4.24. *Data for Exercise 4.19*

Risk	Year					$m(i)$	$ss(i)$
	1	2	3	4	5		
1	2543	1964	2494	2527	2007	2307	346 714
2	2799	2806	2176	2152	2657	2518	432 146
3	2421	2337	1550	1694	2023	2005	587 350

Table 4.25. *Total risk volumes for Exercise 4.19*

Risk	1	2	3
$\sum_{j=1}^{5} P_{ij}$	765	881	535

total risk volume, $\sum_{j=1}^{5} P_{ij}$ for each risk, as recorded in Table 4.25. Using EBCT Model 2, the credibility premium per unit of risk volume for the coming year for risk 1 has been calculated to be 15.53. Calculate the credibility premiums for the coming year for each risk, given that the risk volumes will be 165, 180 and 120 for risks 1, 2 and 3, respectively.

4.20 In EBCT Model 2 show that $\mathbb{E}[X_i^2] = (1/P_i)\mathbb{E}[s^2(\theta)] + \mathbb{E}[m^2(\theta)]$.

4.21 In EBCT Model 2 show that

$$\mathbb{E}\left[\frac{1}{N}\sum_{i=1}^{N}\frac{1}{n-1}\sum_{j=1}^{n}P_{ij}(X_{ij}-\overline{X}_i)^2\right] = \mathbb{E}[s^2(\theta)].$$

Hint: First show that

$$\sum_{j=1}^{n}P_{ij}(X_{ij}-\overline{X}_i)^2 = \sum_{j=1}^{n}P_{ij}X_{ij}^2 - P_{i\bullet}\overline{X}_i^2$$

and $\mathbb{E}[\overline{X}_i^2] = (1/P_{i\bullet})\mathbb{E}[s^2(\theta)] + \mathbb{E}[m^2(\theta)]$.

4.22 In EBCT Model 2 show that, in the case $P_{ij} = 1$ for all i, j, we have $P^* = n(N-1)/(Nn-1)$, and hence show that the estimators of the three structural parameters given in the text revert to those used in Model 1.

4.23 For the past five years an insurance company has insured 12 different small chains of general retailers against loss from, or damage to, their premises by fire, theft or other general insurable cause. For chain

Table 4.26. *Annual claims and numbers of shops for Exercise 4.23*

Chain	Year 1	Year 2	Year 3	Year 4	Year 5
1	4500 ; 4	1990 ; 5	6820 ; 5	7560 ; 6	1440 ; 6
2	940 ; 3	460 ; 3	520 ; 3	2770 ; 3	800 ; 4
3	4010 ; 5	5170 ; 4	2590 ; 3	4380 ; 5	7940 ; 6
4	6150 ; 8	42 140 ; 8	34 730 ; 10	17 110 ; 10	10 890 ; 11

$i = 1, 2, \ldots, 12$ and year $j = 1, 2, \ldots, 5$, the random variable Y_{ij} represents the annual claims and P_{ij} represents the number of shops in the chain. You may assume that the sequence (Y_{ij}, P_{ij}), $i = 1, 2, \ldots, 12$, $j = 1, 2, \ldots, 5$, satisfies the assumptions of EBCT Model 2.

The data from the first four chains in the collective are shown in Table 4.26. The claims are inflation-adjusted and are in £. The cells in the table contain the data in the form Y_{ij} ; P_{ij}.

The credibility premium per shop for the coming year has already been calculated for chains 1 and 2: it is £1077 for chain 1 and £905 for chain 2. Calculate the credibility premium per shop for the coming year for chains 3 and 4.

5

Risk sharing – reinsurance and deductibles

The purpose of risk sharing is to *spread the risk* among those involved. The principal, or direct, insurer may pass on some of the risk to another insurance company, which, in this role, is called the *reinsurer*. In doing so, the direct insurer is purchasing insurance from the reinsurer. In addition, the direct insurer may structure the policy such that the policyholder – the insured party – is responsible for some of the risk, by including a *deductible* or *policy excess* in the conditions of the cover. In this case the insured party has to bear a specified sum whenever a claim is settled – the direct insurer is only responsible for the payment of the amount over and above the excess. The relationship the policyholder has with the direct insurer is parallel to the relationship the direct insurer has with the reinsurer – both the policyholder and the direct insurer are buying insurance to cover part of the risk they are exposed to.

Buying insurance protects the policyholder against the effects of "large" losses. Similarly, the inclusion of a reinsurance arrangement often protects the direct insurer against the effects of "very large" claims. In particular it protects the direct insurer against having sole responsibility (or any responsibility) for the *tails* of the distributions of large claims.

As we shall see later in this chapter, the effects of this are as follows:

- there is a reduction in the mean amount paid out by the direct insurer on claims;
- there is a reduction in the variability of the amount paid out by the direct insurer on claims;
- there is a reduction in the probability that the direct insurer will face a "very large" payout on any particular claim (or collection of claims).

In other words, reinsurance "stabilises" the direct insurer's payouts on claims. One can also argue that the availability of reinsurance arrangements allows smaller companies to become involved in the direct insurance of large risks, thus increasing competition.

There are two principal types of reinsurance arrangement:

- excess of loss reinsurance, which we consider in §5.1;
- proportional reinsurance, which we consider in §5.2.

We will examine the properties of these various arrangements at the *claim level*, so let X be the claim amount (the total amount to be met by all those involved), V the amount of the claim paid by the policyholder, Y the amount of the claim paid by the direct insurer, and Z the amount of the claim paid by the reinsurer. So, in all cases, $X = V + Y + Z$.

We begin by studying the case in which there is no deductible in place, so $V = 0$ and $X = Y+Z$. In §5.3 we consider the case in which there is a deductible in place but no reinsurance arrangement. The case with both a deductible and reinsurance in place is considered in Example 5.9 and as Case study 2 in Chapter 7. In §5.4 we examine the relationship between retention levels (the amounts of risks covered by the direct insurer) and the costs involved in reinsurance contracts – and the consequences for the direct insurer's profit on the business. In §5.5 to §5.8 we consider several selected issues relating to the optimisation of reinsurance contracts seen from the point of view of the direct insurer or the reinsurer.

While we will consider only simple reinsurance arrangements, it must be noted that in practice insurance companies may have in place arrangements with very complicated structures.

First, a note on notation. As each reinsurance contract involves at least the random variables X, Y and Z, we use quantities with suffices to denote which random variable is involved. For example, we write f_X and F_X for the probability density function and the distribution function, respectively, of the claim amount X. Recall that we assume moments are finite as necessary (without saying so explicitly each time).

5.1 Excess of loss reinsurance

Under this arrangement, the direct insurer sets a *retention level M* (>0) and pays in full any claim for which $X \leq M$. The direct insurer retains an amount M of the risk. For claims for which $X > M$, the direct insurer pays M and the reinsurer pays the remaining amount $X - M$. So the amounts paid by the direct insurer and the reinsurer, Y and Z, respectively, are defined as follows:

$$Y = \begin{cases} X & \text{if } X \le M \\ M & \text{if } X > M, \end{cases} \qquad (5.1)$$

$$Z = \begin{cases} 0 & \text{if } X \le M \\ X - M & \text{if } X > M. \end{cases} \qquad (5.2)$$

We can write this conveniently as

$$Y = \min(X, M), \quad Z = \max(0, X - M). \qquad (5.3)$$

In all cases, $X = Y + Z$.

Under this type of reinsurance arrangement, the direct insurer has limited liability (limited to M on each claim) and no exposure to the "risky tail" of the claims distribution. This general feature makes this type of reinsurance attractive to the direct insurer. The reinsurer has unlimited liability (unless there is a cap on the claim amount, and we will see an example of this in Case study 3 in Chapter 7) – the reinsurer has sole responsibility for the "risky tail" of the claims distribution. As a result, this type of reinsurance is, in general, not so attractive to the reinsurer.

The probability that a claim involves the reinsurer is

$$\Pr(X > M) = 1 - F_X(M).$$

To avoid trivialities, we assume that M is such that $F_X(M) < 1$.

It is easy to derive an expression for the reduction in the mean amount paid by the direct insurer on a claim, as follows:

$$\begin{aligned}
\mathbb{E}[Y] &= \int_0^M x f_X(x) dx + \int_M^\infty M f_X(x) dx \\
&= \int_0^\infty x f_X(x) dx - \int_M^\infty x f_X(x) dx + \int_M^\infty M f_X(x) dx \\
&= \mathbb{E}[X] - \int_M^\infty (x - M) f_X(x) dx.
\end{aligned}$$

We can write this as

$$\mathbb{E}[Y] = \mathbb{E}[X] - \int_0^\infty y f_X(y + M) dy. \qquad (5.4)$$

So the expected reduction in payout by the direct insurer on a claim is given by

$$\mathbb{E}[\text{reduction for direct insurer}] = \int_0^\infty y f_X(y + M) dy.$$

This expected reduction in payout on a claim by the direct insurer is of course the expected payout on the claim by the reinsurer, $\mathbb{E}[Z]$, and can also be found directly and easily from the definition of Z:

$$\mathbb{E}[Z] = \int_M^\infty (x - M) f_X(x) dx = \int_0^\infty y f_X(y + M) dy. \qquad (5.5)$$

We note that the distribution of the direct insurer's payout Y, net of reinsurance, is neither purely discrete nor purely continuous. It has an atom at M of size $\Pr(Y = M) = \Pr(X > M)$ and it has a density f_X on $(0, M)$. Hence it is of the form given in (1.11). From the formula for the expectation of such a random variable in (1.13) we obtain

$$\mathbb{E}[Y] = M \Pr(Y = M) + \int_0^M x f_X(x) dx,$$

which gives the same answer for $\mathbb{E}[Y]$ as obtained earlier.

Example 5.1 Suppose we model the claim amount, X, as $X \sim \text{Exp}(\lambda)$ with mean $\mu = 1/\lambda$. The proportion of claims which involve the reinsurer is given by the tail probability (from (2.13))

$$\Pr(X > M) = 1 - F_X(M) = e^{-\lambda M} = e^{-M/\mu}.$$

Using (5.4) we find the direct insurer's expected payout on a claim as follows:

$$\begin{aligned}
\mathbb{E}[Y] &= \mathbb{E}[X] - \int_0^\infty y \lambda e^{-\lambda(y+M)} dy \\
&= \mathbb{E}[X] - e^{-\lambda M} \int_0^\infty y \lambda e^{-\lambda y} dy \\
&= \mathbb{E}[X] - e^{-\lambda M} \mathbb{E}[X] \\
&= (1 - e^{-\lambda M}) \mathbb{E}[X] = (1 - e^{-M/\mu})\mu.
\end{aligned}$$

Hence (or directly) the reinsurer's expected payout on a claim is given by

$$\mathbb{E}[Z] = \mathbb{E}[X] - \mathbb{E}[Y] = e^{-\lambda M} \mathbb{E}[X] = \mu e^{-M/\mu}.$$

As an illustration, suppose that $\mathbb{E}[X] = £1000$ and that the retention level is $M = £2000$. Working in units of £1000, we have $\lambda = \mu = 1$ and $M = 2$, and we find that the reinsurer is involved in about $100e^{-2}\% = 13.5\%$ of claims. In the event of a claim, the expected payouts by the two insurers potentially involved are $\mathbb{E}[Y] = 0.8647$ units and $\mathbb{E}[Z] = 0.1353$ units; that is, the expected amounts paid by the direct insurer and the reinsurer are, respectively, £865 and £135 approximately. However, as the reinsurer will in general only be interested in claims with which they are actually involved, we revisit these calculations later (in Example 5.3).

Example 5.2 Suppose we model the claim amount, X, as a two-parameter Pareto random variable with parameters α (>0) and λ (>0); that is, with probability density function as given in (2.30):

$$f_X(x) = \frac{\alpha \lambda^\alpha}{(\lambda + x)^{\alpha+1}}, \; x > 0.$$

The proportion of claims that involve the reinsurer is given by

$$\Pr(X > M) = 1 - F_X(M) = \left(\frac{\lambda}{\lambda + M}\right)^\alpha.$$

Assume $\alpha > 1$ so that we may consider expectations (see §2.2.7). By (5.5) the reinsurer's expected payout on a claim is given by

$$\begin{aligned}
\mathbb{E}[Z] &= \int_0^\infty y f_X(y + M) dy \\
&= \int_0^\infty y \frac{\alpha \lambda^\alpha}{(\lambda + M + y)^{\alpha+1}} dy \\
&= \left(\frac{\lambda}{\lambda + M}\right)^\alpha \int_0^\infty y \frac{\alpha (\lambda + M)^\alpha}{(\lambda + M + y)^{\alpha+1}} dy.
\end{aligned}$$

The final integral defines the mean of a Pareto random variable with parameters α and $\lambda + M$ and so equals $(\lambda + M)/(\alpha - 1)$, so we have

$$\begin{aligned}
\mathbb{E}[Z] &= \left(\frac{\lambda}{\lambda + M}\right)^\alpha \left(\frac{\lambda + M}{\alpha - 1}\right) \\
&= \frac{\lambda}{\alpha - 1} \left(\frac{\lambda}{\lambda + M}\right)^{\alpha-1} \\
&= \left(\frac{\lambda}{\lambda + M}\right)^{\alpha-1} \mathbb{E}[X].
\end{aligned}$$

It follows that the direct insurer's expected payout on a claim is given by

$$\begin{aligned}
\mathbb{E}[Y] &= \mathbb{E}[X] - \mathbb{E}[Z] \\
&= \left[1 - \left(\frac{\lambda}{\lambda + M}\right)^{\alpha-1}\right] \mathbb{E}[X].
\end{aligned}$$

As an illustration, suppose we work in units of £1000, with Pareto parameters $\alpha = 3$ and $\lambda = 2$, then $\mathbb{E}[X] = 1$ (= £1000). With retention level $M = 2$ (= £2000), the reinsurer is involved in $100(2/4)^3 \% = 12.5\%$ of claims. In the event of a claim, the expected payouts by the two insurers potentially involved are $\mathbb{E}[Y] = 0.75$ units and $\mathbb{E}[Z] = 0.25$ units, that is £750 and £250, respectively – but see Example 5.4.

5.1.1 Reinsurance claims

Let us call a claim which involves the reinsurer (a claim such that $X > M$) a "reinsurance claim". It is important to recognise the information available to the parties involved. While the direct insurer knows all claim amounts, a typical claim record (of actual payouts) for the direct insurer may be

$$x_1, x_2, M_{(1)}, x_3, M_{(2)}, x_4, x_5, x_6, \ldots, M_{(m)}, x_n,$$

in which there are n claim amounts $\leq M$ and m reinsurance claims. A set of data of this form is an example of a "censored sample", in which some observations are not known (or stated) exactly and are replaced by a value which they are know to exceed.

The reinsurer's claims record may involve only the reinsurance claims – the reinsurer may not have any information about the other claims. In this case a typical claims record for the reinsurer will be $z_1^*, z_2^*, \ldots, z_m^*$, where $z_j^* = x_j - M$ is the amount paid out by the reinsurer on a reinsurance claim.

Formally we can introduce a new random variable Z^*, the amount paid by the reinsurer on a reinsurance claim, defined as

$$Z^* \equiv X - M \mid (X > M) \tag{5.6}$$

and read as $Z^* = X - M$, given $X > M$. We can also write $Z^* \equiv Z \mid (Z > 0)$.

The variable Z^* is a very useful addition to the tools we use in this chapter. We can establish its distribution as follows: for $z \geq 0$ we have

$$\begin{aligned}
\Pr(Z^* > z) &= \Pr(X - M > z \mid X > M) \\
&= \Pr(X > z + M \mid X > M) \\
&= \frac{\Pr(X > z + M)}{\Pr(X > M)}.
\end{aligned}$$

We can write this as

$$\Pr(Z^* > z) = \frac{1 - F_X(z + M)}{1 - F_X(M)}. \tag{5.7}$$

Now let I be the random variable which indicates whether or not a claim is a reinsurance claim (that is, $I = 1$ if the claim is a reinsurance claim and $I = 0$ if the claim is not a reinsurance claim). Noting that $\mathbb{E}[Z \mid (X > M)] = \mathbb{E}[Z^*]$ and $\mathbb{E}[Z \mid (X \leq M)] = 0$, we have, using the conditional expectation formula (1.3),

$$\begin{aligned}
\mathbb{E}[Z] &= \mathbb{E}[\mathbb{E}[Z \mid I]] \\
&= \mathbb{E}[Z \mid I = 1] \Pr(I = 1) + \mathbb{E}[Z \mid I = 0] \Pr(I = 0) \\
&= \mathbb{E}[Z \mid (X > M)] \Pr(X > M) + \mathbb{E}[Z \mid (X \leq M)] \Pr(X \leq M),
\end{aligned}$$

which gives

$$\mathbb{E}[Z] = \mathbb{E}[Z^*] \Pr(X > M). \tag{5.8}$$

This is just the mathematics of common sense – for example, if the reinsurer pays out on 25% of claims, and the reinsurer's average payout on such claims is £10 000, then the reinsurer's average payout over all claims is £2500.

Example 5.3 (*Example 5.1 revisited.*) The exponential tail probability is $\Pr(X > x) = 1 - F_X(x) = e^{-\lambda x}$, so

$$\Pr(Z^* > z) = \frac{e^{-\lambda(z+M)}}{e^{-\lambda M}} = e^{-\lambda z}.$$

Hence $Z^* \sim \text{Exp}(\lambda)$, a result which illustrates the "lack of memory" property of the exponential distribution (see (2.17)). So the expected payout by the reinsurer on claims with which the reinsurer is involved is given by $\mathbb{E}[Z^*] = 1/\lambda$. We also note from Example 5.1 that $\mathbb{E}[Z] = (1/\lambda)e^{-\lambda M}$, illustrating result (5.8), namely

$$\mathbb{E}[Z] = \mathbb{E}[Z^*] \Pr(X > M).$$

For the illustration in Example 5.1, $Z^* \sim \text{Exp}(1)$, and

$$\mathbb{E}[Z] = 1 \times \exp(-2) = \mathbb{E}[Z^*] \Pr(X > 2).$$

Example 5.4 (*Example 5.2 revisited.*) The $\text{Pa}(\alpha, \lambda)$ tail probability is given by $1 - F_X(x) = (\lambda/(\lambda + x))^\alpha$, so from (5.7) we have

$$\Pr(Z^* > z) = \frac{\left(\frac{\lambda}{\lambda + M + z}\right)^\alpha}{\left(\frac{\lambda}{\lambda + M}\right)^\alpha} = \left(\frac{\lambda + M}{\lambda + M + z}\right)^\alpha.$$

Hence $Z^* \sim \text{Pa}(\alpha, \lambda + M)$. This is the *conditional Pareto tail* result of (2.36):

$$X \sim \text{Pa}(\alpha, \lambda) \Rightarrow X - M \mid X > M \sim \text{Pa}(\alpha, \lambda + M).$$

So the expected payout by the reinsurer on claims with which the reinsurer is involved is given by $\mathbb{E}[Z^*] = (\lambda + M)/(\alpha - 1)$. We also note that

$$\mathbb{E}[Z] = \frac{\lambda}{\alpha - 1} \left(\frac{\lambda}{\lambda + M}\right)^{\alpha - 1},$$

illustrating result (5.8). For the illustration in Example 5.2, $Z^* \sim \text{Pa}(3, 4)$. We also have $\mathbb{E}[Z^*] = 2$ and $\mathbb{E}[Z] = 0.25 = 2 \times 0.125 = \mathbb{E}[Z^*] \Pr(X > 2)$.

5.1.2 Simulation results

To illustrate the results of Example 5.2 and Example 5.4 involving the Pareto distribution, a simulation of 10 000 claim amounts was carried out (using R). The claim amount distribution used in the simulation is $X \sim \mathrm{Pa}(6, 50)$, for which $\mathbb{E}[X] = 10$ and $\mathrm{Var}[X] = 150$. Excess of loss reinsurance with retention level $M = 25$ is in place.

First, we calculate some theoretical values. We find that

$$\Pr(X > M) = (50/75)^6 = 0.08779.$$

Letting Y, Z and Z^* be as above, we have $\mathbb{E}[Y] = 8.683$, $\mathbb{E}[Z] = 1.317$ and $\mathbb{E}[Z^*] = 15$ (using Example 5.2 and Example 5.4). In addition, let XLO be the amount of a claim which is less than 25, that is $XLO = X \mid X \le 25$, and let XRE be the amount of a reinsurance claim, that is $XRE = X \mid X > 25$. We calculate $\mathbb{E}[XRE]$ by noting

$$\mathbb{E}[XRE] = \mathbb{E}[X \mid X > 25] = \mathbb{E}[X - 25 \mid X > 25] + 25$$
$$= \mathbb{E}[Z^*] + 25 = 15 + 25 = 40.$$

Now, using

$$\mathbb{E}[X] = \mathbb{E}[X \mid X \le 25] \Pr(X \le 25) + \mathbb{E}[X \mid X > 25] \Pr(X > 25)$$
$$= \mathbb{E}[XLO] \Pr(X \le 25) + \mathbb{E}[XRE] \Pr(X > 25),$$

we have

$$10 = \mathbb{E}[XLO] \times \{1 - (2/3)^6\} + 40 \times (2/3)^6,$$

which gives $\mathbb{E}[XLO] = 7.113$.

In the simulation, the claim amounts were generated from a vector of 10 000 random numbers – from a uniform distribution on $(0,1)$ – using the probability integral transformation method (see §2.2.7 for simulation from a Pareto distribution): the R code used was

```
x=50*(runif(10000)^(-1/6) - 1)
```

The claim amounts vector x was then manipulated to produce vectors y, z and zstar containing the values indicated by these vectors' names, and these vectors were also summarised, using the commands (for example) length(y) and summary(y). In the simulation, the claim amounts ranged from 0.002 to 198.8, and 916 of them were reinsurance claims (compared to an expected number of 877.9). Finally, vectors containing the values of variables XLO and XRE were constructed and summarised.

Table 5.1. *Theoretical results behind the simulation in §5.1.2*

	Number	Expected number	Mean	SD
X	10 000	–	10	12.25
Y	10 000	–	8.683	7.706
Z	10 000	–	1.317	6.903
Z^*	–	877.9	15	18.37
XLO	–	9122	7.113	6.083
XRE	–	877.9	40	18.37

Table 5.2. *Simulation results in §5.1.2 (selected from **R** output)*

	Number	Min.	Median	Mean	Max.	SD
x	10 000	0.002	6.121	10.14	198.8	12.65
y	10 000	0.002	6.121	8.718	25.00	7.767
z	10 000	0.000	0.000	1.426	173.8	7.295
zstar	916	0.002	10.04	15.57	173.8	19.00
xlo	9084	0.002	5.361	7.076	25.00	6.082
xre	916	25.00	35.04	40.57	198.8	19.00

The results (theoretical and from the simulation) are summarised in Tables 5.1 and 5.2. In addition to the theoretical means, the relevant standard deviations (SD) have been calculated (the values of the standard deviations of Y and Z can be obtained from results given in Exercise 5.4), and are given for information and for comparison with the results from the simulation.

The histogram in Figure 5.1 gives a display of the simulated claim amounts (up to 140). The strong skew of the amounts is evident – about 50% of the claim amounts are less than 6.1 (about 25% are less than 2.5 and about 75% are less than 13). Three claims for very large amounts (greater than 140) are not included in the histogram. The two histograms in Figure 5.2 give displays of the amounts paid by the direct insurer and the 916 non-zero amounts paid by the reinsurer. The spike at 25 in the insurer's histogram is self-explanatory, and the otherwise general skewness of both displays is again evident.

5.1.3 Aggregate claims model with excess of loss reinsurance

We now consider the aggregate claims model from Chapter 3 with excess of loss reinsurance arrangements (at the claim level) in place. Let X_i be the amount of the ith claim, and let Y_i and Z_i be the amounts of this claim paid by

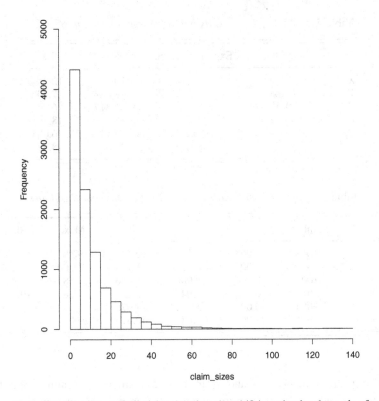

Figure 5.1. Histogram of all claim sizes less than 140 in a simulated sample of 10 000 claim sizes from a Pareto distribution with parameters $\alpha = 6$ and $\lambda = 50$.

the direct insurer and reinsurer, respectively. Let N and N^* be the number of claims and the number of reinsurance claims, respectively. Let S be the aggregate claim amount, and let S_I and S_R be the aggregate claim amounts paid by the insurer and reinsurer, respectively. Then $S = S_I + S_R$, and we have the following representations as compound distributions involving all claims:

$$S = X_1 + X_2 + \cdots + X_N; \quad S_I = Y_1 + Y_2 + \cdots + Y_N; \quad S_R = Z_1 + Z_2 + \cdots + Z_N. \quad (5.9)$$

We have a second representation for S_R in terms of reinsurance claims only:

$$S_R = Z_1^* + Z_2^* + \cdots + Z_{N^*}^*. \quad (5.10)$$

In the important case that $N \sim \text{Poi}(\lambda)$, the random variables S, S_I and S_R above have compound Poisson (CP) distributions (see §3.4.1):

$$S \sim \text{CP}(\lambda, F_X); \quad S_I \sim \text{CP}(\lambda, F_Y); \quad S_R \sim \text{CP}(\lambda, F_Z). \quad (5.11)$$

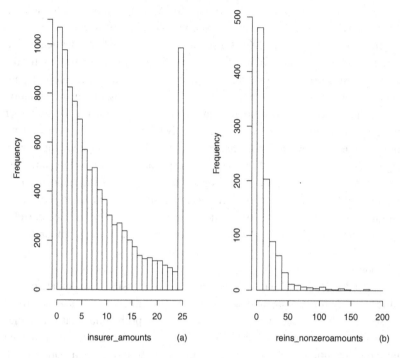

Figure 5.2. Histograms of payouts by direct insurer (a) and payouts by reinsurer on 916 reinsurance claims (b).

We find the distribution of N^* as follows. Note that $N^* \mid (N = n)$ has a bi(n, p) distribution, where $p = \Pr(X > M)$. Using the conditional expectation formula (1.3), and the fact that the probability generating function of a bi(n, p) distribution is $(pz + 1 - p)^n$ (see (2.7)), we have

$$G_{N^*}(z) = \mathbb{E}[z^{N^*}] = \mathbb{E}\big[\mathbb{E}[z^{N^*} \mid N]\big] = \mathbb{E}\big[(pz + 1 - p)^N\big] = G_N(pz + 1 - p).$$

When $N \sim \mathrm{Poi}(\lambda)$, this becomes

$$G_{N^*}(z) = \exp\left(\lambda(pz + (1 - p) - 1)\right) = \exp\left(p\lambda(z - 1)\right).$$

Hence N^* has a Poisson distribution, with mean $p\lambda$. Using the representation for S_R in (5.10) we have

$$S_R \sim \mathrm{CP}(p\,\lambda, F_{Z^*}). \tag{5.12}$$

Equating the means of S_R from its two representations (in (5.11) and (5.12)) gives $\lambda\mathbb{E}[Z] = p\lambda\mathbb{E}[Z^*]$, that is $\mathbb{E}[Z] = \mathbb{E}[Z^*]\Pr(X > M)$, as noted in (5.8).

As discussed in Chapter 3, the model for S as a compound distribution incorporates important structural assumptions. One of these is that the sequence of claim amounts $\{X_i\}$ is a sequence of independent, identically distributed random variables. While this is probably a realistic assumption for the risks in most portfolios, there are particular situations in which the independence of the variables may be questioned. It is quite possible that a situation could exist in which claim amounts tend to be all rather low (or indeed rather high), with the amounts therefore being positively correlated, not independent. For example, if we were to consider claims for vehicle damage arising from a single road traffic accident, there may well be a tendency for the claim amounts to be mostly low (or high), depending on the severity of the incident. This feature could also arise, for example, in the case of claims arising on home insurance policies in an area hit by flooding. Similary, there may be correlations among direct insurer's amounts $\{Y_i\}$, or among reinsurer's amounts $\{Z_i\}$. As with all models, it is wise to keep the assumptions in mind when applying the models.

As also discussed in Chapter 3, it is not usually possible to calculate probabilities exactly for compound distributions. In Example 5.5, S has a compound Poisson distribution and we want to calculate the probability that it assumes a value less than a specified amount. The method adopted here is to assume that the distribution of S can be approximated by a normal distribution (with matching mean and variance) as discussed in §3.6.1. With Poisson parameter $\lambda = 100$, the approximation will be acceptable, but we must be aware that in other circumstances the use of a normal approximation may be unjustified and produce misleading answers. A compound Poisson variable $S \sim \mathrm{CP}(\lambda, F_X)$ has positive skewness (the coefficient of skewness is actually $k/\sqrt{\lambda}$, where $k = \mathbb{E}[X^3]/\{\mathbb{E}[X^2]\}^{3/2}$), and this will have to be recognised in the case that λ is small. In addition, the claim amount variable may well be modelling a distribution which in practice has a fat (or heavy) tail and is strongly positively skewed. When either or both of these situations pertains, the use of a normal distribution to approximate the distribution of S will lead to inaccurate answers – in particular it will underestimate tail probabilities of the form $\Pr(S > c)$ for "large" c.

Example 5.5 Suppose S has a compound Poisson distribution with claim rate (Poisson mean) 100 and individual claim amount variable X with the simple discrete distribution

x	10	20	30	60
$\Pr(X = x)$	0.2	0.4	0.3	0.1

for which $\mathbb{E}[X] = 25$ and $\mathbb{E}[X^2] = 810$. Using the expression for compound Poisson cumulants in (3.17), we find

$$\mathbb{E}[S] = 100 \times \mathbb{E}[X] = 2500,$$

$$\text{Var}[S] = 100 \times \mathbb{E}[X^2] = 81\,000,$$

$$\text{SD}[S] = 284.6.$$

Suppose the direct insurer wants to be 95% sure of making a profit on this business, and the direct insurer decides to set a premium P using the expected value principle with relative security loading α (see (4.1)), so that $P = (1 + \alpha)\mathbb{E}[S]$. Then

$$\text{Pr(insurer makes a profit)} = \text{Pr}(S < P) = \text{Pr}\,(S < (1 + \alpha)\mathbb{E}[S]).$$

Using a normal approximation for the distribution of S, we have

$$\text{Pr(insurer makes a profit)} = \Phi\left(\frac{\alpha\mathbb{E}[S]}{\sqrt{\text{Var}[S]}}\right),$$

where we recall that Φ is the $N(0, 1)$ distribution function. In this case, and for a probability of 0.95, we have $2500\alpha/\sqrt{81\,000} = 1.6449$, which gives $\alpha = 0.1873$. So the direct insurer sets a premium using a relative security loading of 18.73%, and therefore charges a premium of $1.1873 \times 2500 = 2968$.

Suppose the direct insurer enters into an excess of loss reinsurance contract with retention level $M = 28$. The direct insurer's payout on a claim, Y, assumes values 10, 20 and 28 with probabilities 0.2, 0.4 and 0.4, respectively. We have $\mathbb{E}[Y] = 21.2$ and $\mathbb{E}[Y^2] = 493.6$. The reinsurer's payout on a claim, Z, assumes values 0, 2 and 32 with probabilities 0.6, 0.3 and 0.1, respectively. We have $\mathbb{E}[Z] = 3.8$ and $\mathbb{E}[Z^2] = 103.6$.

The distribution of the direct insurer's aggregate claim amount in (5.11) is $S_I \sim CP(100, F_Y)$, for which we have

$$\mathbb{E}[S_I] = 100 \times \mathbb{E}[Y] = 2120,$$

$$\text{Var}[S_I] = 100 \times \mathbb{E}[Y^2] = 49\,360,$$

$$\text{SD}[S_I] = 222.2.$$

The distribution of the reinsurer's aggregate claim amount in (5.11) is $S_R \sim CP(100, F_Z)$, for which we have

$$\mathbb{E}[S_R] = 100 \times \mathbb{E}[Z] = 380,$$

$$\text{Var}[S_R] = 100 \times \mathbb{E}[Z^2] = 10\,360,$$

$$\text{SD}[S_R] = 101.8.$$

Claims greater than $M = 28$ have amounts 30 and 60, and these occur with probabilities in the ratio 3:1, so Z^* assumes values 2 and 32 with probabilities 0.75 and 0.25, respectively. Hence $\mathbb{E}[Z^*] = 9.5$ and $\mathbb{E}[(Z^*)^2] = 259$. We also have Pr(claim is a reinsurance claim) = 0.4, so the rate of reinsurance claims is $0.4 \times 100 = 40$. The alternative representation for S_R (in terms of reinsurance claims only, see (5.12)) is therefore $S_R \sim \mathrm{CP}(40, F_{Z^*})$, from which we find that

$$\mathbb{E}[S_R] = 40 \times \mathbb{E}[Z^*] = 380,$$
$$\mathrm{Var}[S_R] = 40 \times \mathbb{E}[(Z^*)^2] = 10\,360,$$

as found above.

Suppose now that the reinsurer also sets a premium (to be paid by the direct insurer) using the expected value principle, in this case with a relative security loading of 25%. Then the reinsurer's premium, P_R say, is given by $P_R = 1.25 \times \mathbb{E}[S_R] = 1.25 \times 380 = 475$.

The direct insurer's position is now as follows – premium income $P = 2968$, payout on claims S_I with mean 2120 and variance 49 360, and reinsurance premium to pay $P_R = 475$. Let us calculate the quantity

Pr(direct insurer's total payout > 3000)

with and without the reinsurance contract in place (using normal approximations to the distributions of the direct insurer's total payout). With no reinsurance in place, the direct insurer's total payout is S and has mean 2500 and variance 81 000. We find

$$\mathrm{Pr}(S > 3000) = 1 - \Phi(1.757) = 0.0395.$$

With the reinsurance in place, the direct insurer's total payout is $S_I + P_R$ and has mean $2120 + 475 = 2595$ and variance 49 360. We find

$$\mathrm{Pr}(S_I + P_R > 3000) = 1 - \Phi(1.823) = 0.0342,$$

which we note is lower than 0.0395. Going further into the tail of the distributions, we find that, without reinsurance,

Pr(direct insurer's total payout > 3300) = 0.0025,

and is only 0.00075 with reinsurance. The size of the "with reinsurance" probability relative to the corresponding "no reinsurance" probability is sensitive to the parameters of the situation, for example the size of the reinsurer's security loading.

Example 5.6 Suppose S has a compound Poisson distribution with claim rate (Poisson mean) 100 and individual claim amount variable $X \sim \mathrm{Exp}(1)$ with mean 1 (in using this value for the mean we are effectively adopting the expected claim amount as our monetary unit). The direct insurer has an excess of loss reinsurance contract in place with retention level $M = 2$ (that is, twice the expected claim amount). We have $\mathbb{E}[X] = 1$ and $\mathbb{E}[X^2] = 2$, and so we have

$$\mathbb{E}[S] = 100 \times \mathbb{E}[X] = 100,$$
$$\text{Var}[S] = 100 \times \mathbb{E}[X^2] = 200,$$
$$\text{SD}[S] = 14.14.$$

The direct insurer's payout on a claim, Y, has $\mathbb{E}[Y] = 1 - e^{-2} = 0.8647$ (from Example 5.1) and $\mathbb{E}[Y^2] = 2(1 - 3e^{-2}) = 1.188$ (obtainable from the result in Exercise 5.3(a)). The reinsurer's payout on a claim, Z, has $\mathbb{E}[Z] = e^{-2} = 0.1353$ and $\mathbb{E}[Z^2] = 2e^{-2} = 0.2707$ (again from Example 5.1 and Exercise 5.3(a)).

For the direct insurer we have $S_I \sim \text{CP}(100, F_Y)$, and

$$\mathbb{E}[S_I] = 100 \times \mathbb{E}[Y] = 86.47,$$
$$\text{Var}[S_I] = 100 \times \mathbb{E}[Y^2] = 118.8,$$
$$\text{SD}[S_I] = 10.9.$$

For the reinsurer we have $S_R \sim \text{CP}(100, F_Z)$, and

$$\mathbb{E}[S_R] = 100 \times \mathbb{E}[Z] = 13.53,$$
$$\text{Var}[S_R] = 100 \times \mathbb{E}[Z^2] = 27.07,$$
$$\text{SD}[S_R] = 5.20.$$

From Example 5.3, $Z^* \sim \text{Exp}(1)$. Hence $\mathbb{E}[Z^*] = 1$ and $\mathbb{E}[(Z^*)^2] = 2$. We also have that the probability that a claim is a reinsurance claim is e^{-2} so the rate of reinsurance claims is $100e^{-2} = 13.53$ (13.534 to five significant figures). The alternative representation for S_R (in terms of reinsurance claims only, (5.12)) is therefore $S_R \sim \text{CP}(13.534, F_{Z^*})$, from which we obtain

$$\mathbb{E}[S_R] = 13.534 \times \mathbb{E}[Z^*] = 13.53,$$
$$\text{Var}[S_R] = 13.534 \times \mathbb{E}[(Z^*)^2] = 27.07,$$

as found above.

Suppose the expected claim amount is actually £1070. The calculations above are being carried out using £1070 as the monetary unit, so the values of $\mathbb{E}[S_I]$ and $\text{SD}[S_I]$ are actually $86.4665 \times 1070 = £92\,519$ and $\sqrt{118.7988} \times 1070 = £11\,662$, etc.

To illustrate the results, a simulation of 10 000 observations of S, S_I and S_R was carried out, using the claim amount distribution and parameter values as above. In order to generate each observation of $\{S, S_I, S_R\}$, a value for the number of claims, n, was generated (from a Poi(100) distribution) and then n claim amounts were generated (from an exponential distribution with mean 1),

Table 5.3. *Simulation results in Example 5.6 (selected from **R** output)*

	Number	Min.	Median	Mean	Max.	SD
S	10 000	54.62	99.61	100.1	160.4	14.15
S_I	10 000	48.92	86.28	86.57	129.8	10.91
S_R	10 000	0.7794	13.06	13.53	40.33	5.22

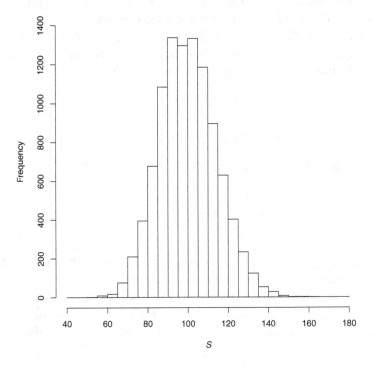

Figure 5.3. Histogram of 10 000 simulated aggregate claim amounts from compound Poisson distribution $CP(100, F_X)$, where $X \sim \text{Exp}(1)$.

say amounts $x_1, x_2, \ldots, x_n$. The amounts of each claim to be paid by the insurer and the reinsurer were calculated, say $y_1, y_2, \ldots, y_n$ and $z_1, z_2, \ldots, z_n$, respectively, and summing the x_i, y_i and z_i gave single simulated values of S, S_I and S_R. The process was repeated 10 000 times, and the results are summarised and given in Table 5.3.

The means and standard deviations of the simulated data are in close agreement with the theoretical results.

The histogram in Figure 5.3 gives a display of the simulated values of the aggregate claim amounts S. The two histograms in Figure 5.4 give displays of

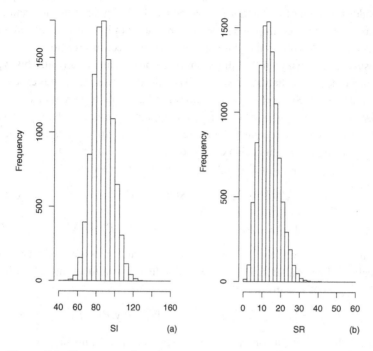

Figure 5.4. Histograms of aggregate payouts by direct insurer (a) and reinsurer (b).

the aggregate amounts paid by the direct insurer and the reinsurer, here denoted SI and SR, respectively. The histograms of SI and SR exhibit approximate symmetry and normality (that of SR less so).

5.2 Proportional reinsurance

Under this arrangement, the direct insurer pays a proportion of each claim (say a proportion β) and the reinsurer pays the remainder of the claim (a proportion $1 - \beta$). The value of β, the proportion of the risk retained by the direct insurer, is agreed in advance by the parties involved.

So, Y and Z, the amounts paid by the direct insurer and the reinsurer, respectively, are defined simply as follows:

$$Y = \beta X, \ Z = (1 - \beta)X, \tag{5.13}$$

where X is the claim amount. As before, we have $X = Y + Z$ (of course).

Under this type of reinsurance arrangement, both the direct insurer and the reinsurer are involved in paying each claim, and both have unlimited liability (unless there is a cap on the claim amount). The direct insurer now has some exposure to the "risky tail" of the claims distribution, and as a result this type of reinsurance is, in general, less attractive to the direct insurer than is excess of loss reinsurance. Since the reinsurer does not have sole responsibility for the "risky tail" of the claims distribution, this type of reinsurance is, in general, more attractive to the reinsurer than is excess of loss reinsurance. This type of reinsurance arrangement is also called "quota share" reinsurance.

From the definitions of Y and Z, we have immediately that

$$\mathbb{E}[Y] = \beta\mathbb{E}[X], \qquad \mathrm{SD}[Y] = \beta\,\mathrm{SD}[X]; \qquad (5.14)$$

$$\mathbb{E}[Z] = (1 - \beta)\mathbb{E}[X], \qquad \mathrm{SD}[Z] = (1 - \beta)\,\mathrm{SD}[X]. \qquad (5.15)$$

The mathematics of the situation are straightforward, involving only a change of scale of the variable X. For example, the distribution function of the direct insurer's payout Y is given by

$$F_Y(y) = \Pr(Y \le y) = \Pr(\beta X \le y) = \Pr(X \le y/\beta) = F_X(y/\beta).$$

The following results on the scaling of variables will be useful:

$X \sim \mathrm{Exp}(\lambda) \Rightarrow \beta X \sim \mathrm{Exp}(\lambda/\beta)$ by (2.16);
$X \sim \mathrm{Pa}(\alpha, \lambda) \Rightarrow \beta X \sim \mathrm{Pa}(\alpha, \beta\lambda)$ by (2.35);
$X \sim \mathrm{gamma}(\alpha, \lambda) \Rightarrow \beta X \sim \mathrm{gamma}(\alpha, \lambda/\beta)$ by (2.22);
$X \sim \mathrm{lognormal}(\mu, \sigma) \Rightarrow \beta X \sim \mathrm{lognormal}(\log\beta + \mu, \sigma)$.

The last of these follows easily from $\log(\beta X) = \log(\beta) + \log(X)$ and then noting that, if $X \sim \mathrm{lognormal}(\mu, \sigma)$, then $\log(X) \sim \mathrm{N}(\mu, \sigma^2)$.

The extension to the aggregate claims model is easy – the proportionality carries through to the aggregate amounts. As before, letting S be the aggregate claim amount and letting S_I and S_R be the aggregate claim amounts paid by the direct insurer and the reinsurer, respectively, we have

$$S = X_1 + X_2 + \cdots + X_N,$$

$$S_I = Y_1 + Y_2 + \cdots + Y_N = \beta X_1 + \beta X_2 + \cdots + \beta X_N = \beta S, \qquad (5.16)$$

and similarly

$$S_R = Z_1 + Z_2 + \cdots + Z_N = (1 - \beta)S. \qquad (5.17)$$

Example 5.7 Suppose an aggregate claim amount variable S has a compound Poisson distribution with claim rate λ and individual claim amount variable X

which is exponentially distributed with mean μ. There is a proportional reinsurance contract in place under which the direct insurer retains (pays) a proportion β of each claim amount.

From (3.18) we have $\mathbb{E}[S] = \lambda\mu$, $\text{Var}[S] = 2\lambda\mu^2$, and we note that S_I and S_R also have compound Poisson distributions. Using (5.14)–(5.17) we have $S_I = \beta S$, $S_R = (1 - \beta)S$, and the mean and variance of S_I and S_R are given by

$$\mathbb{E}[S_I] = \lambda\beta\mu \text{ and } \text{Var}[S_I] = 2\lambda\beta^2\mu^2;$$

$$\mathbb{E}[S_R] = \lambda(1 - \beta)\mu \text{ and } \text{Var}[S_R] = 2\lambda(1 - \beta)^2\mu^2.$$

5.3 Deductibles (policy excesses)

This arrangement will probably be familiar to anyone with a general insurance policy – certainly to those with a private motor policy. The policyholder agrees to bear the first amount, say D, of any loss (and so only submits a claim when the loss exceeds D). In this arrangement D is called a *deductible* or the *policy excess*.

An obvious benefit to the insurer of including deductibles in policies is the removal of small claims from the insurer's record, leading to savings in administrative effort and costs. The overall reduction in the number and average amount of potential claims to be settled opens up the possibility of a lowering of premiums, with consequent market advantages.

The policyholder and the insurer are effectively in the same relative positions as the direct insurer and reinsurer, respectively, in an excess of loss reinsurance arrangement as described in §5.1. There is an obvious difference, however, in that the policyholder does not receive any money from another party (unlike a direct insurer, who receives a premium). The policyholder's financial involvement is all "outgo" – consisting of a premium payment to the insurer and an amount D or less of each loss.

Let X represent the loss, and now let Y and Z represent the amounts of the loss paid by the *policyholder* and the *insurer*, respectively. Then

$$Y = \begin{cases} X & \text{if } X \leq D \\ D & \text{if } X > D, \end{cases} \tag{5.18}$$

$$Z = \begin{cases} 0 & \text{if } X \leq D \\ X - D & \text{if } X > D; \end{cases} \tag{5.19}$$

that is,

$$Y = \min(X, D), \quad Z = \max(0, X - D). \tag{5.20}$$

Losses up to D are uninsured; losses in excess of D are insured, and the insured loss is given by $Z^* \equiv X - D \mid X > D$ (that is, the loss incurred by the insurer on losses with which the insurer is involved). The insurer is, of course, only exposed to risk from insured losses.

From (2.17) and (2.36), we note that

for $X \sim \text{Exp}(\lambda)$, the insured loss $\sim \text{Exp}(\lambda)$, and
for $X \sim \text{Pa}(\alpha, \lambda)$, the insured loss $\sim \text{Pa}(\alpha, \lambda + D)$.

Increasing the value of D increases the policyholder's risk exposure and reduces the insurer's exposure. Whether or not this is acceptable to the policyholder depends on whether or not they receive a fair reduction in premium to compensate for the increased exposure. Whether or not it is worthwhile for the insurer depends on whether the balance between risk exposure and profits is acceptable. Some effects of introducing deductibles, and aspects of the relationship between the positions of the policyholder and the insurer and the size of the deductible, are explored in Exercises 5.18 and 5.19. In Case study 2 in Chapter 7 we examine a situation in which a deductible and a reinsurance contract are both in place.

Example 5.8 Suppose we model the individual loss, X, as $X \sim \text{Exp}(1/\mu)$ (with mean μ). Then, using results from Examples 5.1 and 5.3, and averaging over all losses (potential claims), we find the expected payments for the policyholder and the insurer to be, respectively,

$$\mathbb{E}[Y] = (1 - e^{-D/\mu})\mu \text{ and } \mathbb{E}[Z] = \mu e^{-D/\mu}.$$

For insured losses, $Z^* \sim \text{Exp}(1/\mu)$.

In the case that μ = £3000 and D = £150, we find that $\Pr(X > 150)$ = 0.9512, and so the insurer is involved in about 95.1% of all losses. Over all losses, the expected payouts by the policyholder and the insurer on an individual loss are £146.31 and £2853.69, respectively. About 4.9% of losses are for amounts less than or equal to £150 and do not lead to claims – the expected size of such losses, which are borne by the policyholder, is £74.38. For insured losses, the insurer's expected payout is £3000.

Suppose now we consider the aggregate model in which S has a compound Poisson distribution, losses occur at rate λ, and the individual loss distribution is as above. Let S_P and S_I represent the aggregate payouts by the policyholder and the insurer, respectively. Then we have

$$S_P \sim \text{CP}(\lambda, F_Y) \text{ and } \mathbb{E}[S_P] = \lambda\mu(1 - e^{-D/\mu}),$$

and

$$S_I \sim \text{CP}(\lambda, F_Z) \text{ and } \mathbb{E}[S_I] = \lambda\mu e^{-D/\mu}.$$

For insured losses, we have $S_I \sim \text{CP}(p\lambda, F_Z^*)$, where $p = \Pr(X > D) = e^{-D/\mu}$, giving again $\mathbb{E}[S_I] = \lambda\mu e^{-D/\mu}$.

Example 5.9 Consider a situation in which individual losses, X, are modelled by a Pareto random variable $X \sim \text{Pa}(\lambda + 1, \lambda)$, with parameters chosen so that the monetary unit is the expected loss ($\mathbb{E}[X] = 1$) for convenience. An insurer writes policies to cover such losses with an individual excess of D per loss; the policyholder submits a claim on any loss which exceeds D. Let Y and Z be the amount of a loss paid by the policyholder and the insurer, respectively. We find expresssions for $\mathbb{E}[Z]$ and $\mathbb{E}[Y]$.

We note that $Z = 0$ for $X \leq D$ and $Z = X - D$ for $X > D$. Then, using the conditional tail property of the Pareto distribution (2.36), the insured loss $Z^* \equiv X - D \mid X > D \sim \text{Pa}(\lambda + 1, \lambda + D)$. The probability that the insurer is involved in a loss is

$$\Pr(X > D) = \left(\frac{\lambda}{\lambda + D}\right)^{\lambda+1},$$

and the expected payout by the insurer on a claim is $\mathbb{E}[Z^*] = (\lambda + D)/\lambda$. Then we obtain

$$\mathbb{E}[Z] = \mathbb{E}[Z \mid X \leq D]\Pr(X \leq D) + \mathbb{E}[Z \mid X > D]\Pr(X > D)$$
$$= 0 + \left(\frac{\lambda + D}{\lambda}\right)\left(\frac{\lambda}{\lambda + D}\right)^{\lambda+1}$$
$$= \left(\frac{\lambda}{\lambda + D}\right)^{\lambda}.$$

Hence

$$\mathbb{E}[Y] = \mathbb{E}[X] - \mathbb{E}[Z] = 1 - \left(\frac{\lambda}{\lambda + D}\right)^{\lambda}.$$

Suppose now that aggregate losses S have a compound Poisson distribution $S \sim \text{CP}(\pi, F_X)$ with X as above, and that the insurer has arranged proportional reinsurance with another company, with retained proportion β of each claim amount. Both insurers set their premiums using the expected value principle (4.1): the direct insurer uses a loading factor (relative security loading) of θ_1 and the reinsurer uses a loading factor of $\theta_2(>\theta_1)$ – we will see later in §5.4 why we must have $\theta_2 > \theta_1$. We consider the direct insurer's expected profit on this business.

Using the results of §5.1.3 and working with insured losses only, the aggregate claim amount, S_{CL} say, has distribution $S_{CL} \sim CP(\pi^*, F_{Z^*})$, where

$$\pi^* = \pi \left(\frac{\lambda}{\lambda + D} \right)^{\lambda+1} \text{ and } Z^* \sim Pa(\lambda + 1, \lambda + D),$$

giving

$$\mathbb{E}[S_{CL}] = \pi \left(\frac{\lambda}{\lambda + D} \right)^{\lambda}.$$

Note that we can also represent the distribution of S_{CL} as $S_{CL} \sim CP(\pi, F_Z)$, and we find the same formula for $\mathbb{E}[S_{CL}]$.

The premium charged by the direct insurer is

$$(1 + \theta_1)\mathbb{E}[S_{CL}] = (1 + \theta_1)\pi \left(\frac{\lambda}{\lambda + D} \right)^{\lambda}.$$

With the reinsurance in place, the direct insurer pays βS_{CL} and the reinsurer pays $(1 - \beta)S_{CL}$. So the insurer's expected claim amount is

$$\mathbb{E}[\beta S_{CL}] = \beta \pi \left(\frac{\lambda}{\lambda + D} \right)^{\lambda}.$$

The reinsurer's premium is

$$(1 + \theta_2)\mathbb{E}[(1 - \beta)S_{CL}] = (1 + \theta_2)(1 - \beta)\pi \left(\frac{\lambda}{\lambda + D} \right)^{\lambda}.$$

It follows that the direct insurer's expected profit is

$$(1 + \theta_1)\pi \left(\frac{\lambda}{\lambda + D} \right)^{\lambda} - \beta \pi \left(\frac{\lambda}{\lambda + D} \right)^{\lambda} - (1 + \theta_2)(1 - \beta)\pi \left(\frac{\lambda}{\lambda + D} \right)^{\lambda}$$

$$= \pi \left(\frac{\lambda}{\lambda + D} \right)^{\lambda} [\theta_1 - (1 - \beta)\theta_2].$$

5.4 Retention levels and reinsurance costs

We consider now the relationship between retention levels and the costs involved in reinsurance contracts (in this section we will ignore deductibles).

Suppose the aggregate claim amount faced by a direct insurer, S, has a compound Poisson distribution with claim rate λ and individual claim amount variable X. Let Y and Z represent the payouts on a claim by the direct insurer and the reinsurer, respectively. For simplicity, let us assume that both the direct insurer and the reinsurer set their premiums using the expected value principle, with relative security loadings of θ and ψ, respectively.

Note that we must have $\psi > \theta$. If this were not so, the direct insurer could reinsure the entire risk and in so doing make a guaranteed profit (by "pocketing the difference" $(\theta - \psi)$ per unit of risk) – the transaction would produce a risk-free profit to the direct insurer and would constitute an arbitrage opportunity. The case $\psi = \theta$ corresponds to a risk sharing arrangement for which all retention levels are possible and equivalent – this is not a realistic scenario.

Before any considerations of reinsurance, the direct insurer has an expected payout of $\mathbb{E}[S] = \lambda\mathbb{E}[X]$ (by (3.18)), and so, with the security loading θ there is premium income from the policyholders of $(1 + \theta)\lambda\mathbb{E}[X]$.

The reinsurer has an expected payout of $\lambda\mathbb{E}[Z]$, and so, with security loading ψ the reinsurer has premium income from the direct insurer of $(1 + \psi)\lambda\mathbb{E}[Z]$.

So the direct insurer's premium income, net of reinsurance costs, is

$$(1 + \theta)\lambda\mathbb{E}[X] - (1 + \psi)\lambda\mathbb{E}[Z].$$

The direct insurer's expected payout, net of reinsurance, is $\lambda\mathbb{E}[Y]$. To ensure that the insurer has a positive expected profit, we therefore require

$$(1 + \theta)\lambda\mathbb{E}[X] - (1 + \psi)\lambda\mathbb{E}[Z] > \lambda\mathbb{E}[Y],$$

which, with $\mathbb{E}[Y] = \mathbb{E}[X] - \mathbb{E}[Z]$, gives the condition

$$\theta\mathbb{E}[X] > \psi\mathbb{E}[Z]. \qquad (5.21)$$

We note therefore that we require the two relative security loadings to satisfy the following inequalities, which give lower and upper bounds on the reinsurer's security loading for contracts with the direct insurer to be practicable:

$$\theta < \psi < \frac{\mathbb{E}[X]}{\mathbb{E}[Z]}\theta.$$

In the case of proportional reinsurance with retained proportion β, the condition which ensures positive expected profit for the insurer, namely (5.21), becomes $\theta\mathbb{E}[X] > \psi(1-\beta)\mathbb{E}[X]$, which in turn gives the condition $\beta > 1-(\theta/\psi)$. The condition imposes a *lower bound* on the retained proportion β. The direct insurer must retain a certain proportion of the risk (for example, with $\theta = 0.1$ and $\psi = 0.2$, the direct insurer must retain at least 50% of the risk) – it is not possible to pass over too much of the risk exposure while maintaining profitability. The more risk the direct insurer passes over, the more secure the direct insurance business is, but there is a price to pay in reduced profitability (less direct insurance business is being conducted, and reinsurance is expensive).

In the case of excess of loss reinsurance with $X \sim \text{Exp}(1/\mu)$ (with mean μ) and retention level M, we have, from Example 5.1, that $\mathbb{E}[Z] = \mu e^{-M/\mu}$

and condition (5.21) becomes $M > \mu \log(\psi/\theta)$. The condition imposes a *lower bound* on the retention level M (for example, with $\theta = 0.1$ and $\psi = 0.2$, the direct insurer's retention level M must be at least 0.693μ) – again we see that the direct insurer must retain at least a specified level of exposure to the risks involved.

The direct insurer cannot make profits if there is not enough direct insurance business, that is if the direct insurer does not maintain enough risk exposure. The results above are good illustrations of the trade-off between security and profitability.

5.5 Optimising the reinsurance contract

In the remaining sections of this chapter we examine selected issues concerning optimising reinsurance arrangements seen from the point of view of the direct insurer or the reinsurer as appropriate. In §5.6 we establish optimal retention levels for the direct insurer based on maximising the expected utility of the direct insurer's assets at the end of the period of insurance – we do this separately for excess of loss contracts and proportional contracts. In §5.7 we look at the problem of minimising the variance of the aggregate claims for each party involved. In §5.8 we consider the optimisation of reinsurance contracts for a portfolio of independent risks, based on minimising the uncertainty of the direct insurer's net profit.

We continue to consider only simple reinsurance structures, noting again that in practice insurance companies may have in place much more complex arrangements.

5.6 Optimising reinsurance contracts based on maximising expected utility

Consider a direct insurer who has a known initial wealth and a set of possible actions which lead to different financial gains. We will suppose that the direct insurer has adopted a utility function $u(x)$ (see Appendix A) and chooses which action to take using the expected utility criterion, that is the action that maximises the expected utility of wealth is chosen.

In particular, we suppose the direct insurer has initial wealth W and uses an exponential utility function $u(x) = -e^{-\alpha x}$ for some $\alpha > 0$. Then, if action A leads to financial gain $G(A)$, the direct insurer chooses the action A which maximises

$$\mathbb{E}[u(W + G(A))] = \mathbb{E}[-\exp\{-\alpha(W + G(A))\}]. \qquad (5.22)$$

We suppose that the direct insurer is exposed to a risk with aggregate claims variable S, where $S \sim CP(\lambda, F_X)$, and where the claim amount variable X has probability density function $f_X(x)$ and distribution function $F_X(x)$.

5.6.1 Excess of loss reinsurance

Suppose the direct insurer effects excess of loss reinsurance with retention level M, and let the direct insurer's premium be P. We assume that the reinsurer's premium is calculated using the expected value principle (4.1), with relative security loading ψ. We will denote this premium by P_R, which is, of course, a function of M. Then P_R is given by

$$P_R = (1 + \psi)\lambda\mathbb{E}[Z] - (1 + \psi)\lambda \int_M^\infty (x - M)f_X(x)dx,$$

where Z is the reinsurer's payment on a claim.

The aggregate claims S_I net of reinsurance paid by the direct insurer has a $CP(\lambda, F_Y)$ distribution, where Y is the direct insurer's payout on a claim. By (3.16) the random variable S_I has a moment generating function $M_{S_I}(t) = \mathbb{E}[\exp(tS_I)] = \exp[\lambda\{M_Y(t) - 1\}]$, which is also a function of M because the distribution of Y depends on M. The direct insurer's wealth at the end of the insurance period is $W + P - P_R - S_I$, and M is chosen such that

$$\mathbb{E}[u(W + P - P_R - S_I)] = \mathbb{E}[-\exp\{-\alpha(W + P - P_R - S_I)\}]$$
$$= -e^{-\alpha(W+P)}\exp(\alpha P_R)\mathbb{E}[\exp(\alpha S_I)],$$

is maximised, which is equivalent to minimising

$$\exp(\alpha P_R)\mathbb{E}[\exp(\alpha S_I)] = \exp(\alpha P_R)\exp[\lambda\{M_Y(\alpha) - 1\}].$$

Taking logarithms, the problem can now be expressed as follows: minimise with respect to M the function

$$h(M) = \alpha P_R + \lambda(M_Y(\alpha) - 1). \qquad (5.23)$$

We have

$$P_R = (1 + \psi)\lambda\left(\int_0^\infty xf_X(x)dx - \int_0^M xf_X(x)dx - M(1 - F_X(M))\right).$$

Assume that f_X is well-behaved, so that we may differentiate with respect to M to obtain

$$\frac{\partial P_R}{\partial M} = (1 + \psi)\lambda\left(-Mf_X(M) - (1 - F_X(M)) + Mf_X(M)\right)$$
$$= -(1 + \psi)\lambda(1 - F_X(M)).$$

By (1.15), the moment generating function of Y is

$$M_Y(\alpha) = \int_0^M e^{\alpha x} f_X(x)dx + e^{\alpha M}(1 - F_X(M)),$$

and so

$$\frac{\partial M_Y(\alpha)}{\partial M} = e^{\alpha M} f_X(M) + \alpha e^{\alpha M}(1 - F_X(M)) - e^{\alpha M} f_X(M)$$
$$= \alpha e^{\alpha M}(1 - F_X(M)).$$

Hence, from (5.23) we have

$$h'(M) = \alpha\frac{\partial P_R}{\partial M} + \lambda\frac{\partial M_Y(\alpha)}{\partial M}$$
$$= \lambda\alpha(1 - F_X(M))\left(e^{\alpha M} - (1 + \psi)\right).$$

It follows that $h'(M) = 0$ when M satisfies

$$F_X(M) = 1 \text{ or } e^{\alpha M} = 1 + \psi.$$

The first of these corresponds to the "no reinsurance" case, while the second gives $M = (1/\alpha)\log(1 + \psi)$. When $M < $ (resp. $>$) $(1/\alpha)\log(1 + \psi)$, it is easy to check that $h'(M) < 0$ (resp. > 0), so that $M = (1/\alpha)\log(1 + \psi)$ does indeed give a minimum. We conclude that the direct insurer's expected utility of wealth at the end of the insurance period is maximised by setting the retention level M at

$$M = \frac{1}{\alpha}\log(1 + \psi). \tag{5.24}$$

This result is interesting and instructive in several ways: we see that the direct insurer's optimal retention level

- does not depend on the direct insurer's initial wealth (this is to be expected in the presence of the exponential utility function);
- does not depend on the individual claim size distribution;
- is a decreasing function of the direct insurer's utility function parameter α (α is related to the direct insurer's risk aversion – the more "safety/security conscious" is the direct insurer, the higher will be the value of α) – increasing α corresponds to setting a lower retention level and passing more of the risk to the reinsurer;
- is an increasing function of the reinsurer's premium security loading ψ – the more expensive is the reinsurance, the less the direct insurer will

purchase – this corresponds to the direct insurer setting a higher retention level and passing less of the risk to the reinsurer.

5.6.2 Proportional reinsurance

Suppose the direct insurer effects proportional reinsurance with retained proportion β, then the aggregate claim paid by the direct insurer, S_I, is given by $S_I = \beta S$ and $S_I \sim \text{CP}(\lambda, F_Y)$, where $Y = \beta X$. The aggregate claim paid by the reinsurer, S_R, is given by $S_R = (1 - \beta)S$ and $S_R \sim \text{CP}(\lambda, F_Z)$, where $Z = (1 - \beta)X$. The moment generating function of S_R is given by

$$M_{S_R}(t) = \exp[\lambda\{M_Z(t) - 1\}] = \exp[\lambda\{M_X((1 - \beta)t) - 1\}].$$

Let P denote the direct insurer's premium, and, in this case, let us assume that the reinsurer sets an "exponential premium" with parameter η; that is, an exponential utility function $u(x) = -e^{-\eta x}$ is used and the premium is set at P_R using the exponential premium principle (4.5), so that, assuming $M_X((1 - \beta)M) < \infty$ and using (3.16) we have

$$P_R = \frac{1}{\eta} \log M_{S_R}(\eta)$$
$$= \frac{1}{\eta}[\lambda\{M_X((1 - \beta)\eta) - 1\}]$$
$$= \frac{\lambda}{\eta}\left[\int_0^\infty \exp((1 - \beta)\eta x)f_X(x)dx - 1\right].$$

The direct insurer's wealth at the end of the insurance period is $W + P - P_R - \beta S$, and β is chosen to maximise

$$\mathbb{E}[u(W + P - P_R - \beta S)] = \mathbb{E}[-\exp\{-\alpha(W + P - P_R - \beta S)\}]$$
$$= -e^{-\alpha(W+P)} \exp(\alpha P_R)\mathbb{E}[\exp(\alpha\beta S)].$$

This is equivalent to minimising

$$\exp(\alpha P_R)\mathbb{E}[\exp(\alpha\beta S)] = \exp(\alpha P_R)\exp[\lambda\{M_X(\alpha\beta) - 1\}],$$

where we assume $M_X(\alpha\beta) < \infty$. Taking logarithms, the problem can now be expressed as follows: minimise with respect to β the function

$$h(\beta) = \alpha P_R + \lambda\left[\int_0^\infty \exp(\alpha\beta x)f_X(x)dx\right]. \tag{5.25}$$

Assuming that f_X is well-behaved, we find that the derivatives with respect to β of

$$P_R = \frac{\lambda}{\eta}\left[\int_0^\infty \exp((1 - \beta)\eta x)f_X(x)dx - 1\right]$$

and

$$\int_0^\infty \exp(\alpha\beta x) f_X(x) dx$$

are, respectively,

$$\frac{\lambda}{\eta}(-\eta)\int_0^\infty x\exp((1-\beta)\eta x) f_X(x) dx = -\lambda\int_0^\infty x\exp((1-\beta)\eta x) f_X(x) dx$$

and

$$\alpha\int_0^\infty x\exp(\alpha\beta x) f_X(x) dx.$$

Hence, from (5.25) we have

$$h'(\beta) = -\alpha\lambda\int_0^\infty [\exp((1-\beta)\eta x) - \exp(\alpha\beta x)]x f_X(x) dx.$$

It follows that $h'(\beta) = 0$ if and only if $(1-\beta)\eta = \alpha\beta$, that is if and only if $\beta = \eta/(\eta+\alpha)$. The reader can verify that $h'' > 0$, thus verifying that we have found a minimum for $h(\beta)$.

We conclude that the direct insurer's expected utility of wealth at the end of the insurance period is maximised by setting the retained proportion of the risk, β, at

$$\beta = \frac{\eta}{\eta+\alpha}. \tag{5.26}$$

As in §5.6.1, the result is interesting and instructive in several ways: we see that the direct insurer's optimal retained proportion

- does not depend on the direct insurer's initial wealth (this is to be expected in the presence of the exponential utility function);
- does not depend on the individual claim size distribution;
- is a decreasing function of the direct insurer's utility function parameter α – increasing α corresponds to setting a lower retained proportion and passing more of the risk to the reinsurer;
- is an increasing function of the reinsurer's utility function parameter η – the more risk averse the reinsurer is, the higher the reinsurer's premium will be and the less reinsurance the direct insurer will purchase – this corresponds to the direct insurer setting a higher retained proportion and passing less of the risk to the reinsurer.

Example 5.10 Suppose the annual aggregate claims for a portfolio of risks, S, is modelled as $S \sim N(\mu, \sigma^2)$ and P denotes the annual premium charged by the direct insurer to cover the overall risk. Suppose the direct insurer (who has initial wealth W) effects proportional reinsurance, with retained proportion β

of each claim (so $S_I = \beta S$ and $S_R = (1 - \beta)S$), where β is to be chosen such that the insurer's expected utility of wealth at the end of the year is maximised with respect to the exponential utility function $u(x) = -\exp(-\alpha x)$. Suppose also that the reinsurer sets the premium P_R using the expected value principle with relative security loading ψ, where $\psi < \alpha\sigma^2/\mu$.

From (2.10), S has the moment generating function

$$M_S(t) = \mathbb{E}[\exp(tS)] = \exp(t\mu + t^2\sigma^2/2).$$

The direct insurer's expected utility at the end of the year is given by

$$\mathbb{E}[-\exp\{-\alpha(W + P - P_R - S_I)\}] = -k\mathbb{E}[\exp\{\alpha(P_R + S_I)\}],$$

where $k = \exp\{-\alpha(W + P)\}$. The reinsurer's premium is

$$P_R = (1 + \psi)\mathbb{E}[S_R] = (1 + \psi)(1 - \beta)\mu,$$

and

$$\mathbb{E}[\exp(\alpha S_I)] = \mathbb{E}[\exp(\alpha\beta S)] = M_S(\alpha\beta) = \exp(\alpha\beta\mu + \alpha^2\beta^2\sigma^2/2).$$

Hence the direct insurer's expected utility at the end of the year can be written as $-k\exp(h(\beta))$, where

$$h(\beta) = \alpha\mu(1 + \psi)(1 - \beta) + \alpha\beta\mu + \alpha^2\beta^2\sigma^2/2,$$

where $k > 0$ and does not depend on β. We maximise the expected utility by finding the value of β which minimises $h(\beta)$. Differentiating with respect to β gives

$$h'(\beta) = -\alpha\mu(1 + \psi) + \alpha\mu + \alpha^2\sigma^2\beta = 0 \Rightarrow \beta = \psi\mu/(\alpha\sigma^2).$$

Noting that $h''(\beta) = \alpha^2\sigma^2 > 0$ confirms that we have a minimum.

The optimal value of the retained proportion β is an increasing function of ψ – this reflects the fact that, as the price of reinsurance increases, the direct insurer will purchase less reinsurance and retain more of the risk (higher β). The optimal β is a decreasing function of α – this reflects the role of α in the utility function as a measure of the direct insurer's attitude to risk – a higher value of α reflects a more risk-averse insurer, one who will therefore pass over more of the risk – and will purchase more reinsurance and retain less of the risk (lower β). Note also that increasing the direct insurer's uncertainty (as measured by the risk variance σ^2) while keeping the expected aggregate claims (μ) the same produces a lower value of β – other things being equal, the insurer faced with a more uncertain overall commitment will pass over more of the risk.

Suppose instead that the reinsurer sets an "exponential premium" with parameter η, so that

$$P_R = \frac{1}{\eta} \log M_{S_R}(\eta).$$

The moment generating function of S_R is given by

$$M_{S_R}(\eta) = \mathbb{E}[\exp(\eta S_R)] = \mathbb{E}[\exp(\eta(1 - \beta)S)]$$
$$= M_S[\eta(1 - \beta)] = \exp\{\eta(1 - \beta)\mu + \eta^2(1 - \beta)^2\sigma^2/2\},$$

so

$$P_R = (1 - \beta)\mu + \eta(1 - \beta)^2\sigma^2/2.$$

In this case we have to minimise

$$h(\beta) = \alpha(1 - \beta)\mu + \alpha\eta(1 - \beta)^2\sigma^2/2 + \alpha\beta\mu + \alpha^2\beta^2\sigma^2/2.$$

Taking the derivative, we find

$$h'(\beta) = -\alpha\mu - \alpha\eta(1 - \beta)\sigma^2 + \alpha\mu + \alpha^2\sigma^2\beta,$$

and setting $h'(\beta) = 0$ gives $\alpha\beta = \eta(1 - \beta)$, and so

$$\beta = \frac{\eta}{\eta + \alpha}.$$

Again we have $h''(\beta) > 0$, confirming a minimum. We note that this is the same result as we have in (5.26) when S is modelled as a compound Poisson random variable.

5.7 Optimising reinsurance contracts based on minimising the variance of aggregate claims

Maximising expected income, profits or wealth, or minimising expected pay-out, are not necessarily the only desirable outcomes insurers may want to consider – they may also be interested in reducing the *uncertainty* inherent in their situation. Statistically, the insurer may want to consider not only the *level*, but also the *spread* of the insurer's payout – in other words to consider second order moments (in addition to first order moments) of the payout distribution.

Many investors would choose to gamble for a probabilistic return X_1 distributed $X_1 \sim N(2, 1)$ rather than for a return X_2 distributed $X_2 \sim N(3, 4)$, despite the fact that the second gamble has a higher *expected* return. The probability that the first gamble produces a positive return is 0.977, whereas for the second it is only 0.933.

In this section we consider five situations in which our optimality criterion is expresed in terms of minimising the variance of the aggregate payout of the insurer (S_I) or the reinsurer (S_R), or the sum of the variances of the two payouts, subject to some appropriate constraint.

It is convenient now to consider reinsurance arrangements in which the amount paid by the direct insurer can be defined directly as a function, say $g(\cdot)$, of the aggregate claims S, rather than being defined in terms of the individual claim amounts. We define two principal types of such a reinsurance arrangement:

- stop loss reinsurance (with retention level M): $S_I = g(S) = \min(S, M)$ (and $S_R = \max(0, S - M)$);
- proportional reinsurance (with retained proportion β): $S_I - g(S) = \beta S$ (and $S_R = (1 - \beta)S$).

Note that, while "stop loss" reinsurance is not the same as "excess of loss" reinsurance (which is based on individual claim amounts), proportional reinsurance as defined above *is* equivalent at the aggregate claims level to the proportional reinsurance introduced earlier.

We note that the function $g(\cdot)$ satisfies $0 \le g(x) \le x$ for $x \ge 0$; in this context we may refer to it as a *reinsurance function*. See Case study 3 in Chapter 7 for other properties of reinsurance functions.

5.7.1 Minimising $\mathrm{Var}[S_I]$ subject to fixed $\mathbb{E}[S_I]$

Suppose the insurer wants to reduce the uncertainty of the aggregate payout S_I. Trivially, the whole risk could be reinsured, but this is not realistic. What the insurer needs to look for is an arrangement that minimises the uncertainty in S_I, but is subject to some sensible constraint. An appropriate such constraint is given by setting a value for the expected payout, so let us fix the expected payout at $\mathbb{E}[S_I] = c$ for some c ($<\mathbb{E}[S]$).

We show in the following that, among all reinsurance arrangements with $\mathbb{E}[S_I] = c$, it is stop loss reinsurance, with retention level M such that $\mathbb{E}[\min(S, M)] = c$, that minimises $\mathrm{Var}[S_I]$. To show this, let $g(\cdot)$ be any reinsurance function such that $S_I = g(S)$ and $\mathbb{E}[g(S)] = c$. Recall that $0 \le g(x) \le x$. We now consider $\mathrm{Var}[g(S)]$, given by

$$\mathrm{Var}[g(S)] = \mathbb{E}[(g(S) - c)^2] = \mathbb{E}[(g(S) - M + M - c)^2]$$
$$= \mathbb{E}[(g(S) - M)^2] - (M - c)^2.$$

This holds for any reinsurance function g such that $\mathbb{E}[g(S)] = c$, and in particular it holds for the above stop loss reinsurance, so we have

$$\text{Var}\left[\min(S, M)\right] = \mathbb{E}[(\min(S, M) - M)^2] - (M - c)^2.$$

We know that

$$\min(S, M) - M = \begin{cases} S - M & \text{if } S \le M \\ 0 & \text{otherwise,} \end{cases}$$

so

$$\mathbb{E}[(\min(S, M) - M)^2] = \mathbb{E}[(S - M)^2 1(S \le M)],$$

where we recall that $1(A)$ is the indicator function of the event A. If $S \le M$, then $0 \le g(S) \le S \le M$, so that

$$(S - M)^2 1(S \le M) \le (g(S) - M)^2 1(S \le M) \le (g(S) - M)^2.$$

This means that

$$\mathbb{E}[(\min(S, M) - M)^2] \le \mathbb{E}[(g(S) - M)^2],$$

and this implies that

$$\begin{aligned} \text{Var}[\min(S, M)] &= \mathbb{E}[(\min(S, M) - M)^2] - (M - c)^2 \\ &\le \mathbb{E}[(g(S) - M)^2] - (M - c)^2 \\ &= \text{Var}[g(S)]. \end{aligned}$$

Hence $\text{Var}[S_I]$ is minimised, subject to $\mathbb{E}[S_I] = c$, by using stop loss reinsurance $S_I = \min(S, M)$ with retention M determined by $\mathbb{E}[S_I] = c$. The result reflects the fact that the direct insurer will find attractive a reinsurance arrangement where the direct insurer has limited liability (and no exposure to the "risky tail" of the aggregate claims distribution).

5.7.2 Minimising $\text{Var}[S_R]$ subject to fixed $\text{Var}[S_I]$

Suppose the reinsurer wants to reduce the uncertainty of the aggregate payout S_R. Trivially, the reinsurer could accept no risk at all, but this scenario is of no interest here. What the reinsurer needs to look for is an arrangement that minimises the uncertainty in S_R, but is subject to some sensible constraint. An appropriate such constraint is given by setting a value for the uncertainty faced by the direct insurer, so let us fix the variance of S_I at $\text{Var}[S_I] = c$ for some $c > 0$. Then we have

$$\begin{aligned} \text{Var}[S_R] &= \text{Var}[S - S_I] \\ &= \text{Var}[S] + \text{Var}[S_I] - 2 \, \text{Cov}[S, S_I]. \end{aligned}$$

Now Var[S] is fixed and Var[S_I] is fixed ($= c$) by our constraint. So Var[S_R] is minimised when Cov[S, S_I] is maximised.

The correlation coefficient between S and S_I is given by Cov[S, S_I] divided by $\{\text{Var}[S]\,\text{Var}[S_I]\}^{1/2}$, which is fixed. So Cov[S, S_I] is maximised when the correlation coefficient between S and S_I is maximised, which occurs (with value $+1$) when S and S_I are linearly related in a positive sense, that is when $S_I = \beta S$, with $\beta > 0$.

Hence Var[S_R] is minimised, subject to Var[S_I] $= c$, by using proportional reinsurance with $S_I = \beta S$ (and where β is given by $\beta = \{\text{Var}[S_I]/\,\text{Var}[S]\}^{1/2}$). The result reflects the fact that the reinsurer will find attractive a reinsurance arrangement in which the reinsurer does not have sole responsibility for the "risky tail" of the aggregate claims distribution.

5.7.3 Comparing stop loss and equivalent proportional reinsurance arrangements

Suppose the direct insurer effects stop loss reinsurance with retention level M, so that $S_I = \min(S, M)$. Consider a proportional arrangement which is equivalent as regards uncertainty, in the sense that the direct insurer's payouts under both arrangements have the same variance. So let $S_I^* = \beta S$, where $\beta = \{\text{Var}[S_I]/\,\text{Var}[S]\}^{1/2}$ (>0), giving Var[S_I^*] = Var[S_I].

We introduce a proportional reinsurance arrangement which gives the direct insurer the same expected payout as the stop loss arrangement above – so let $S_I^{\dagger} = \gamma S$, where $\gamma = \mathbb{E}[S_I]/\mathbb{E}[S]$, $0 \leq \gamma \leq 1$, and giving $\mathbb{E}[S_I^{\dagger}] = \mathbb{E}[S_I]$.

In §5.7.1, we found that, for fixed expected payout, the direct insurer achieves minimum variance for this payout by using stop loss reinsurance. It follows that Var[S_I] $\leq$ Var[$S_I^{\dagger}$]. But

$$\text{Var}[S_I] = \text{Var}[S_I^*] = \beta^2\,\text{Var}[S] \quad \text{and} \quad \text{Var}[S_I^{\dagger}] = \gamma^2\,\text{Var}[S],$$

from which it follows that $\beta^2\,\text{Var}[S] \leq \gamma^2\,\text{Var}[S]$ and hence $\beta \leq \gamma$. Hence

$$\mathbb{E}[S_I^*] = \beta\mathbb{E}[S] \leq \gamma\mathbb{E}[S] = \mathbb{E}[S_I].$$

We conclude that, if we compare stop loss and proportional reinsurance arrangements which give the direct insurer the same uncertainty (equal variance of payout), the direct insurer's expected payout under the proportional arrangement is less than or equal to that under the stop loss arrangement.

We next consider the above reinsurance contracts with regard to the *coefficient of variation* of the payouts for the direct insurer and for the reinsurer. The coefficient of variation (c.v.) of a random variable is the standard deviation expressed as a multiple of the mean – it is a dimensionless measure of uncertainty which takes into account the level (the expected value) of the variable. So for a random variable with mean μ and standard deviation σ, the c.v. is given by σ/μ. In most circumstances lower values of a c.v. are preferred to higher values. It is interesting to examine the implications of the results of §5.7.2 and §5.7.3 for the c.v.s of the payouts by the direct insurer and the reinsurer under stop loss and proportional contracts.

Define a stop loss arrangement such that $S_I = \min(S, M)$ and $S_R = \max(0, S - M)$, and a proportional arrangement such that $S_I^* = \beta S$ and $S_R^* = (1 - \beta)S$, subject to the constraint $\mathrm{Var}[S_I] = \mathrm{Var}[S_I^*]$.

The result of §5.7.2 implies that $\mathrm{Var}[S_R] \geq \mathrm{Var}[S_R^*]$. The result of the current subsection implies that $\mathbb{E}[S_I] \geq \mathbb{E}[S_I^*]$, which in turn implies that $\mathbb{E}[S_R] \leq \mathbb{E}[S_R^*]$. We deduce from these results that, for "equivalent" contracts we have the following.

For the direct insurer

$$
\begin{array}{cc}
\text{stop loss} & \text{proportional} \\
\dfrac{\mathrm{SD}[S_I]}{\mathbb{E}[S_I]} & \leq \dfrac{\mathrm{SD}[S_I^*]}{\mathbb{E}[S_I^*]};
\end{array}
$$

for the reinsurer

$$
\begin{array}{cc}
\text{stop loss} & \text{proportional} \\
\dfrac{\mathrm{SD}[S_R]}{\mathbb{E}[S_R]} & \geq \dfrac{\mathrm{SD}[S_R^*]}{\mathbb{E}[S_R^*]}.
\end{array}
$$

So, for the direct insurer's payout, c.v.(stop loss) $\leq$ c.v.(proportional), and for the reinsurer's payout, c.v.(stop loss) $\geq$ c.v.(proportional). These considerations show that, using the criterion of "low c.v.", stop loss arrangements are desirable for the direct insurer, while proportional arrangements are desirable for the reinsurer.

5.7.4 Minimising $\mathrm{Var}[S_I] + \mathrm{Var}[S_R]$

Consider an aggregate claims variable S with $S = S_I + S_R$ under a general, unspecified, reinsurance arrangement. Consider a proportional reinsurance arrangement with the same variance for the direct insurer's payout. This arrangement has $S_I^* = \beta S$ and $S_R^* = (1 - \beta)S$, where $\beta = \{\mathrm{Var}[S_I]/\mathrm{Var}[S]\}^{1/2}$, giving $\mathrm{Var}[S_I^*] = \mathrm{Var}[S_I]$.

For the general arrangement we have

$$
\begin{aligned}
\mathrm{Var}[S_I] + \mathrm{Var}[S_R] &= \mathrm{Var}[S_I] + \mathrm{Var}[S - S_I] \\
&= \mathrm{Var}[S_I] + \mathrm{Var}[S] + \mathrm{Var}[S_I] - 2\,\mathrm{Cov}[S, S_I].
\end{aligned}
$$

Similarly, for the proportional arrangement we have

$$\mathrm{Var}[S_I^*] + \mathrm{Var}[S_R^*] = \mathrm{Var}[S_I^*] + \mathrm{Var}[S - S_I^*]$$
$$= \mathrm{Var}[S_I^*] + \mathrm{Var}[S] + \mathrm{Var}[S_I^*] - 2\,\mathrm{Cov}[S, S_I^*].$$

Noting $\mathrm{Var}[S_I^*] = \mathrm{Var}[S_I]$ and the relationship between the correlation coefficient (Corr) and the covariance, $\mathrm{Corr}[X, Y] = \mathrm{Cov}[X, Y]/\{\mathrm{Var}[X]\,\mathrm{Var}[Y]\}^{1/2}$, we have

$$\mathrm{Var}[S_I^*] + \mathrm{Var}[S_R^*] - \{\mathrm{Var}[S_I] + \mathrm{Var}[S_R]\}$$
$$= 2\,\mathrm{Cov}[S, S_I] - 2\,\mathrm{Cov}[S, S_I^*]$$
$$= 2\,\{\mathrm{Var}[S]\,\mathrm{Var}[S_I]\}^{1/2}\{\mathrm{Corr}[S, S_I] - \mathrm{Corr}[S, S_I^*]\}.$$

But we know that $\mathrm{Corr}[S, S_I] \leq 1$ and $\mathrm{Corr}[S, S_I^*] = 1$, so that $\mathrm{Corr}[S, S_I] - \mathrm{Corr}[S, S_I^*] \leq 0$. This gives

$$\mathrm{Var}[S_I^*] + \mathrm{Var}[S_R^*] \leq \mathrm{Var}[S_I] + \mathrm{Var}[S_R].$$

Hence the sum of the variances of the payouts by both parties is minimised by using proportional reinsurance.

Continuing, we note that $\mathrm{Var}[\beta S] + \mathrm{Var}[(1 - \beta)S] = (2\beta^2 - 2\beta + 1)\,\mathrm{Var}[S]$, which is minimised by setting $\beta = 0.5$. Hence the sum of the variances of the payouts by both parties is minimised by using proportional reinsurance with retained proportion 0.5, that is when the aggregate risk is shared equally between the direct insurer and the reinsurer.

5.7.5 Minimising the sum of variances when two independent risks are shared between two insurers

Consider two insurance companies A and B which are exposed to aggregate claims S_A and S_B, respectively, where S_A and S_B are independent random variables. Let us suppose that the two companies share their risks with each other as follows:

company A retains $g_1(S_A)$ and reinsures $S_A - g_1(S_A)$ with company B;
company B retains $g_2(S_B)$ and reinsures $S_B - g_2(S_B)$ with company A,

where $g_i(\cdot)$, $i = 1, 2$, are reinsurance functions (so they satisfy $0 \leq g_i(x) \leq x$, $i = 1, 2$). Then the total payouts for company A and company B are T_A and T_B, respectively, where

$$T_A = g_1(S_A) + \{S_B - g_2(S_B)\} \quad \text{and} \quad T_B = g_2(S_B) + \{S_A - g_1(S_A)\}.$$

Hence, since S_A and S_B are independent, we have

$$\text{Var}[T_A] = \text{Var}[g_1(S_A)] + \text{Var}[S_B - g_2(S_B)]$$

and

$$\text{Var}[T_B] = \text{Var}[g_2(S_B)] + \text{Var}[S_A - g_1(S_A)].$$

Suppose the function $g_1(\cdot)$ does *not* define a proportional reinsurance arrangement. If we now choose $\beta_1 = \{\text{Var}[g_1(S_A)]/\text{Var}[S_A]\}^{1/2}$ and define g_1^* by $g_1^*(S_A) = \beta_1 S_A$, then $g_1^*(\cdot)$ *does* define a proportional arrangement and we have $\text{Var}[g_1^*(S_A)] = \text{Var}[g_1(S_A)]$. We know that the variance of the reinsurance payout, subject to fixed variance for the direct insurance payout, is minimised by proportional reinsurance (see §5.7.2), hence

$$\text{Var}[S_A - g_1^*(S_A)] \leq \text{Var}[S_A - g_1(S_A)].$$

A similar argument holds for $g_2(\cdot)$. Hence $\text{Var}[T_A] + \text{Var}[T_B]$ is minimised by both companies using proportional reinsurance.

In Exercise 5.31 you are asked to show that the optimal arrangement is in fact that in which each company shares its risk equally with the other.

Example 5.11 Suppose an aggregate claims variable S is distributed as $S \sim \text{CP}(6, F_X)$, where the individual claim amount is fixed at $X = £1000$. We consider four cases, which will illustrate the results of §5.7.1–§5.7.4. Note that $\mathbb{E}[X] = 1000$, $\mathbb{E}[X^2] = 10^6$, $\mathbb{E}[S] = 6000$ and $\text{Var}[S] = 6 \times 10^6$.

Case 1 Suppose the direct insurer requires a fixed expected payout and, subject to this, wants to minimise the variance of the payout, $\text{Var}[S_I]$. We know from the result in §5.7.1 that a stop loss reinsurance contract must be arranged, with retention level M set to produce the required $\mathbb{E}[S_I]$.

The number of claims, N, has distribution $N \sim \text{Poi}(6)$, so that

$$p_k = \text{Pr}(N = k) = e^{-6}6^k/k!, \quad k = 0, 1, 2, \ldots.$$

We note that $N = 0$ implies that $S = S_I = S_R = 0$. If $N = r$, then $S = 1000r$, $r = 1, 2, \ldots$, and the situation can be displayed as in Table 5.4. Suppose we fix M between $1000r$ and $1000(r + 1)$. The direct insurer's expected payout is given by

$$\mathbb{E}[S_I] = \sum_{k=0}^{r} 1000k \times p_k + M\left(1 - \sum_{k=0}^{r} p_k\right).$$

For example, with $r = 3$,

$$\mathbb{E}[S_I] = 1000p_1 + 2000p_2 + 3000p_3 + M(1 - p_0 - p_1 - p_2 - p_3),$$

Table 5.4. *Stop loss arrangement with claim size fixed at £1000 in Example 5.11*

N	Probability	S	S_I	S_R
0	p_0	0	0	0
1	p_1	1000	1000	0
2	p_2	2000	2000	0
.	.	.	.	.
r	p_r	1000r	1000r	0
r + 1	p_{r+1}	1000(r + 1)	M	1000(r + 1) − M
r + 2	p_{r+2}	1000(r + 2)	M	1000(r + 2) − M
.	.	.	.	.

Table 5.5. *Stop loss arrangement of Table 5.4 for which the direct insurer's payout has minimum variance, subject to having fixed expected value in Example 5.11*

N	Probability	S	S_I	S_R	
0	p_0	0	0	0	
1	p_1	1000	1000	0	
2	p_2	2000	2000	0	
3	p_3	3000	3000	0	
4	p_4	4000	3685	315	(= 4000 − 3685)
5	p_5	5000	3685	1315	(= 5000 − 3685)
6	p_6	6000	3685	2315	(= 6000 − 3685)
.	.	.	.	.	

which gives

$$\mathbb{E}[S_I] = 150\,000e^{-6} + M(1 - 61e^{-6}) = 371.81 + 0.84880M.$$

For $3000 \leq M \leq 4000$, this gives $2918 \leq \mathbb{E}[S_I] \leq 3767$.

Suppose then the direct insurer requires $\mathbb{E}[S_I] = 3500$. We achieve this by taking $r = 3$ and solving $371.81 + 0.84880M = 3500$, which gives $M = 3685.4$. We take $M = 3685$, so

$$S_I = \min(S, 3685) \quad \text{and} \quad S_R = \max(0, S - 3685).$$

So the direct insurer pays a maximum of the first three claims in full and £685 towards the fourth claim. The situation is now as shown in Table 5.5. Continuing, we find

$$\mathbb{E}[S_I^2] = 1000^2 p_1 + 2000^2 p_2 + 3000^2 p_3 + 3685^2 \Pr(\text{four or more claims})$$
$$= 12.52245 \times 10^6,$$

and so

$$\text{Var}[S_I] = 12.52245 \times 10^6 - 3500^2 = 27.25 \times 10^4.$$

The mean and variance of the reinsurer's payout are given by

$$\mathbb{E}[S_R] = 6000 - 3500 = 2500$$

and

$$\text{Var}[S_R] = \sum_{n=4}^{\infty}(1000n - 3685)^2 \times p_n - 2500^2,$$

which evaluates to 480.0×10^4. The standard deviation of the reinsurer's payout (£2191) is more than four times as high as that of the direct insurer's payout (£522). Note that (to four significant figures) $\text{Var}[S_I] + \text{Var}[S_R] = 507.2 \times 10^4$.

Case 2 Consider now the case in which the direct insurer effects proportional reinsurance with retained proportion chosen so that the direct insurer's expected payout is the same as in Case 1, namely £3500. Let $S_I^{(2)} = \beta S$, where β satisfies $\beta \mathbb{E}[S] = 3500$. Then $\beta = 7/12$, so that $S_I^{(2)} = (7/12)S$ and $S_R^{(2)} = (5/12)S$ are the direct insurer's and reinsurer's payouts, respectively.

The mean and variance of the direct insurer's payout are given by

$$\mathbb{E}[S_I^{(2)}] = 3500 \quad \text{and} \quad \text{Var}[S_I^{(2)}] = (7/12)^2 \text{Var}[S] = 204.17 \times 10^4.$$

This greatly exceeds $\text{Var}[S_I]$ in Case 1. The mean and variance of the reinsurer's payout are given by

$$\mathbb{E}[S_R^{(2)}] = 2500 \quad \text{and} \quad \text{Var}[S_R^{(2)}] = (5/12)^2 \text{Var}[S] = 104.17 \times 10^4,$$

which is very much less than $\text{Var}[S_R]$ in Case 1. Note that $\text{Var}[S_I^{(2)}] + \text{Var}[S_R^{(2)}] = 308.3 \times 10^4$.

Case 3 Consider now the case in which the direct insurer effects proportional reinsurance with retained proportion 0.5. Let $S_I^{(3)} = S_R^{(3)} = 0.5S$ denote the direct insurer's and reinsurer's payouts, respectively. The mean and variance of the direct insurer's payout are given by

$$\mathbb{E}[S_I^{(3)}] = 0.5\mathbb{E}[S] = 3000 \quad \text{and} \quad \text{Var}[S_I^{(3)}] = 0.5^2 \text{Var}[S] = 150 \times 10^4,$$

which again greatly exceeds $\text{Var}[S_I]$ in Case 1. Similarly, $\mathbb{E}[S_R^{(3)}] = 3000$ and $\text{Var}[S_R^{(3)}] = 150 \times 10^4$, which again is very much less than $\text{Var}[S_R]$ in Case 1. Note that $\text{Var}[S_I^{(3)}] + \text{Var}[S_R^{(3)}] = 300 \times 10^4$.

Table 5.6. *Summary of results for Cases 1–4 in Example 5.11*

	$\mathbb{E}[S_I]$	$\mathrm{Var}[S_I]$ $\times 10^4$	$\mathbb{E}[S_R]$	$\mathrm{Var}[S_R]$ $\times 10^4$	$\mathrm{Var}[S_I] + \mathrm{Var}[S_R]$ $\times 10^4$
Stop loss ($M = 3685$)	3500	27.25	2500	480.0	507.2
Proportional ($\beta = 7/12$)	3500	204.2	2500	104.2	308.3
Proportional ($\beta = 0.5$)	3000	150	3000	150	300
Proportional ($\beta = 0.2131$)	1279	27.25	4721	371.5	398.8

Case 4 Consider now a further case, with direct insurer's and reinsurer's payouts denoted $S_I^{(4)}$ and $S_R^{(4)}$, respectively, for which $\mathrm{Var}[S_I^{(4)}] = \mathrm{Var}[S_I] = 27.25 \times 10^4$ as in Case 1, and with $\mathrm{Var}[S_R^{(4)}]$ minimised. The result in §5.7.2 informs us that we need a proportional arrangement, so let $S_I^{(4)} = \beta S$, with

$$\beta = \sqrt{\frac{\mathrm{Var}[S]}{\mathrm{Var}[S_I^{(4)}]}} = 0.2131.$$

The mean of the direct insurer's payout is given by $\mathbb{E}[S_I^{(4)}] = \beta \times 6000 = 1279$. The mean and variance of the reinsurer's payout are given by

$$\mathbb{E}[S_R^{(4)}] = (1 - \beta) \times 6000 = 4721 \text{ and}$$
$$\mathrm{Var}[S_R^{(4)}] = (1 - \beta)^2 \times 6 \times 10^6 = 371.5 \times 10^4,$$

which is less than the value of $\mathrm{Var}[S_R]$ in Case 1. Note that (to four significant figures) $\mathrm{Var}[S_I^{(4)}] + \mathrm{Var}[S_R^{(4)}] = 398.8 \times 10^4$.

The results of all four cases in this example are summarised in Table 5.6.

Example 5.12 Consider two insurance companies, A and B. Company A issues one-year term assurances to a large group of independent lives aged 30 – the sum assured in each case is £600 000 and the number of deaths has a Poi(5) distribution. Company B issues one-year term assurances to a different and independent large group of independent lives aged 40 – the sum assured in each case is £800 000 and the number of deaths has a Poi(7) distribution. Each company reinsures with the other. We consider three cases below which will illustrate the result of §5.7.5.

Let SA and SB denote the aggregate claims for companies A and B respectively, in units of £100 000, so that SA $\sim$ CP(5, F_{XA}), where $XA = 6$ and

$SB \sim CP(7, F_{XB})$, where $XB = 8$. The means and variances of SA and SB are given by

$$\mathbb{E}[SA] = 5 \times 6 = 30, \quad \text{Var}[SA] = 5 \times 6^2 = 180,$$

and

$$\mathbb{E}[SB] = 7 \times 8 = 56, \quad \text{Var}[SB] = 7 \times 8^2 = 448.$$

Case 1 Suppose company A retains the first 36 units of its claims ($= £3.6$ million), that is it covers the first six claims, while company B retains the first 56 units of its claims ($= £5.6$ million), that is it covers the first seven claims. Let SA_I and $SA_R = SA - SA_I$ denote the retained and reinsured amounts for company A; similarly for SB_I and SB_R. The calculations below were carried out first "by hand" and were then verified using *R*.

For company A, the number of claims is $N_A \sim \text{Poi}(5)$. The mean and variance of SA_I are given by

$$\mathbb{E}[SA_I] = \sum_{n=0}^{5} 6n \times \Pr(N_A = n) + 36 \Pr(N_A \geq 6) = 27.04$$

and

$$\text{Var}[SA_I] = \sum_{n=0}^{5} (6n)^2 \times \Pr(N_A = n) + 36^2 \Pr(N_A \geq 6) - \mathbb{E}^2[SA_I] = 84.35.$$

The mean and variance of SA_R are given by

$$\mathbb{E}[SA_R] = \mathbb{E}[SA - SA_I] = 30 - 27.04 = 2.96$$

and

$$\text{Var}[SA_R] = \text{Var}[SA - SA_I] = \sum_{n=7}^{\infty} (6n - 36)^2 \Pr(N_A = n) - \mathbb{E}^2[SA_R] = 42.61.$$

For company B, the number of claims is $N_B \sim \text{Poi}(7)$. The mean and variance of SB_I are given by

$$\mathbb{E}[SB_I] = \sum_{n=0}^{6} 8n \times \Pr(N_B = n) + 56 \Pr(N_B \geq 7) = 47.66$$

and

$$\text{Var}[SB_I] = \sum_{n=0}^{6} (8n)^2 \times \Pr(N_B = n) + 56^2 \Pr(N_B \geq 7) - \mathbb{E}^2[SB_I] = 131.85.$$

The mean and variance of SB_R are given by

$$\mathbb{E}[SB_R] = \mathbb{E}[SB - SB_I] = 56 - 47.66 = 8.34$$

Table 5.7. *Summary of each company's payouts in Case 1 in Example 5.12*

	Expected payout
Company *A*	27.04 + 8.34 = 35.38 (= £3 538 000)
Company *B*	47.66 + 2.96 = 50.62 (= £5 062 000)
	Variance of payout
Company *A*	84.35 + 176.90 = 261.25
Company *B*	131.85 + 42.61 = 174.46

and

$$\text{Var}[SB_R] = \text{Var}[SB - SB_I] = \sum_{n=8}^{\infty}(8n - 56)^2 \Pr(N_B = n) - \mathbb{E}^2[SB_R] = 176.90.$$

A summary of each company's payouts is presented in Table 5.7. The sum of the variances of the two payouts is 435.71.

Case 2   Suppose instead that each company retains 60% of its aggregate claims. Let $SA_I^{(2)}$ and $SA_R^{(2)}$ denote the retained and reinsured amounts, respectively, for company *A*; similarly for $SB_I^{(2)}$ and $SB_R^{(2)}$. Then we have

$$SA_I^{(2)} = 0.6SA, \ SA_R^{(2)} = SA - SA_I^{(2)} = 0.4SA$$

and

$$SB_I^{(2)} = 0.6SB, \ SB_R^{(2)} = SB - SB_I^{(2)} = 0.4SB.$$

For company *A*, the means and variances are

$$\mathbb{E}[SA_I^{(2)}] = 0.6 \times 30 = 18, \ \text{Var}[SA_I^{(2)}] = 0.6^2 \times 180 = 64.8$$

and

$$\mathbb{E}[SA_R^{(2)}] = 0.4 \times 30 = 12, \ \text{Var}[SA_R^{(2)}] = 0.4^2 \times 180 = 28.8.$$

For company *B*, the means and variances are

$$\mathbb{E}[SB_I^{(2)}] = 0.6 \times 56 = 33.6, \ \text{Var}[SB_I^{(2)}] = 0.6^2 \times 448 = 161.28$$

and

$$\mathbb{E}[SB_R^{(2)}] = 0.4 \times 56 = 22.4, \ \text{Var}[SB_R^{(2)}] = 0.4^2 \times 448 = 71.68.$$

A summary of each company's payouts is presented in Table 5.8. The sum of the variances of the two payouts is 326.56.

Table 5.8. *Summary of each company's payouts*
in Case 2 in Example 5.12

	Expected payout
Company A	18 + 22.4 = 40.4 (= £4 040 000)
Company B	33.6 + 12 = 45.6 (= £4 560 000)
	Variance of payout
Company A	64.8 + 71.68 = 136.48
Company B	161.28 + 28.8 = 190.08

Case 3 Suppose now that each company retains 50% of its aggregate claims. Let $SA_I^{(3)}$ and $SA_R^{(3)}$ denote the retained and reinsured amounts, respectively, for company A; similarly for $SB_I^{(3)}$ and $SB_R^{(3)}$. Then we have

$$SA_I^{(3)} = SA_R^{(3)} = 0.5SA, \quad SB_I^{(3)} = SB_R^{(3)} = 0.5SB.$$

For company A, the means and variances are

$$E[SA_I^{(3)}] = 0.5 \times 30 = 15, \quad Var[SA_I^{(3)}] = 0.5^2 \times 180 = 45$$

and

$$E[SA_R^{(3)}] = 15, \quad Var[SA_R^{(3)}] = 45.$$

For company B, the means and variances are

$$E[SB_I^{(3)}] = 0.5 \times 56 = 28, \quad Var[SB_I^{(3)}] = 0.5^2 \times 448 = 112$$

and

$$E[SB_R^{(3)}] = 28, \quad Var[SB_R^{(3)}] = 112.$$

For each company, the expected payout is $15 + 28 = 43$ (= £4 300 000) and the variance of the payout is $45 + 112 = 157$. The sum of the variances of the two payouts is 314.

We summarise the results of all three cases in Table 5.9. The rows indicated $A + B$ give the sums of the means (= 86) and the sums of the variances of the payments by the two companies. We note that the sum of the variances of the payouts is lower for the proportional reinsurance arrangements than for the stop loss arrangement (as indicated by the result in §5.7.5). The sum of the variances is lowest for the "equal share of the risk" proportional arrangement (as indicated by the result of Exercise 5.31).

Table 5.9. *Summary of each company's payouts for Cases*
1–3 in Example 5.12

Arrangement	Company	Mean	Variance
No reinsurance	A	30.00	180.00
	B	56.00	448.00
	A + B	86.00	628.00
Stop loss (A retains 36, B 56)	A	35.38	261.25
	B	50.62	174.46
	A + B	86.00	435.71
Proportional (each retains 60%)	A	40.40	136.48
	B	45.60	190.08
	A + B	86.00	326.56
Proportional (each retains 50%)	A	43.00	157.00
	B	43.00	157.00
	A + B	86.00	314.00

5.8 Optimising reinsurance contracts for a group of independent risks based on minimising the variance of the direct insurer's net profit – finding the optimal relative retentions

In this final section on reinsurance contracts, we consider a portfolio consisting of n independent risks. The direct insurer will arrange a reinsurance contract (of the same type, excess of loss or proportional) for each risk, and the problem is to decide the relative retentions that should be used for the risks in the portfolio.

The criterion we will use is that of minimising the variance of the direct insurer's net profit after reinsurance costs, which we will assume are calculated on the basis of the expected value principle (that is, with a simple security loading being applied to the reinsurer's expected payout for each risk). The minimisation of variance will be performed subject to a suitable constraint, namely that the expected value of the direct insurer's net profit is fixed, say at some value c.

The results which follow are due to the influential Italian probabilist and statistician Bruno de Finetti (1906–1985).

5.8.1 Optimal relative retentions in the case of excess of loss reinsurance

Suppose we have n independent risks, with aggregate claims for risk i being denoted $S_i, i = 1, \ldots, n$. For this derivation, we assume that each S_i has a compound Poisson distribution, so let $S_i \sim CP(\lambda_i, F_{X_i}), i = 1, \ldots, n$, where X_i is the claim amount variable for risk i. Let X_i have probability density function f_i and distribution function F_i, where $F_i(x) = 0$ for $x \le 0$ and $F_i(x) < 1$ for all finite x (that is, the claim amount is not bounded above).

For each risk i, assume there is an excess of loss reinsurance arrangement with retention $M_i, i = 1, \ldots, n$. For risk i, let the premium charged by the direct insurer to cover the risk, the direct insurer's payout, the reinsurer's payout and the reinsurer's security loading be denoted P_i, S_i^I, S_i^R and ψ_i respectively. Let $\underline{M} = (M_1, \ldots, M_n)$.

The reinsurer's premium for risk i is given by

$$(1 + \psi_i) \times \mathbb{E}[S_i^R] = (1 + \psi_i) \times \lambda_i \mathbb{E}[\text{reinsurer's payout on claim } X_i]$$

$$= (1 + \psi_i)\lambda_i \int_{M_i}^{\infty} (x - M_i) f_i(x) dx.$$

The direct insurer's expected payout on risk i is

$$\mathbb{E}[S_i^I] = \lambda_i \mathbb{E}[\text{direct insurer's payout on claim } X_i]$$

$$= \lambda_i \left(\int_0^{M_i} x f_i(x) dx + M_i (1 - F_i(M_i)) \right).$$

The direct insurer's net profit, say $IP(\underline{M})$, is given by

$$IP(\underline{M}) = \sum_{i=1}^{n} \left\{ P_i - (1 + \psi_i) \mathbb{E}[S_i^R] - S_i^I \right\}.$$

We want to minimise $\text{Var}[IP(\underline{M})]$, subject to the constraint $\mathbb{E}[IP(\underline{M})] = c$.

First, we note expressions for the required quantities:

$$\mathbb{E}[IP(\underline{M})] = \sum_{i=1}^{n} \left\{ P_i - (1 + \psi_i)\lambda_i \int_{M_i}^{\infty} (x - M_i) f_i(x) dx \right.$$

$$\left. - \lambda_i \left(\int_0^{M_i} x f_i(x) dx + M_i (1 - F_i(M_i)) \right) \right\}$$

and

$$\text{Var}[IP(\underline{M})] = \sum_{i=1}^{n} \text{Var}[S_i^I]$$

$$= \sum_{i=1}^{n} \lambda_i \mathbb{E}[(\text{direct insurer's payout on claim } X_i)^2]$$

$$= \sum_{i=1}^{n} \lambda_i \left\{ \int_0^{M_i} x^2 f_i(x) dx + M_i^2 (1 - F_i(M_i)) \right\}.$$

We use the method of Lagrange multipliers to perform the constrained optimisation. Let

$$h(\underline{M}) = \mathrm{Var}[IP(\underline{M})] - \gamma\{\mathbb{E}[IP(\underline{M})] - c\}.$$

We set $\partial h/\partial M_i = 0$, $i = 1,\ldots,n$. The derivatives are given by

$$\frac{\partial}{\partial M_i}\mathrm{Var}[IP(\underline{M})] = \lambda_i\left[M_i^2 f_i(M_i) + 2M_i(1 - F_i(M_i)) - M_i^2 f_i(M_i)\right]$$
$$= 2\lambda_i M_i(1 - F_i(M_i))$$

and

$$\frac{\partial}{\partial M_i}\mathbb{E}[IP(\underline{M})] = \psi_i\lambda_i(1 - F_i(M_i)).$$

These expressions yield

$$\frac{\partial h}{\partial M_i} = 2\lambda_i M_i(1 - F_i(M_i)) - \gamma\psi_i\lambda_i(1 - F_i(M_i)),$$

and hence

$$\frac{\partial h}{\partial M_i} = 0 \Leftrightarrow 2M_i(1 - F_i(M_i)) = \gamma\psi_i(1 - F_i(M_i)).$$

The claim amount is unbounded, which means that $1 - F_i(M_i) \neq 0$. Hence

$$\frac{\partial h}{\partial M_i} = 0 \Leftrightarrow M_i = \gamma\psi_i/2 = \theta\psi_i \text{ for some } \theta\ (= \gamma/2).$$

It is easy to show that $\partial^2 h/\partial M_i^2 > 0$, $i = 1,\ldots,n$, at the turning point; for a function of the form of $h(\cdot)$, this is sufficient to confirm that we have a minimum. Hence the optimal relative retentions are given by

$$M_i = \theta\psi_i\,, \quad i = 1,\ldots,n, \text{ for some } \theta. \tag{5.27}$$

The result is very simple indeed – the relative retentions for the risks are simply proportional to the reinsurer's security loadings and do not depend on the P_i or the distributions of the S_i. The actual values of the M_i are obtained from the constraint $\mathbb{E}[IP(\underline{M})] = c$, using a specified value for c. We note from the result $M_i \propto \psi_i$ (from (5.27)) that higher reinsurance costs (higher ψ_i) correspond to higher retentions for the direct insurer.

It may be convenient to use an alternative form for $IP(\underline{M})$, which comes from writing $S_i^I = S_i - S_i^R$ and which gives

$$IP(\underline{M}) = \sum_{i=1}^{n}\left\{P_i - (1 + \psi_i)\mathbb{E}[S_i^R] - (S_i - S_i^R)\right\}.$$

From this we get

$$\mathbb{E}[IP(\underline{M})] = \sum_{i=1}^{n} \left\{ P_i - \mathbb{E}[S_i] - \psi_i \mathbb{E}[S_i^R] \right\}. \tag{5.28}$$

It may be easier to calculate the value of this expression, which involves $\mathbb{E}[S_i^R]$, than the original expression, which also involves $\mathbb{E}[S_i^I]$.

We note from this expression that if the direct insurer wants a higher expected profit, this corresponds to having a lower $\mathbb{E}[S_i^R]$ and hence a higher $\mathbb{E}[S_i^I]$, which in turn corresponds to higher retentions, exposure to more risk and reduced security.

Example 5.13   Consider three risks:

$$S_1 \sim \text{CP}(100, F_{X_1}), \qquad S_2 \sim \text{CP}(200, F_{X_2}), \qquad S_3 \sim \text{CP}(100, F_{X_3}),$$

where X_1, X_2 and X_3 have exponential distributions with means 1, 2 and 3, respectively. The direct insurers premium is calculated using a 20% security loading on the expected aggregate claim amount, while the reinsurer's premiums are calculated using loadings of 30%, 40% and 50%, respectively, on the reinsurer's expected payouts for the three risks.

We want to arrange excess of loss reinsurance such that $\text{Var}[IP(\underline{M})]$ is minimised, subject to the requirement that the direct insurer's expected profit be 40. We find immediately from the above result that the optimal retentions satisfy $M_1 = 0.3\theta$, $M_2 = 0.4\theta$ and $M_3 = 0.5\theta$, for some θ.

Now (see Example 5.3) for $X \sim \text{Exp}(1/\mu)$ (with mean μ), and with retention M, we have $\mathbb{E}[Z^*] = \mu$, $\Pr(X > M) = e^{-M/\mu}$ and $\mathbb{E}[Z] = \mu e^{-M/\mu}$. Here we have $\mathbb{E}[S_1] = 100$, $\mathbb{E}[S_2] = 400$ and $\mathbb{E}[S_3] = 300$, and the direct insurer's premiums are $P_1 = 120$, $P_2 = 480$ and $P_3 = 360$. We also have

$$\mathbb{E}[S_1^R] = 100 \times 1 \times e^{-0.3\theta/1} = 100e^{-0.3\theta},$$
$$\mathbb{E}[S_2^R] = 200 \times 2e^{-0.4\theta/2} = 400e^{-\theta/5},$$
$$\mathbb{E}[S_3^R] = 100 \times 3e^{-0.5\theta/3} = 300e^{-\theta/6}.$$

Hence, using (5.28) we have

$$\begin{aligned}
\mathbb{E}[IP(\underline{M})] &= (120 - 100 - 0.3 \times 100e^{-0.3\theta}) \\
&\quad + (480 - 400 - 0.4 \times 400e^{-0.2\theta}) \\
&\quad + (360 - 300 - 0.5 \times 300e^{-\theta/6}) \\
&= 160 - 30e^{-0.3\theta} - 160e^{-0.2\theta} - 150e^{-\theta/6}.
\end{aligned}$$

Setting this equal to 40, and solving numerically (by simple computer evaluation and search, or the use of a mathematics package, or a formal iterative solving technique), we find $\theta = 5.455$, from which we find the optimal retention levels to be $M_1 = 1.64$, $M_2 = 2.18$ and $M_3 = 2.73$.

Suppose the direct insurer had required an expected profit of 30 instead of 40. We find $\theta = 5.030$, and the optimal retention levels are lower, at $M_1 = 1.51$, $M_2 = 2.01$ and $M_3 = 2.52$ (lower profit corresponds to reduced exposure to risk and greater security).

5.8.2 Optimal relative retentions in the case of proportional reinsurance

Suppose we have n independent risks, with aggregate claims for risk i being denoted S_i, $i = 1, \ldots, n$. Suppose that for each i there is a proportional reinsurance arrangement in place. For risk i, let the premium charged by the direct insurer to cover the risk, the retention level (direct insurer's retained proportion), the direct insurer's payout, the reinsurer's payout and the reinsurer's security loading be denoted $P_i, \beta_i, S_i^I, S_i^R$ and ψ_i, respectively. Thus we have $S_i^I = \beta_i S_i$ and $S_i^R = (1 - \beta_i)S_i$.

The reinsurer's premium for risk i is

$$(1 + \psi_i) \times \mathbb{E}[S_i^R] = (1 + \psi_i)(1 - \beta_i)\mathbb{E}[S_i].$$

The direct insurer's expected payout on risk i is $E[S_i^I] = \beta_i \mathbb{E}[S_i]$. Write $\underline{\beta}$ for $(\beta_1, \ldots, \beta_n)$. Then the direct insurer's net profit, $IP(\underline{\beta})$ say, is given by

$$IP(\underline{\beta}) = \sum_{i=1}^{n} \left\{ P_i - (1 + \psi_i)(1 - \beta_i)\mathbb{E}[S_i] - S_i^I \right\}.$$

We want to minimise $\mathrm{Var}[IP(\underline{\beta})]$ subject to the constraint $\mathbb{E}[IP(\underline{\beta})] = c$.

First, we note expressions for the required quantities:

$$\mathbb{E}[IP(\underline{\beta})] = \sum_{i=1}^{n} \{P_i - (1 + \psi_i)(1 - \beta_i)\mathbb{E}[S_i] - \beta_i \mathbb{E}[S_i]\}$$

$$= \sum_{i=1}^{n} \{P_i - (1 + \psi_i - \psi_i \beta_i)\mathbb{E}[S_i]\}$$

and

$$\mathrm{Var}[IP(\underline{\beta})] = \sum_{i=1}^{n} \mathrm{Var}[S_i^I] = \sum_{i=1}^{n} \beta_i^2 \, \mathrm{Var}[S_i].$$

We again use the method of Lagrange multipliers to perform the constrained optimisation. Let

$$h(\underline{\beta}) = \mathrm{Var}[IP(\underline{\beta})] - \gamma \left\{ \mathbb{E}[IP(\underline{\beta})] - c \right\}.$$

We set $\partial h/\partial \beta_i = 0$, $i = 1, \ldots, n$. The derivatives are given by

$$\frac{\partial h}{\partial \beta_i} = 2\beta_i \, \text{Var}[S_i] - \gamma \psi_i \mathbb{E}[S_i],$$

and hence

$$\frac{\partial h}{\partial \beta_i} = 0 \Leftrightarrow 2\beta_i \, \text{Var}[S_i] = \gamma \psi_i \mathbb{E}[S_i] \Leftrightarrow \beta_i = \frac{1}{2}\gamma \psi_i \frac{\mathbb{E}[S_i]}{\text{Var}[S_i]}.$$

Hence the optimal relative retentions are given by

$$\beta_i = \theta \psi_i \frac{\mathbb{E}[S_i]}{\text{Var}[S_i]}, \quad i = 1, \ldots, n, \text{ for some } \theta. \tag{5.29}$$

Clearly $\partial^2 h/\partial \beta_i^2 > 0$, $i = 1, \ldots, n$; this is sufficient for a function of the form of $h(\cdot)$ to confirm that we have a minimum. The actual values of the β_i are obtained from the constraint $\mathbb{E}[IP(\beta)] = c$, using a specified value for c.

The relative retentions for the risks are proportional to the reinsurer's security loadings and do not depend on the P_i. We note from (5.29) that higher reinsurance costs (higher ψ_i, higher $\psi_i \mathbb{E}[S_i]$) correspond to higher retentions for the direct insurer, while higher uncertainty in the aggregate claims (higher $\text{Var}[S_i]$) corresponds to lower retentions. We note also (from the expression for $\mathbb{E}[IP(\beta)]$ above) that if the direct insurer wants a higher expected profit, this corresponds to having higher retentions, and thus exposure to more risk and reduced security.

Note that it is possible that this approach to optimising the proportions to be retained by the direct insurer may result in one or more of them turning out to exceed 1. If we do get $\beta_1 > 1$ (say) then what we would do in practice is set $\beta_1 = 1$ and use this value in the calculations based on $\mathbb{E}[IP(\beta)] = c$.

Example 5.14 Consider three risks with the same means and variances as the risks in Example 5.13, so $\mathbb{E}[S_1] = 100$, $\text{Var}[S_1] = 200$, $\mathbb{E}[S_2] = 400$, $\text{Var}[S_2] = 1600$, $\mathbb{E}[S_3] = 300$ and $\text{Var}[S_3] = 1800$.

Suppose, also as in Example 5.13, that the direct insurers premium is calculated using a 20% security loading on the expected aggregate claim amount, while the reinsurer's premiums are calculated using loadings of 30%, 40% and 50%, respectively on the reinsurer's expected payouts for the three risks.

We want to arrange proportional reinsurance using the above criterion, and also requiring that the direct insurer's expected profit is 30. We find immediately from (5.29) that the optimal retentions satisfy

$$\beta_1 = 0.3 \times (100/200)\theta = 3\theta/20,$$
$$\beta_2 = 0.4 \times (400/1600)\theta = \theta/10,$$
$$\beta_3 = 0.5 \times (300/1800)\theta = \theta/12.$$

Then

$$\mathbb{E}[IP(\underline{\beta})] = 120 - (1.3 - 0.3 \times 3\theta/20) \times 100$$
$$+ 480 - (1.4 - 0.4 \times \theta/10) \times 400$$
$$+ 360 - (1.5 - 0.5 \times \theta/12) \times 300$$
$$= 33\theta - 180.$$

Setting this equal to 30, we find $\theta = 70/11$, from which we find the optimal retentions to be $\beta_1 = 0.955, \beta_2 = 0.636$ and $\beta_3 = 0.530$.

Suppose the direct insurer had required an expected profit of 40 instead of 30. We find the optimal retentions are higher, at $\beta_1 = 1$ (the insurer now retains 100% of risk 1), $\beta_2 = 0.667$ and $\beta_3 = 0.556$ (higher expected profit corresponds to higher exposure to risk and reduced security).

Exercises

5.1 Claim amounts from a general insurance portfolio are lognormally distributed with mean £1000 and standard deviation £2000. Excess of loss reinsurance with retention level £1750 is in place. Calculate the probability that the reinsurer is involved in a claim.

5.2 Claim amounts have a Pa(5, 4) distribution, and excess of loss reinsurance with retention level M is in place. Calculate the value of M such that the mean amounts paid by the direct insurer and the reinsurer on a claim are equal.

5.3 Suppose claim amounts are distributed as $X \sim \text{Exp}(1/\mu)$ and an excess of loss contract with retention level M is in place. Let Y and Z be the amounts paid out on a claim by the direct insurer and the reinsurer, respectively. Let $p = \Pr(X > M) = e^{-M/\mu}$.

(a) Show that the moment generating function $M_Y(t)$ of Y is given by

$$M_Y(t) = \frac{1}{1 - \mu t}\{1 - \mu p t e^{Mt}\},$$

and hence, or otherwise, show that $\mathbb{E}[Y^2]$ is given by

$$\mathbb{E}[Y^2] = 2\mu\{\mu(1 - p) - Mp\}.$$

(b) Show that $\mathbb{E}[Z^2]$ is given by

$$\mathbb{E}[Z^2] = 2\mu^2 p.$$

5.4   Suppose claim amounts are distributed as $X \sim Pa(\alpha, \lambda)$ and an excess of loss contract with retention level M is in place. Let Y and Z be the amounts paid out on a claim by the direct insurer and the reinsurer, respectively.

Show that $\mathbb{E}[Y^2]$ and $\mathbb{E}[Z^2]$ are given by

$$\mathbb{E}[Y^2] = \frac{2\lambda}{(\alpha - 1)(\alpha - 2)} \left[\lambda(1 - \eta^{\alpha-2}) - (\alpha - 2)M\eta^{\alpha-1} \right]$$

and

$$\mathbb{E}[Z^2] = \frac{2\lambda^2\eta^{\alpha-2}}{(\alpha - 1)(\alpha - 2)},$$

where $\eta = \lambda/(\lambda + M)$.

5.5   The claim amounts arising from a risk have a $Pa(\alpha, \lambda)$ distribution, where the value of λ is known. The direct insurer has arranged an excess of loss reinsurance contract with retention level M. In the past year, there were a total of m reinsurance claims, that is there were m claims which exceeded M and hence involved the reinsurer. The amounts of these claims are known to the reinsurer – suppose the amounts paid by the reinsurer on these claims are $r_1, r_2, \ldots, r_m$, and this is all the information the reinsurer has.

(a) Show that the maximum likelihood estimator of α, based on the information available to the reinsurer, is given by

$$\hat{\alpha} = m \left(\sum_{i=1}^{m} \log(\lambda + M + r_i) - m\log(\lambda + M) \right)^{-1}.$$

(b) In a particular situation, with $\lambda = 5000$ and $M = £6000$, a total of 176 claims exceeded £6000 last year, and for these claims

$$\sum_{i=1}^{176} \log(r_i + 11\,000) = 1693.3,$$

where the $\{r_i\}$ are defined as above.

(1) Show that the maximum likelihood estimate of α is 3.171.
(2) Using the fitted model, calculate the probability that a claim will involve the reinsurer, show that the mean amount of reinsurance claims is £11 067, and hence, or otherwise, find the mean amount of claims which do not involve the reinsurer.

5.6   An insurer covers an individual loss X with excess of loss reinsurance in place with retention level M. The insurer pays Y and the reinsurer pays $Z = X - Y$. Show that

$$YZ = \begin{cases} 0 & \text{if } X \le M \\ M(X - M) & \text{if } X > M. \end{cases}$$

Hence show that $\mathbb{E}[YZ] = M\mathbb{E}[Z]$ and $\text{Cov}[Y, Z] \ge 0$. Deduce that

$$\text{Var}[X] \ge \text{Var}[Y] + \text{Var}[Z].$$

5.7 Let X be a random variable with a lognormal distribution with parameters μ and σ and probability density function $f(x)$. Verify the following result, which is needed for some reinsurance calculations: for any real number $c > 0$ and $k = 0, 1, 2, \ldots$

$$\int_0^c x^k f(x)dx = \exp\left(k\mu + \frac{k^2\sigma^2}{2}\right) \Phi\left(\frac{\log c - \mu - k\sigma^2}{\sigma}\right).$$

5.8 The aggregate claims for a risk S, in units of £1000, has a compound Poisson distribution $S \sim \text{CP}(100, F_X)$, where $X \sim \text{Pa}(5, 4)$. The direct insurer has an excess of loss reinsurance contract in place, with retention level M, where M is the upper 5% point of the individual claims distribution.

(a) Show that $M = £3282$ and specify the distribution of the amount paid by the reinsurer on a reinsurance claim (that is, on a claim which involves the reinsurer).

(b) Specify the distribution of S_R, the aggregate annual claim amount paid by the reinsurer, and find the mean and standard deviation of S_R.

(c) Find the mean of S_I, the aggregate annual claim amount paid by the direct insurer.

5.9 Suppose aggregate claims $S \sim \text{CP}(100, F_X)$, where the claim amount X is measured in units of £1000 and modelled as a $\text{Pa}(6, 10)$ random variable. The direct insurer is considering entering into one of two different types of reinsurance contract – excess of loss with retention M, and proportional with retained proportion β. The direct insurer wants an expected payout of £1500 on a claim. The reinsurer's premium is to be chosen such that there is a probability of 0.95 that the reinsurer makes a profit on the business, and you may assume that the reinsurer's aggregate payout can be approximated by a normal distribution.

(a) Show that $\beta = 0.75$ and $M = 3.195$.

(b) Calculate the reinsurer's premium for each arrangement; comment on the answers.

5.10 The distribution of total annual claims S on a general insurance portfolio is to be modelled as a compound Poisson distribution $S \sim \text{CP}(1000, F_X)$. The direct insurer has an excess of loss reinsurance contract in place, with retention level £4000 on each claim. Consider the following models (all with the same mean) for a loss X (in units of £1000).

Model 1: X has an exponential distribution with mean 2.

Model 2: X has a Pareto distribution with parameters $\alpha = 4$ and $\lambda = 6$.

Model 3: X has a lognormal distributon with parameters $\mu = 0.2877$ and $\sigma = 0.9005$.

(a) State, or derive, the distribution of $Z^* \equiv X - 4 \mid X > 4$ using Model 1.

(b) State, or derive, the distribution of $Z^* \equiv X - 4 \mid X > 4$ using Model 2.

(c) Calculate the values of the mean and standard deviation of the total annual claims paid by the reinsurer under each of the three models. (*Note:* the calculations for Model 3 require the result of Exercise 5.7 above and are the most onerous.)

5.11 Consider the aggregate claims model $S \sim CP(\lambda, F_X)$, where $X = Y + Z$ and $S = S_I + S_R$, in the usual notation for the amounts paid out by the direct insurer and the reinsurer, respectively, in the presence of a reinsurance contract. In particular, suppose there is an excess of loss reinsurance contract in place with retention level M on each claim. Let $p = \Pr(X > M)$, and let Z^* denote the amount paid by the reinsurer on a reinsurance claim.

From (5.8) we know that $\mathbb{E}[Z^*] = (1/p)\mathbb{E}[Z]$. Show that

$$\text{Var}[Z^*] = \frac{1}{p}\text{Var}[Z] - \frac{1-p}{p^2}\{\mathbb{E}[Z]\}^2.$$

5.12 Suppose the annual aggregate claims S from a portfolio of risks has a compound Poisson distribution, with Poisson parameter 100. Each claim which arises comes from one of two separate sub-portfolios and is of one of two types: the amounts of type 1 claims have an exponential distribution with mean 1, while those of type 2 claims have an exponential distribution with mean 2, where 60% of claims are of type 1. Let X represent a randomly selected claim, so we have $S \sim CP(100, F_X)$.

Excess of loss reinsurance is in place, with retention level 1.8 on each claim. Let $X = Y + Z$ and $S = S_I + S_R$ in the usual notaion.

(a) By using conditional expectation arguments with X, Y and Z, verify the following values:

$\mathbb{E}[X] = 1.4$　　　$\mathbb{E}[X^2] = 4.4,$

$\mathbb{E}[S] = 140$　　　$\text{Var}[S] = 440,$

$\mathbb{E}[Y] = 0.9756$　　$\mathbb{E}[Y^2] = 1.3727,$

$\mathbb{E}[Z] = 0.4244$　　$\mathbb{E}[Z^2] = 1.4994,$

$\mathbb{E}[S_I] = 97.56$　　$\text{Var}[S_I] = 137.27,$

$\mathbb{E}[S_R] = 42.44$　　$\text{Var}[S_R] = 149.94.$

(b) Suppose the heterogeneity of the portfolio is not recognised, and the distribution of claim amounts is represented by the single variable X which is exponentially distributed with mean 1.4. Recalculate all the moments in part (a) and comment on the consequences for the reinsurer of being unaware of the heterogeneity present in the system.

5.13 Consider a portfolio of private motor policies. In the event of an accident or other incident covered by the policy, the loss (cost of repairs and/or replacement of parts or the whole vehicle) has a $Pa(\alpha, \lambda)$ distribution. A deductible (excess) of £150 is applied to all losses – no claim is made if the loss is less than £150; otherwise a claim is always made (for the loss less the deductible). A sample of 100 claims has mean £1210 and standard deviation £1790.

(a) Calculate method of moments estimates of α and λ.

(b) Estimate the proportion of losses that do not lead to claims.

(c) Excess of loss insurance is to be arranged with another company so that the direct insurer's mean payout on claims is reduced to £1000. Find the retention limit the direct insurer must set on individual claims to achieve the required reduction in the direct insurer's mean payout.

5.14 The annual aggregate claims S from a risk has a compound Poisson distribution $S \sim CP(200, F_X)$, where the individual claim X is modelled crudely as taking the value £1000 or £5000, with probabilities 0.75 and 0.25, respectively. The direct insurer's premium is calculated using the expected value principle, with relative security loading factor 0.3.

The direct insurer wants to arrange excess of loss reinsurance for this risk. The reinsurer's premiums are calculated using the expected value principle, but with a loading factor 0.5.

Let $IP(M)$ denote the direct insurer's annual profit, net of reinsurance costs, under a contract with retention level for a claim set at M. Suppose that M is a value between £1000 and £5000 chosen such that the variable $IP(M)$ has coefficient of variation 1/3, that is

$$\frac{\text{standard deviation}(IP(M))}{\text{mean}(IP(M))} = \frac{1}{3}.$$

(a) Show that $IP(M)$ satisfies

$$IP(M) = 145\,000 + 75M - S_I,$$

where S_I is the aggregate claims paid out by the direct insurer, and hence show that that $M = £3557$.

(b) Using the value of M found in (a), and assuming normality as required, calculate the approximate probability of each of the following two events, quoting your calculated probabilities to three decimal places:

(1) the direct insurer's annual profit is positive;

(2) the reinsurer's annual profit is positive.

5.15 Consider the aggregate claims model $S \sim CP(\lambda, F_X)$. A reinsurance arrangment is in place, defined at the level of individual claims. Let Y and S_I represent, respectively, the amount of an individual claim and the aggregate amount, paid by the direct insurer; similarly for Z and S_R for the reinsurer. So $X = Y + Z$ and $S = S_I + S_R$.

(a) By considering $\text{Var}[S_I + S_R]$, show that the covariance between S_I and S_R is given by $\text{Cov}[S_I, S_R] = \lambda \mathbb{E}[YZ]$ and hence find an expression for the correlation coefficient between S_I and S_R in terms of moments of Y and Z.

(b) Calculate the correlation coefficient between the direct insurer's and the reinsurer's aggregate claim amounts in Example 5.5.

(c) Calculate the correlation coefficient between the direct insurer's and the reinsurer's aggregate claim amounts in Example 5.6.

5.16 A life insurance company covers 1000 lives for one-year term insurance in amounts (in units of £100 000) as shown below:

Benefit amount	1	2
Number of insured lives	600	400

The insured lives can be assumed to be independent, with a single probability of a claim of 0.025 applying to all lives.

(a) Find the mean and standard deviation of the total claim amount, S, and hence calculate the (approximate) probability that the direct insurer's total payout on this business exceeds £4.5 million.

(b) A reinsurance contract is arranged with the aim of reducing both the uncertainty in the direct insurer's payout and the probability calculated in (a). The direct insurer sets a retention level of £160 000 and purchases the necessary cover at a cost of £0.0275 per £1 of cover.

(1) Calculate the cost of the reinsurance.

(2) Find the mean and standard deviation of the direct insurer's total payout on claims with the reinsurance in place, and hence calculate the revised (approximate) probability that the direct insurer's total payout on this business exceeds £4.5 million; comment briefly on the effect of the reinsurance.

5.17 Suppose that S has a compound Poisson distribution with claim rate λ and that individual claim amount variable X has a Pareto distribution with parameters α (>2) and γ. The direct insurer has a proportional reinsurance contract in place under which a proportion β of each claim amount is retained. Find expressions for the mean and variance of the aggregate claim amounts paid by the direct insurer and the reinsurer.

5.18 An insurer sells 2000 one-year policies covering independent and identical risks. A maximum of one claim is permitted under each policy. The probability of a loss arising under each policy is 0.15, and the loss X has an exponential distribution with mean 1 (we are taking the expected loss as our monetary unit). One thousand of the policies sold have no excess, while 1000 of them each have an excess of size D imposed, where D has been set at the lower 30% point of the loss distribution. For policies with an excess, a claim arises if and only if the loss exceeds the excess.

Let S_1 denote the aggregate claims which arise on the 1000 policies with no excess and let S_2 denote the aggregate claims which arise on the 1000 policies with an excess. Let $S = S_1 + S_2$.

(a) (1) Show that $D = 0.3567$, and specify the distribution of $X - D \mid X > D$.

(2) Calculate the mean and variance of S_1 and S_2, and hence calculate the mean and variance of S.

(b) Assume that the insurer sets premiums using the expected value principle with a relative security loading of 20%. Using the normal distribution, calculate approximately the probabilities that the insurer makes a profit on the business consisting of:

(1) the 1000 policies with no excess;
(2) the 1000 policies with an excess;
(3) all 2000 policies;

comment briefly on the results.

5.19 Suppose S has a compound Poisson distribution, with losses occurring at rate λ and with individual loss variable X exponentially distributed with mean μ. A policyholder has cover against this risk by means of a policy that incorporates a deductible D on each loss. The insurer's premium is calculated using the expected value principle with relative security loading θ. Let P, IP and PTC denote the premium charged by the insurer, the insurer's profit and the policyholder's total costs/payout, respectively.

(a) Show that

$$P = (1 + \theta)\lambda\mu e^{-D/\mu},$$

$$\mathbb{E}[IP] = \theta\lambda\mu e^{-D/\mu},$$

$$\text{Var}[IP] = 2\lambda\mu^2 e^{-D/\mu},$$

$$\mathbb{E}[PTC] = \lambda\mu(1 + \theta e^{-D/\mu}).$$

(b) In the case $\lambda = 0.3, \theta = 0.2$ and $\mu = 1$ (in units of £1000), calculate the premium charged by the insurer, the insurer's expected
profit and the policyholder's expected total costs, for each of the
following cases:

(1) $D = 0$,
(2) $D = 0.1$ (= £100),
(3) $D = 0.2$ (= £200),
(4) $D = 0.3$ (= £300);

comment on the results. Calculate the policyholder's relative savings
in premium (in percentage terms relative to the case $D = 0$) of having
a policy with a deductible as given in cases (2)–(4) above.

Calculate the policyholder's relative savings in expected total costs
(in percentage terms relative to the case $D = 0$) of having a policy
with a deductible as given in cases (2)–(4) above.

(c) Suppose S represents the aggregate losses for a portfolio with $\lambda = 200$. Again suppose $\theta = 0.2$ and $\mu = £1000$. Using a normal approximation, calculate the probability that the insurer makes a profit on
this business for each of the cases (1)–(4) given in part (b) above;
comment on the results.

5.20 Suppose S has a compound Poisson distribution with claim rate λ and
individual claim amount variable X with mean μ. The insurer's premium
is calculated using the expected value principle with relative security
loading θ. The insurer enters into a proportional reinsurance contract
with retained proportion β, where the reinsurer's premium is calculated
according to the same principle to that used by the direct insurer, but
with a loading factor of ψ.

Show that the direct insurer's expected profit is given by

$$\lambda\mu[\psi\beta - (\psi - \theta)],$$

and comment briefly on the relationship between the expected profit and
the value of β.

5.21 Claims occur on a portfolio of insurance policies according to a Poisson process. Individual claims have value 1 or 2, each value occurring
with probability 0.5. The insurer adopts an exponential utility function $u(x) = -e^{-\alpha x}$ and makes decisions on the basis of maximising the

insurer's expected utility of wealth. The insurer effects excess of loss reinsurance with retention level M, where $1 < M < 2$, where the reinsurer's premium is calculated using a simple security loading factor of ψ on the reinsurer's expected payout.

Using the approach described in §5.6.1, calculate the values of the retention level M for which the direct insurer's expected utility at the end of the year is maximised in the following cases:

(a) $\psi = 0.3$, $\alpha = 0.15$,

(b) ψ is reduced by one-third to 0.2, $\alpha = 0.15$,

(c) $\psi = 0.3$, α is increased by one-third to 0.2;

comment on the results.

5.22 Claims occur on a portfolio of insurance policies according to a Poisson process. All individual claims have the same value. The insurer adopts an exponential utility function $u(x) = -e^{-\alpha x}$ and makes decisions on the basis of maximising the insurer's expected utility of wealth, and effects proportional reinsurance with retained proportion β. The reinsurer sets an exponential premium with parameter η (see §4.1.6 and the example of proportional reinsurance in §5.6.2).

Using the approach of §5.6.2, calculate the values of the retained proportion β for which the direct insurer's expected utility at the end of the year is maximised in the following cases:

(a) $\alpha = 0.4$, $\eta = 1$,

(b) $\alpha = 0.4$, η is increased by one-half to 1.5,

(c) α is doubled to 0.8, $\eta = 1$;

comment on the results.

5.23 The total annual claims from an insurer's portfolio have a compound Poisson distribution with rate parameter λ and with individual claim sizes which have an exponential distribution with mean μ.

The insurer receives a total premium P in respect of this portfolio (where $P > \lambda\mu$) and arranges a quota share (proportional) reinsurance contract under which the two parties share the risk and the premium. The direct insurer retains a proportion β of the risk where $0 \le \beta \le 1$ and passes on $(1 - \beta)P$ as premium to the reinsurer.

The direct insurer's net profit for the year is the premium income retained less the proportion of claims paid out by the direct insurer. The direct insurer adopts the utility function $u(x) = -e^{-\alpha x}$.

(a) Show that the expected utility of the direct insurer's net profit for the year is given by

$$ -\exp[-\alpha\beta P + \lambda\{(1 - \alpha\beta\mu)^{-1} - 1\}], $$

and hence show that this expected utility is maximised when the direct insurer chooses β to satisfy

$$\beta = \frac{1}{\alpha\mu}\left[1 - \left(\frac{\lambda\mu}{P}\right)^{1/2}\right].$$

(b) Comment on the dependence of the direct insurer's optimum β on (i) the degree of the direct insurer's risk aversion and (ii) the premium received P.

(c) Evaluate the optimum value of β in the case $\lambda = 100$, $\mu = £150$, $P = £20\,000$ and $\alpha = 0.005$.

5.24 An individual is considering buying insurance to cover a loss, $X(£)$, which is modelled as having a uniform distribution on $(0, 200)$. The insurance on offer has a compulsory excess of £50, so the individual pays $\min(X, 50)$ and the insurer pays $\max(0, X - 50)$.

(a) By considering the expected outgoings of the individual (or of the insurer), justify the fact that the individual will have to be prepared to pay a premium of at least £56.25 for the cover on offer.

(b) Suppose the individual has current wealth £300, that the exponential utility function $u(x) = -e^{-0.02x}$ is adopted, and that the individual acts rationally in utility terms. Decide whether or not the individual will purchase the cover if it is on offer for a premium of £100.

(c) In the general case with the individual having current wealth £W and adopting a utility function $u(x) = -e^{-\alpha x}$, show that the maximum premium P the individual will be prepared to pay for the cover satisfies the inequality

$$e^{\alpha P}[(150\alpha + 1)e^{50\alpha} - 1] < e^{200\alpha} - 1,$$

and hence verify the result found in (b).

5.25 Suppose aggregate claims are distributed $S \sim N(\mu, \sigma^2)$. A stop loss reinsurance contract is in place, with retention level M, so that the payments by the direct insurer and the reinsurer are, respectively, $S_I = \min(S, M)$ and $S_R = \max(0, S - M)$. The reinsurer sets an exponential premium with parameter η (see §4.1.6 and §5.6.2).

Show that the reinsurer's premium is given by

$$\frac{1}{\eta}\log\left\{\Phi\left(\frac{M - \mu}{\sigma}\right) + g(M, \mu, \sigma, \eta)\right\},$$

where

$$g(M,\mu,\sigma,\eta) =$$

$$\exp\left(\frac{1}{2}\eta^2\sigma^2 - \eta(M - \mu)\right)\left[1 - \Phi\left(\frac{M - \mu - \eta\sigma^2}{\sigma}\right)\right].$$

5.26 In Example 5.10, suppose the reinsurer sets the premium P_R using the standard deviation principle as $P_R = \mathbb{E}[S_R] + \psi\mathrm{SD}[S_R]$, where $\psi < \alpha\sigma$.

Show that the direct insurer's optimal retention is $\beta = \psi/(\alpha\sigma)$, and comment on the nature of the dependence of the optimal retention on ψ, α and σ.

5.27 Let S represent the total annual claims from an insurer's portfolio. Assume we adopt a simple model for S, namely that S has an exponential distribution with mean £10 million.

(a) The insurer wishes to enter into a reinsurance contract with the requirement $\mathbb{E}[S_I] = $ £8 million. Calculate and compare the variances of the direct insurer's claims under a stop loss arrangement and under a proportional arrangement. Comment on your results.

(b) The insurer wishes to enter into a reinsurance contract with the requirement $\mathrm{SD}[S_I] = $ £3 million. Calculate and compare the variances of the reinsurer's claims under a stop loss arrangement and under a proportional arrangement. Comment on your results.

(c) Calculate and compare $\mathrm{Var}[S_I] + \mathrm{Var}[S_R]$, the sum of the variances of the direct insurer's claims and the reinsurer's claims under

(1) a stop loss arrangement, with retention chosen so that the sum of variances is minimised;

(2) a proportional arrangement with retention $\beta = 0.75$;

(3) a proportional arrangement with retention $\beta = 0.5$.

Comment on the results.

5.28 Consider the aggregate claims model $S \sim \mathrm{CP}(\lambda, F_X)$, where $F_X(0) = 0$ and $F_X(x) < 1$ for all finite $x > 0$, that is the values of the claim amount variable X are positive and not bounded above. An excess of loss reinsurance arrangement is in place, with retention level M on each claim. The parties want to find the value of M which will minimise $\mathrm{Var}[S_I] + \mathrm{Var}[S_R]$, the sum of the variances of the direct insurer's claims and the reinsurer's claims.

(a) Show that $\mathrm{Var}[S_I] + \mathrm{Var}[S_R]$ has a turning point at M, where M satisfies

$$\int_M^\infty (x - M)f_X(x)dx = M\{1 - F_X(M)\}.$$

(b) In each of the following cases, derive an expression for the optimal value of M in terms of the parameters of the distribution of the claim amount variable X (in each case show explicitly that a minimum value for the sum of the variances has been found):

(1) X has a Pareto distribution with parameters α (>2) and γ;

(2) X has an exponential distribution with mean μ.

(c) Suppose we drop the assumption that the values of X are unbounded and we allow for a cap on the claim amount – in particular suppose X has a uniform distribution on $(0, 1)$. Find the optimal value of M in this case.

5.29 In this exercise we minimise the variance of the direct insurer's payout, subject to a fixed expectation – a parallel result to that of §5.7.1 in the case that aggregate claims have a compound Poisson distribution and we compare excess of loss (rather than stop loss) and proportional arrangements.

Suppose S has a compound Poisson distribution with individual claim amount variable X with probability density function $f(x)$. Let $\min(X, M)$ and βX be the direct insurer's payouts on a claim under an excess of loss arrangement and a proportional arrangement, respectively, with

$$\mathbb{E}[\min(X, M)] = \mathbb{E}[\beta X] = c,$$

for some c. Let S_I and S_I^* denote the direct insurer's aggregate payouts, under these excess of loss and proportional arrangements, respectively.

By considering

$$\int_0^\infty (\beta x - M)^2 f(x)dx$$

and using an argument similar to that used in §5.7.1, deduce that $\text{Var}[S_I] \leq \text{Var}[S_I^*]$; that is, that the variance of the direct insurer's aggregate payout is minimised (for the two competing arrangements) by using excess of loss reinsurance.

5.30 In this exercise we compare excess of loss and equivalent proportional reinsurance arrangements – a parallel result to that of §5.7.3.

Suppose aggregate claims S are distributed $S \sim CP(\lambda, F_X)$, with $F_X(0) = 0$, and the insurer has an excess of loss reinsurance contract in place with retention level M on each claim amount. Let S_I denote the direct insurer's aggregate payout under this arrangement.

Consider a proportional arrangement which is equivalent as regards uncertainty, in the sense that the direct insurer's payouts under both arrangements have the same variance. So define $S_I^* = \beta S$, where $\beta = \{\text{Var}[S_I]/\text{Var}[S]\}^{1/2}$, giving $\text{Var}[S_I^*] = \text{Var}[S_I]$. By applying the result in §5.7.1 to individual claims, and arguing in a similar way to that used in §5.7.3, show that $\mathbb{E}[S_j^*] \le \mathbb{E}[S_I]$.

5.31 Consider the situation explored in §5.7.5, in which we seek to minimise the sum of the variances of the payouts when two independent risks are shared between two insurers.

Extend the argument to show that the optimal arrangement is that in which each company shares its risk equally with the other.

5.32 Consider two insurance companies A and B, which are exposed to independent risks and for which the aggregate claims variables are denoted SA and SB, respectively. Consider the situation for which SA ~ $CP(200, F_X)$ and SB ~ $CP(400, F_X)$, and where, in both cases, individual claim amounts X have an exponential distribution with mean 1 (we are taking the expected claim amount as our monetary unit). Each company reinsures with the other. Let T_A and T_B represent the resulting total payouts by companies A and B, respectively.

(a) Suppose each company enters into an excess of loss reinsurance agreement with the other with retention level 1.6. Calculate the value of $\text{Var}[T_A] + \text{Var}[T_B]$.

(b) Explain clearly why the minimum possible value for $\text{Var}[T_A] + \text{Var}[T_B]$ under any reinsurance arrangement is 600.

5.33 Let $S_1, \ldots, S_n$ be independent and identically distributed random variables representing the annual claims from each of n (>1) insurance companies. Suppose the companies effect a mutual reinsurance agreement under which company j ($j = 1, \ldots, n$) pays an amount $R(S_i)$ of the claims of each other company i, where $0 \le \text{Var}[R(S_i)] \le \text{Var}[S_i]$. So company j pays S_j less the amounts paid towards risk j by the other companies, namely $(n-1)R(S_j)$, plus the amount to reinsure parts of the other risks, namely $\sum_{i \ne j} R(S_i)$.

We can therefore represent the amount paid out by company j, T_j, as follows:

$$T_j = S_j - (n-1)R(S_j) + \sum_{i \ne j} R(S_i).$$

Show that $\text{Var}[T_j]$ is minimised in the case that the function $R(\cdot)$ is of the form $R(x) = x/n$.

Hint: Find an expression for $\text{Var}[T_j]$ in terms of n, $\text{Var}[S_j]$, $\text{Var}[R(S_j)]$ and $\text{Cov}[S_j, R(S_j)]$. Noting that, given any rule $R(x)$, we can select a rule $R^*(x) = \beta x$ with $\text{Var}[R^*(S)] = \text{Var}[R(S)]$, find an expression for the minimum value of $\text{Var}[T_j]$ in terms of n, β and $\text{Var}[S_j]$ and hence deduce the result.

5.34 Consider a portfolio consisting of three independent risks, for which the aggregate claims are modelled as $S_1 \sim N(10, 1)$, $S_2 \sim N(20, 2^2)$ and $S_3 \sim N(30, 3^2)$. The insurer charges premiums for the three risks of 11, 24, and 39, respectively, and wants to arrange proportional reinsurance for each risk. The reinsurer's premiums are calculated using the expected value principle, with relative security loadings on the reinsurer's expected payouts of 20%, 30% and 40%, respectively.

The direct insurer wants to determine which the retentions to set so that the variance of the direct insurer's net profit is minimised, subject to the constraint that the direct insurer's expected net profit is 5.

(a) (1) Show that the expected value of the direct insurer's net profit can be expressed in terms of the retentions (β_1, β_2, β_3, respectively) as $2\beta_1 + 6\beta_2 + 12\beta_3 - 6$.

(2) Determine the percentage of each risk the direct insurer should retain.

(b) Examine, and comment on, the way in which the results in (a) change if:

(1) the direct insurer's expected net profit requirement is increased from 5 to 7 (with all other original values remaining unchanged);

(2) the direct insurer's combined premium is reduced from 74 to 70.5 (with all other original values remaining unchanged);

(3) the reinsurer's security loading factor is increased for the third risk from 40% to 50% (with all other original values remaining unchanged).

6

Ruin theory for the classical risk model

In Chapter 3 we considered risk models for a fixed time period. Now we widen our focus and consider risk models that evolve over time. Our basic model, the *classical risk model*, is introduced in §6.1, and in this chapter our aim is to study *ruin theory* for this model. Ruin theory is concerned with quantities related to the event that the insurance company's capital becomes negative at some point in time. These ruin quantities are defined in §6.1 and §6.2, and their properties and behaviour are studied in later sections of this chapter. Numerical calculation of, and statistical inference for, these ruin quantities are discussed in §6.6 and §6.7.

6.1 The classical risk model

We aim to build a time-evolving risk model that captures the evolution of the reserves of an insurance company. We follow the principle of putting together separate models for the arrivals of claims and for the claim sizes, an approach that we found to be useful and tractable in Chapter 3. In addition we incorporate an extra ingredient describing the inflow of premium income.

In the *classical risk model* we assume the following.

(i) The claim sizes $X_1, X_2, \ldots$ are iid positive random variables with distribution function F_X and *finite* mean μ.

(ii) The claims arrive in a Poisson process with rate λ (>0).

(iii) The claim sizes $X_1, X_2 \ldots$ are independent of the claim-arrival process.

(iv) Premium income accrues linearly in time at rate c (>0), so that by time t the total amount of premiums received is ct.

(v) At time $t = 0$ the insurance company has (non-negative) initial capital u.

Note that in the above assumptions for the classical risk model, the claim sizes have finite mean μ. We write $M_X(r)$ for the moment generating function of the X_i. This is not necessarily finite for all r. Recall that, when we write expressions involving moment generating functions, we assume, unless stated otherwise, that r is such that the moment generating function is finite.

There are various important stochastic processes arising from the classical risk model. The first is the Poisson process $\{N(t) : t \geq 0\}$, where $N(t)$ is the *number of claims* that arrive up to and including time t.

The second stochastic process is $\{S(t) : t \geq 0\}$, where

$$S(t) = \sum_{i=1}^{N(t)} X_i, \tag{6.1}$$

so that $S(t)$ is the *total amount claimed* by time t.

Thirdly we have the *risk reserve* (or *surplus*) process $\{U(t) : t \geq 0\}$, where $U(t)$ is the risk reserve at time t given by

$$U(t) = u + ct - S(t) = u + ct - \sum_{i=1}^{N(t)} X_i. \tag{6.2}$$

Here $U(t)$ is the amount of surplus at time t, taking account of the inflow of premiums and the outflow of claim payments, starting with capital u at time $t = 0$. Note that in (6.2) $u + ct$ is deterministic, but $S(t)$ is a random variable, so that $U(t)$ is a random variable.

We can summarise various properties of $N(t)$ and $S(t)$. From §2.1.1 and §2.2.3, we have $N(t) \sim \text{Poi}(\lambda t)$, and the times between successive claim arrivals are iid exponential random variables with mean $1/\lambda$. In addition, the sample paths of $\{N(t)\}$ are non-decreasing step functions. We also have that, for fixed t, the random variable $S(t)$ has a compound Poisson distribution given by

$$S(t) \sim \text{CP}(\lambda t, F_X),$$

and the results of Chapter 3 may be applied to $S(t)$. In particular, from our equations for the mean and variance of a compound Poisson distribution in (3.18), we have

$$\mathbb{E}[S(t)] = \lambda \mu t \text{ and } \text{Var}[S(t)] = \lambda t \mathbb{E}[X_1^2]. \tag{6.3}$$

Recall that we assume that moments of X_1 are finite as required, so that if we write $\mathbb{E}[X_1^2]$, then we are tacitly assuming that it is finite.

We also have, from the convolution series formula for a compound distribution in Theorem 3.8,

$$\Pr(S(t) \le x) = \sum_{n=0}^{\infty} \frac{e^{-\lambda t}(\lambda t)^n}{n!} F_X^{\star n}(x),$$

and, from the expression for a compound Poisson moment generating function in (3.16), we have

$$M_{S(t)}(r) = \exp(\lambda t(M_X(r) - 1)).$$

In the above formulation, $U(t)$ is the risk reserve of an insurance company. However, this model is also often used for the "risk reserve" corresponding to a particular portfolio of policies. In this second context, u is the capital allocated to this portfolio by the insurance company, and ruin theory for this portfolio is one way of assessing its riskiness.

We remark here that the classical risk model involves many simplifications of how an insurance business operates in practice. For example, the model assumes that the claim-arrival rate λ remains constant over time, that no interest is earned on the surplus, that there is no inflation, that premium income is received continuously in time, that claims are paid out immediately, and in addition there are many independence assumptions. It is easy to think of real life situations where these assumptions do not hold. Nevertheless, this model forms the basis of many models used in insurance mathematics. As we shall see, analysis of this "simple" model is relatively tractable, although it does call upon rather heavier mathematical machinery than that used for the fixed-time models in Chapter 3. This mathematical machinery involves, for example, results from renewal theory, properties of Laplace transforms, and techniques of integration theory. In the text of this chapter, we quote and use such results without proof, referring the interested reader to other sources.

6.1.1 The relative safety loading

From (6.2) and (6.3), the expected surplus at time t is

$$\mathbb{E}[U(t)] = u + ct - \mathbb{E}[S(t)] = u + ct - \lambda \mu t, \tag{6.4}$$

and the expected profit per unit time in $(0, t]$ is

$$\frac{\mathbb{E}[U(t) - U(0)]}{t} = \frac{(u + ct - \lambda \mu t) - u}{t} = c - \lambda \mu.$$

In general, we hope to make a profit, and this motivates the *net profit condition*:

$$c > \lambda \mu. \tag{6.5}$$

Given λ and μ, we aim to set the premium rate c so that (6.5) is satisfied (although bear in mind that there may be other considerations in the setting of premiums, for example forces of competition between insurance companies). For the classical risk model, $c - \lambda\mu$ is the expected net income in each unit time interval. In view of (6.5), we introduce the *relative safety loading* (or *premium loading factor* or *relative security loading*) θ, defined by

$$\theta = \frac{c - \lambda\mu}{\lambda\mu},$$

so that the premium rate c may be written $c = (1 + \theta)\lambda\mu$. The net profit condition is satisfied if and only if the relative safety loading θ is positive. We note that the relative safety loading θ is the same as the relative security loading for the expected value premium principle in §4.1.1.

6.1.2 Ruin probabilities

The expected surplus in (6.4) may be written in terms of the relative safety loading,

$$\mathbb{E}[U(t)] = u + \theta\lambda\mu t,$$

and the expected profit over $(0, t]$ is $\theta\lambda\mu t$. We assume that we have positive relative safety loading, so that the expectation of $U(t)$ is positive and increases steadily with time t. However, $\{U(t)\}$ is a stochastic process and has random fluctuations about its expectation, and so there could still be a possibility that $U(t)$ becomes negative at some point. Indeed, from (6.3), we know that $\text{Var}[U(t)] = \lambda t\mathbb{E}[X_1^2]$, so that the variability of $U(t)$ is increasing in t.

More formally, if $U(t) < 0$ for some $t \geq 0$ then ruin is said to occur. An obvious concern is with the probability that a ruin event occurs. Intuitively, we see that this probability depends on the initial capital u, in the sense that we might expect that a higher value of u is safer, leading to a reduction in the probability of ruin. We shall see later that this is indeed so.

We define the *probability of ruin* given initial capital u to be

$$\psi(u) = \Pr(U(t) < 0 \text{ for some } t > 0). \tag{6.6}$$

This is also known as the probability of eventual ruin, or the infinite-time horizon ruin probability.

A related quantity is the *finite-time ruin probability* with time horizon t_0 (>0), given by

$$\psi(u, t_0) = \Pr(U(t) < 0 \text{ for some } t \text{ in } (0, t_0]).$$

There are several logical relationships that hold between the various ruin probabilities. We have already mentioned that we expect $\psi(u)$ to be non-increasing in u, that is for $0 \leq u_1 \leq u_2$ we expect

$$\psi(u_1) \geq \psi(u_2). \tag{6.7}$$

This can be seen formally as follows. For $u \geq 0$, let A_u be the event that ruin occurs in $(0, \infty)$ when the initial capital is u. Suppose that $u_2 \geq u_1$, and consider a ruin sample path for initial capital u_2. Translating this sample path downwards, we see that this will also be a ruin sample path when it starts at u_1. Thus we have $A_{u_2} \subseteq A_{u_1}$. This implies $\Pr(A_{u_2}) \leq \Pr(A_{u_1})$, and we obtain (6.7). We can see by a similar argument that, for $t_0 \geq 0$,

$$\psi(u_1, t_0) \geq \psi(u_2, t_0).$$

Define $A_{u,t}$ to be the event that ruin occurs in $(0, t]$ when the initial capital is u. Then, for $0 < t_1 \leq t_2 < \infty$, we have $A_{u,t_1} \subseteq A_{u,t_2} \subseteq A_u$, and so

$$\psi(u, t_1) \leq \psi(u, t_2) \leq \psi(u).$$

Further, if $t_n \to \infty$ as $n \to \infty$, and if $0 < t_1 \leq t_2, \leq \cdots$, then $A_{u,t_1} \subseteq A_{u,t_2}, \subseteq \cdots$ and $\cup_{n=1}^{\infty} A_{u,t_n} = A_u$. By continuity of probability (see, for example, chap. 1 of Grimmett and Stirzaker (2001)), we obtain

$$\psi(u, t_n) \to \psi(u) \text{ as } n \to \infty,$$

and further

$$\psi(u, t) \to \psi(u) \text{ as } t \to \infty.$$

We also define the *survival probability*

$$\varphi(u) = 1 - \psi(u) = \Pr(U(t) \geq 0 \text{ for all } t \geq 0), \tag{6.8}$$

so that $\varphi(u)$ is the probability of never being ruined with initial capital u.

When $U(t)$ is interpreted as the surplus for a particular portfolio, then ruin probabilities may be used as measures of "dangerousness" for that portfolio, with, of course, higher ruin probabilities corresponding to more dangerous portfolios.

When $U(t)$ is interpreted as the surplus for a whole insurance company, then ruin corresponds to insolvency of the insurance company. In real life, insurance companies do fail, but we note that in practice the solvency/insolvency of an insurance company involves very many different and more complicated considerations. Insurance companies fail for many reasons (for example, because of factors affecting the whole financial market, or because of fraud). In addition, solvency arrangements are subject to legal requirements. These wider aspects of solvency/insolvency, while very interesting and topical, are beyond the scope of this book and are not considered further here.

6.2 Lundberg's inequality and the adjustment coefficient

Ruin probabilities can be difficult to evaluate explicitly except for certain special cases for the claim-size distribution. This is a familiar situation in applied probability modelling, and one which we met in Chapter 3. As in Chapter 3, this means that we often consider approximations. Another common approach is to look for simple bounds for the relevant quantity, and, in this section, we do this for the ruin probability $\psi(u)$ in the classical risk model with positive relative safety loading.

The upper bound given in this section is called Lundberg's bound, and it applies when the claim-size distribution satisfies a condition relating to its moment generating function $M_X(r)$.

Theorem 6.1 *(Lundberg's inequality.) In the classical risk model with Poisson rate $\lambda > 0$, premium income rate $c > 0$ and positive relative safety loading, suppose there exists a unique $R > 0$ that solves*

$$M_X(r) - 1 = \frac{cr}{\lambda},\qquad(6.9)$$

where $M_X(r)$ is the claim-size moment generating function. Then the probability of ruin $\psi(u)$, with initial capital $u(\geq 0)$, satisfies

$$\psi(u) \leq e^{-Ru} \text{ for all } u \geq 0.$$

Note that the Lundberg inequality gives an easy upper bound for the ruin probability. This upper bound is less than 1 for $u > 0$, and is decreasing in u.

The positive number R is the *adjustment coefficient* (or the *Lundberg exponent*) and is an important characteristic of a risk model. We look at various of its properties in §6.2.1; we give a proof of Theorem 6.1 in §6.2.2; and in §6.2.3 we consider when the adjustment coefficient exists.

6.2.1 Properties of the adjustment coefficient

The adjustment coefficient is often used as a single-number summary of riskiness, since larger R corresponds to a smaller upper bound for the ruin probability in the Lundberg inequality in Theorem 6.1. Intuitively we might prefer a portfolio with a larger adjustment coefficient.

Note also that the defining equation (6.9) for the adjustment coefficient may be written

$$M_X(r) - 1 = (1 + \theta)\mu r.\qquad(6.10)$$

This (and (6.9)) may be solved explicitly in certain special cases, as Example 6.2 shows.

Example 6.2 In a classical risk model with positive relative safety loading θ, claims are exponentially distributed with mean μ, so that the claim-size moment generating function is $(1 - \mu r)^{-1}$ for $r < \mu^{-1}$. By (6.10), in order to find the adjustment coefficient R, we seek a positive solution to

$$\frac{1}{1 - \mu r} - 1 = (1 + \theta)\mu r.$$

Because $M_X(r)$ is only defined for $r < \mu^{-1}$, this means that the adjustment coefficient must lie in $(0, \mu^{-1})$. We find that R solves

$$1 - (1 - \mu r) = (1 + \theta)\mu r(1 - \mu r).$$

This simplifies to

$$(1 + \theta)(\mu r)^2 - \theta\mu r = 0,$$

which is

$$(1 + \theta)\mu^2 r\left(r - \frac{\theta}{(1 + \theta)\mu}\right) = 0.$$

This equation has two solutions, $r = 0$ and $r = \theta/(1 + \theta)\mu$. We know $R > 0$, so that the second value is the only candidate for R, and the second value is indeed in the range $(0, \mu^{-1})$ (because $\theta > 0$). Thus the adjustment coefficient for exponentially distributed claims is

$$R = \frac{\theta}{(1 + \theta)\mu} = \frac{1}{\mu} - \frac{\lambda}{c},$$

where λ and c are the Poisson rate and the premium income rate, respectively. Theorem 6.1 yields that for exponential claims the probability of ruin satisfies

$$\psi(u) \leq \exp\left(-\frac{\theta}{(1 + \theta)\mu}u\right) \text{ for all } u \geq 0,$$

completing this example.

How does the adjustment coefficient depend on the other parameters in the risk model? For example, how would you expect it to depend on the relative safety loading θ? On an intuitive level, if θ increases, then we interpret this in terms of the portfolio being "safer", and so we might expect R to be larger. We can see that this is indeed so for exponentially distributed claims as follows. From (6.10), we can find R graphically by plotting the curve $y = M_X(r) - 1$ and the line $y = (1 + \theta)\mu r$. The r-coordinate of the point where the curve and

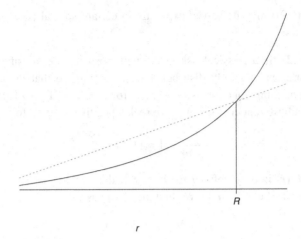

Figure 6.1. Graphs of $y = M_X(r) - 1$ (solid line) and $y = (1 + \theta)\mu r$ (dashed line). The adjustment coefficient is the r-coordinate of the point of intersection.

the line cross is R; see Figure 6.1. Now suppose that θ is increased, but that λ, μ and the claim-size distribution all stay unchanged. Then the gradient of the straight line increases while the curved line remains unchanged, which means that the point of intersection moves further from the origin and so R increases.

The equation satisfied by R in (6.10) does not involve λ, except via θ. So, if the claim-size distribution is unchanged, but λ and c change in such a way that the relative safety loading θ is unchanged, then the value of the adjustment coefficient remains the same.

Another observation is that R depends on the units used for the claim sizes, as in the following example.

Example 6.3   In a classical risk model with claim-arrival rate λ, suppose that the premium rate is c and that the claim amounts $X_1, X_2, \ldots$ (£) are independent exponentially distributed random variables with mean μ. Now consider the same model, but with all monetary quantities given in euros. Let β be the exchange rate, so that the new premium rate is $c' = \beta c$, and the new claim amounts are $X'_1, X'_2, \ldots$, where $X'_i = \beta X_i$. You can check for yourself that both models have the same relative safety loading $\theta = (c - \lambda\mu)/(\lambda\mu)$. For the new claim-amount distribution, we have

$$1 - F_{X'}(y) = \Pr(X'_1 > x) = \Pr(X_1 > x/\beta) = e^{-x/(\beta\mu)},$$

which shows that X_1' has an exponential distribution with mean $\beta\mu$. From Example 6.2, the adjustment coefficient R' for the euro model is

$$R' = \frac{\theta}{(1 + \theta)\mu\beta},$$

and this is $R' = R/\beta$, so that the adjustment coefficient is not the same in the two different currencies. Note, however, that the initial capital will also be in euros, so that the resulting Lundberg upper bound will be the same in both currencies.

When claims are exponentially distributed (and in some other special cases) the adjustment coefficient has an easy explicit expression in terms of the parameters of the model. However, in general the equations for the adjustment coefficient in (6.9) or (6.10) have to be solved numerically, for example using the Newton–Raphson method; see §6.6. In such situations, we often need to have some idea of the size of R in order to start off a computational search or iterative procedure. Lemma 6.4 gives a bound on the size of R which may be useful in such contexts.

Lemma 6.4 *In a classical risk model with positive relative safety loading θ and claim sizes $X_1, X_2, \ldots$, suppose that $\mathbb{E}[X_1] = \mu$ and $\mu_2 = \mathbb{E}[X_1^2] < \infty$. Then the adjustment coefficient R satisfies*

$$R \le \frac{2\mu\theta}{\mu_2}.$$

Observe that this simple upper bound involves the claim-size distribution only via its first two moments, whereas the equations (6.9) and (6.10) for the adjustment coefficient involve the whole claim-size distribution via its moment generating function.

Proof Observe from (6.10) that R satisfies

$$\int_{(0,\infty)} e^{Rx} F_X(dx) - 1 = (1 + \theta)\mu R. \tag{6.11}$$

By taking the first three terms of a Taylor expansion, the left-hand side satisfies

$$\int_{(0,\infty)} e^{Rx} F_X(dx) - 1 \ge \int_{(0,\infty)} \left(1 + Rx + \frac{R^2 x^2}{2}\right) F_X(dx) - 1$$

$$= 1 + R\mu + \frac{R^2}{2}\mu_2 - 1$$

$$= R\mu + \frac{R^2}{2}\mu_2.$$

Combining this with (6.11), we have

$$R\mu + \frac{R^2}{2}\mu_2 \le (1 + \theta)\mu R,$$

and rearrangement gives the upper bound in the lemma. □

By this lemma, we know that $R \in (0, 2\mu\theta/\mu_2)$. See Exercise 6.3 for a positive lower bound on R when the claim sizes are bounded.

6.2.2 Proof of Lundberg's inequality

In this section we go into more mathematical depth and we prove Lundberg's inequality (Theorem 6.1).

Proof of Theorem 6.1 From the definition of the probability of ruin $\psi(u)$ in (6.6) it seems that to check for ruin we must look at $U(t)$ for all $t \ge 0$. However, between claims, the sample path of the surplus process $\{U(t) : t \ge 0\}$ increases linearly at rate $c > 0$ and the path jumps down at the time of a claim. Thus ruin can only occur at the time of a claim, and this means that we only need to look at the surplus process at the claim times. With initial capital u, define, for $n = 1, 2, \ldots$,

$$\psi_n(u) = \Pr(\text{ruin occurs on or before the } n\text{th claim}).$$

Since the occurrence of ruin on or before the nth claim implies ruin somewhere in $(0, \infty)$, we have, for $n = 1, 2, \ldots$,

$$\psi_n(u) \le \psi(u) \text{ for all } u \ge 0.$$

We also have

$$\psi_n(u) \to \psi(u) \quad \text{as } n \to \infty; \tag{6.12}$$

see Exercise 6.4. It follows that

$$\psi(u) \le e^{-Ru} \Leftrightarrow \psi_n(u) \le e^{-Ru} \text{ for all } n = 1, 2, \ldots. \tag{6.13}$$

Hence in order to prove the theorem, it is enough to show that

$$\psi_n(u) \le e^{-Ru} \text{ for all } n = 1, 2, \ldots. \tag{6.14}$$

We prove (6.14) by induction on n. For $n = 1$, note that $\psi_1(u)$ is the probability that ruin occurs at the first claim. Suppose the first claim arrives at $T_1 = t$, so that, just before the first claim, the surplus is $u + ct$. If the first claim X_1

causes ruin, then we must have $X_1 > u + ct$. Recall that claims arrive in a Poisson process so T_1 has an exponential distribution with mean λ^{-1}. Conditioning on the time of the first claim, we find

$$\psi_1(u) = \Pr(\text{ruin occurs at the first claim})$$

$$= \int_0^\infty \Pr(\text{ruin occurs at the first claim} \mid T_1 = t)\lambda e^{-\lambda t}dt$$

$$= \int_0^\infty \Pr(X_1 > u + ct)\lambda e^{-\lambda t}dt$$

$$= \int_0^\infty \lambda e^{-\lambda t} \int_{(u+ct,\infty)} F_X(dx)dt.$$

In the inner integral, we have $x \geq u + ct$, so that, for these x-values,

$$1 \leq \exp(-R(u + ct - x)),$$

where we have also used that $R > 0$. This gives

$$\psi_1(u) \leq \int_0^\infty \lambda e^{-\lambda t} \int_{(u+ct,\infty)} e^{-R(u+ct-x)} F_X(dx)dt$$

$$\leq \int_0^\infty \lambda e^{-\lambda t} \int_{(0,\infty)} e^{-R(u+ct-x)} F_X(dx)dt, \tag{6.15}$$

where the last line follows because we have extended the range of integration of the inner integral and the extra part is non-negative. We rewrite (6.15) to get

$$\psi_1(u) \leq e^{-Ru} \int_0^\infty e^{-(\lambda+cR)t} \lambda \int_{(0,\infty)} e^{Rx} F_X(dx)dt. \tag{6.16}$$

Note that $\lambda \int_{(0,\infty)} e^{Rx} F_X(dx)$ is $\lambda M_X(R)$, and, from the equation (6.9) for R, we know that $\lambda M_X(R) = \lambda + cR$. Then (6.16) becomes

$$\psi_1(u) \leq e^{-Ru} \int_0^\infty (\lambda + cR)e^{-(\lambda+cR)t} \, dt.$$

The integrand is the density of an exponential distribution with mean $(\lambda + cR)^{-1}$, so the integral is 1. Thus we have shown

$$\psi_1(u) \leq e^{-Ru} \text{ for all } u \geq 0,$$

and we have proved (6.14) for $n = 1$.

Now suppose that $n \geq 1$ and assume that $\psi_n(u) \leq \exp(-Ru)$ for all $u \geq 0$. Conditioning on the time and size of the first claim, we find that $\psi_{n+1}(u)$ is

$$\int_0^\infty \int_{(0,\infty)} \lambda e^{-\lambda t} \Pr(\text{ruin by claim } (n+1) \mid T_1 = t, X_1 = x) F_X(dx)dt,$$

and this is

$$\int_0^\infty \int_{(0,u+ct]} \lambda e^{-\lambda t} \Pr(\text{ruin by claim } (n+1) \mid T_1 = t, X_1 = x) F_X(dx) dt$$

$$+ \int_0^\infty \int_{(u+ct,\infty)} \lambda e^{-\lambda t} \Pr(\text{ruin by claim } (n+1) \mid T_1 = t, X_1 = x) F_X(dx) dt.$$

If $X_1 = x$, where $x > u + ct$, then ruin occurs at the first claim, and

$$\Pr(\text{ruin by claim } (n+1) \mid T_1 = t, X_1 = x) = 1.$$

This means that

$$\int_0^\infty \int_{(u+ct,\infty)} \lambda e^{-\lambda t} Pr(\text{ruin by claim } (n+1) \mid T_1 = t, X_1 = x) F_X(dx) dt$$

$$= \int_0^\infty \int_{(u+ct,\infty)} \lambda e^{-\lambda t} F_X(dx) dt.$$

If the first claim is $X_1 = x$, with $x \le u + ct$, then ruin does not occur at the first claim. In this case, immediately after the first claim, the surplus is $u + ct - x$, and, given the independence and distributional assumptions of the model, the evolution of the model from this point onwards is probabilistically the same as the evolution of a new risk model with the same Poisson arrival rate λ, the same premium income rate c, and the same claim-size distribution function F_X, but starting with initial capital $u + ct - x$. The $(n+1)$st claim in the original model corresponds to the nth claim in the new model, so that the probability of ruin on or before the $(n+1)$st claim in the original model is the same as $\psi_n(u + ct - x)$. This means that

$$\int_0^\infty \int_{(0,u+ct]} \lambda e^{-\lambda t} Pr(\text{ruin by claim } (n+1) \mid T_1 = t, X_1 = x) F_X(dx) dt$$

$$= \int_0^\infty \int_{(0,u+ct]} \lambda e^{-\lambda t} \psi_n(u + ct - x) F_X(dx) dt.$$

Putting all this together we obtain

$$\psi_{n+1}(u) = \int_0^\infty \lambda e^{-\lambda t} \int_{(0,u+ct]} \psi_n(u + ct - x) F_X(dx) dt$$

$$+ \int_0^\infty \lambda e^{-\lambda t} \int_{(u+ct,\infty)} F_X(dx) dt.$$

In the second term on the right-hand side above, we have

$$1 \le \exp(-R(u + ct - x)) \text{ for } x > u + ct$$

(as in the $n = 1$ case). For the first term, the inductive hypothesis implies that $\psi_n(u + ct - x) \leq \exp(-R(u + ct - x))$. Therefore we have

$$\psi_{n+1}(u) \leq \int_0^\infty \lambda e^{-\lambda t} \int_{(0, u+ct]} e^{-R(u+ct-x)} F_X(dx) dt$$
$$+ \int_0^\infty \lambda e^{-\lambda t} \int_{(u+ct, \infty)} e^{-R(u+ct-x)} F_X(dx) dt$$
$$= \int_0^\infty \lambda e^{-\lambda t} \int_{(0, \infty)} e^{-R(u+ct-x)} F_X(dx) dt.$$

The last expression is the same as the right-hand side of (6.15), and so we follow the same steps as in the $n = 1$ case to obtain

$$\psi_{n+1}(u) \leq e^{-Ru} \text{ for all } u \geq 0.$$

Hence (6.14) holds by induction, and thus, by (6.13), we have proved Theorem 6.1. □

6.2.3 When does the adjustment coefficient exist?

We now turn to the question of whether or not the adjustment coefficient R exists for a particular risk model. From the equation (6.10) for R, we consider whether or not there is a unique positive solution of

$$M_X(r) - 1 = (1 + \theta)\mu r. \tag{6.17}$$

This equation only makes sense for those r-values for which the claim-size moment generating function is finite.

Fortunately there are relatively simple conditions which imply the existence of R. We write $r \to a^-$ to mean that r converges to a from the left.

Lemma 6.5 *In the classical risk model with positive relative safety loading, suppose there exists r_∞, $0 < r_\infty \leq \infty$, such that the moment generating function $M_X(r)$ of the claim-size distribution satisfies (i) and (ii) below:*

(i) $M_X(r) < \infty$ *for all $r < r_\infty$,*
(ii) $M_X(r) \to \infty$ *as $r \to r_\infty^-$.*

Then there exists a unique positive solution to (6.17).

Proof Let $h(r) = M_X(r) - 1 - (1 + \theta)\mu r$, which is defined for $r < r_\infty$ by condition (i) in the statement of the lemma. By (6.17), we want to show that there is a unique positive solution of $h(r) = 0$ in $(0, r_\infty)$. We use several properties of the moment generating function $M_X(r)$ of the positive random variable X_1 for $0 \leq r < r_\infty$:

(a) $M_X(0) = 1$,

(b) $M_X(r)$ is continuous, differentiable and strictly convex,

(c) $M_X'(0) = \mu$.

Property (a) follows from the definition of $M_X(r) = \mathbb{E}[e^{rX_1}]$. For (b) and (c), we know that M_X is differentiable infinitely many times for $|r| < r_\infty$ with

$$M_X^{(n)}(r) = \mathbb{E}[X_1^n e^{rX_1}]; \tag{6.18}$$

see, for example, the proof of th. 4.8.3 in Gut (2005). Thus M_X is differentiable and continuous on $|r| < r_\infty$, and also (c) holds.

Further, $M_X''(r) = \mathbb{E}[X_1^2 e^{rX_1}]$, and we now show that this is strictly positive. Since $X_1 > 0$ with probability 1, there exists $\eta > 0$ such that $p = \Pr(X_1 > \eta)$ is strictly positive. Hence

$$M_X''(r) \geq p\mathbb{E}[X_1^2 e^{rX_1} \mid X_1 > \eta] \geq p\eta^2 e^{r\eta} > 0.$$

Thus M_X is strictly convex, and we have shown (b).

By (a) we have $h(0) = 0$. By (b) we have that $h(r)$ is continuous and convex on $(0, r_\infty)$, and also continuous at zero. By (c), $h'(0) = \mu - (1 + \theta)\mu = -\theta\mu$, and this is negative (because the relative safety loading θ is positive). So we know that the graph of h starts at zero by going downwards, and that it is continuous and convex. Consider now its behaviour as r approaches r_∞ from the left. If $r_\infty < \infty$, then

$$h(r) \to \left(\lim_{r \to r_\infty^-} M_X(r)\right) - 1 - (1 + \theta)\mu r_\infty,$$

so that, by condition (ii) in the statement of the lemma, we have

$$h(r) \to \infty \text{ as } r \to r_\infty^-.$$

This means that the graph of $h(r)$ is as shown in Figure 6.2, and so there must be a unique R in $(0, r_\infty)$ such that $h(R) = 0$.

If $r_\infty = \infty$, then we need a more careful argument to show that $h(r) \to \infty$, because in this case the term $-(1 + \theta)\mu r$ converges to $-\infty$ as $r \to r_\infty^-$, while $M_X(r) \to \infty$, so we need to establish which of these convergences "wins". With p and η as above, we have, for $r > 0$,

$$M_X(r) = \mathbb{E}[e^{rX_1}] \geq p e^{\eta r}.$$

Hence we have

$$h(r) \geq p e^{r\eta} - 1 - (1 + \theta)\mu r.$$

The convergence of $e^{r\eta}$ to ∞ will be much faster than the convergence of $-(1 + \theta)\mu r$ to $-\infty$, so that the right-hand side converges to ∞, and hence $h(r) \to \infty$ as

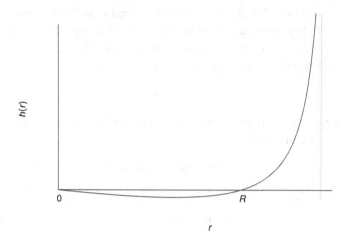

Figure 6.2. Graph of $y = h(r) = M_X(r) - 1 - (1 + \theta)\mu r$ when r_∞ is finite.

$r \to r_\infty^-$. A similar argument to the $r_\infty < \infty$ case then shows there is a unique positive solution to $h(r) = 0$ when $r_\infty = \infty$. □

Example 6.6 (i) Suppose that X_1 has a Pareto distribution Pa(1, 1) with density $f_X(x) = (1 + x)^{-2}$, $x > 0$. For every $r > 0$, the moment generating function $M_X(r)$ is not finite (see Exercise 6.8), so (i) of Lemma 6.5 is not satisfied. Indeed, since $M_X(r) = \infty$ for all $r > 0$, there is no positive r satisfying the adjustment coefficient equation (6.17), and hence there is no adjustment coefficient.

(ii) If X_1 has an exponential distribution with mean μ, then $M_X(r) = (1 - \mu r)^{-1}$ for $r < \mu^{-1}$. Conditions (i) and (ii) of Lemma 6.5 are satisfied with $r_\infty = \mu^{-1}$, so that Lemma 6.5 implies the existence of R (and we have already found it in Example 6.2).

(iii) Suppose X_1 has a uniform distribution on $(0, a)$ for some $a > 0$, so that X_1 has probability density function

$$f_X(x) = \begin{cases} 1/a & \text{for } 0 < x < a \\ 0 & \text{otherwise.} \end{cases}$$

Then

$$M_X(r) = \frac{e^{ra} - 1}{ra},$$

and conditions (i) and (ii) of Lemma 6.5 are satisfied with $r_\infty = \infty$. Hence the adjustment coefficient exists in this case.

The conditions in Lemma 6.5 are often referred to as "small claims" conditions, and they imply a certain bound on the decrease of the tail of the claim-size distribution, which we now explain. By condition (i) of Lemma 6.5, we can find $r_0 > 0$ such that $M_X(r_0) < \infty$. We have, for all $x \geq 0$,

$$e^{r_0 x} 1(X_1 > x) \leq e^{r_0 X_1},$$

where $1(X_1 > x)$ is 1 if the event $X_1 > x$ occurs and is zero otherwise. Taking the expectation of both sides gives

$$e^{r_0 x} \Pr(X_1 > x) \leq M_X(r_0),$$

which tells us that

$$\Pr(X_1 > x) \leq K e^{-r_0 x}, \tag{6.19}$$

where $K = M_X(r_0) < \infty$. Thus any claim-size distribution that satisfies condition (i) of Lemma 6.5 has tail $1 - F_X(x)$ that decreases at least exponentially fast as x tends to infinity. This rules out distributions with heavy tails such as the Pareto distribution.

6.3 Equations for $\psi(u)$ and $\varphi(u)$: the ruin probability and the survival probability

In §6.2 our focus was on the Lundberg inequality, which gives an upper bound for the probability of ruin. In this section we derive equations for the probability of ruin $\psi(u)$ and for the survival probability $\varphi(u)$. By now we have had experience of working with applied probability quantities, and so it comes as no surprise that these equations only have easy explicit solutions for certain special choices for the claim-size distribution. It is also not unexpected that the exponential distribution is one of these special cases. We give examples using the exponential distribution below.

Throughout the rest of this chapter we assume, unless otherwise stated, that the claim-size distribution has a density f_X. This is for ease of exposition and to avoid a few minor complications. For ruin theory in the general case, the interested reader is referred to sects. 5.3 and 5.4 in Rolski *et al.* (1999).

We first show that $\varphi(u)$ and $\psi(u)$ are continuous.

Lemma 6.7 *In the classical risk model with positive relative safety loading, the survival probability $\varphi(u)$ and the ruin probability $\psi(u)$ are continuous at $u > 0$ and right-continuous at $u = 0$.*

Proof In this proof we write $\text{Pr}_u(\cdot)$ to show explicitly the dependence on the initial capital u. Recall that the Poisson rate is λ and the premium income rate is c. Let T_1 be the time of the first claim, which has an $\text{Exp}(\lambda)$ distribution, by results for the inter-event times in a Poisson process in §2.2.3. For all $u \geq 0$ and $h > 0$ we have

$$\varphi(u) = \text{Pr}_u(U(t) \geq 0 \text{ for all } t > 0)$$
$$= \text{Pr}_u(U(t) \geq 0 \text{ for all } t > 0 \mid T_1 > h)e^{-\lambda h}$$
$$+ \int_0^h \text{Pr}_u(U(t) \geq 0 \text{ for all } t > 0 \mid T_1 = s)\lambda e^{-\lambda s} \, ds. \qquad (6.20)$$

If $T_1 > h$ then no claims occur in $(0, h]$ and so $U(h) = u + ch$. Because of the memoryless property of the exponential distribution (see (2.17)), we know that $T_1 - h \mid (T_1 > h) \sim T_1$, and this means that, given $T_1 > h$, then the time from $t = h$ to the first claim has an exponential distribution with mean $1/\lambda$. Hence, when $T_1 > h$, the probabilistic behaviour of the surplus process after time $t = h$ is the same as that of the original process but with initial capital $u + ch$. Thus (6.20) becomes

$$\varphi(u) = \varphi(u+ch)e^{-\lambda h} + \int_0^h \text{Pr}_u(U(t) \geq 0 \text{ for all } t > 0 \mid T_1 = s)\lambda e^{-\lambda s} \, ds. \quad (6.21)$$

The integral term satisfies

$$0 \leq \int_0^h \text{Pr}_u(U(t) \geq 0 \text{ for all } t > 0 \mid T_1 = s)\lambda e^{-\lambda s} \, ds \leq \int_0^h \lambda e^{-\lambda s} \, ds,$$

and this tends to zero as $h \downarrow 0$. Hence, taking the limit as $h \downarrow 0$ in (6.21), we obtain

$$\varphi(u) = \lim_{h \downarrow 0} \varphi(u + ch),$$

so that φ is right-continuous at u for $u \geq 0$.

For $u > 0$, let $h > 0$ be such that $u - ch > 0$. Then (6.21) implies that

$$\varphi(u - ch) = \varphi(u)e^{-\lambda h} + \int_0^h \text{Pr}_{u-ch}(U(t) \geq 0 \text{ for all } t > 0 \mid T_1 = s)\lambda e^{-\lambda s} \, ds.$$
$$(6.22)$$

Letting $h \downarrow 0$, we find that

$$\lim_{h \downarrow 0} \varphi(u - ch) = \varphi(u),$$

so that $\varphi(u)$ is left-continuous at $u > 0$. Thus $\varphi(u)$ is continous at all $u > 0$ and right-continous at $u = 0$. Since $\psi(u) = 1 - \varphi(u)$, the same is true for $\psi(u)$. $\qquad \square$

Lemma 6.8 shows that the survival probability $\varphi(u)$ and the ruin probability $\psi(u)$ each satisfy an *integro-differential equation*. It is a side result of the proof of the lemma that (at least when F_X has a density) the functions $\varphi(u)$ and $\psi(u)$ are differentiable at $u > 0$. For the general case, see sect. 5.3.1 in Rolski *et al.* (1999).

Lemma 6.8 *In the classical risk model with Poisson rate λ, premium rate c, claim-size distribution function F_X (with density f_X) and positive relative safety loading, the survival probability $\varphi(u)$ satisfies, for $u > 0$,*

$$\varphi'(u) = \frac{\lambda}{c}\varphi(u) - \frac{\lambda}{c}\int_0^u \varphi(u - x)f_X(x)dx, \qquad (6.23)$$

and the ruin probability $\psi(u)$ satisfies, for $u > 0$,

$$\psi'(u) = \frac{\lambda}{c}\psi(u) - \frac{\lambda}{c}(1 - F_X(u)) - \frac{\lambda}{c}\int_0^u \psi(u - x)f_X(x)dx. \qquad (6.24)$$

Proof Condition on the time T_1 and size X_1 of the first claim to see that

$$\varphi(u) = \Pr(U(t) \geq 0 \text{ for all } t \geq 0)$$
$$= \int_0^\infty \int_0^\infty \Pr(U(t) \geq 0 \, \forall t \geq 0 \mid X_1 = x, T_1 = s)f_X(x)dx\lambda e^{-\lambda s}\,ds.$$

If the size $X_1 = x$ of the first claim is greater than the surplus $u + cs$ at the time $T_1 = s$ of the first claim, then the first claim causes ruin, and so

$$\Pr(U(t) \geq 0 \, \forall t \geq 0 \mid X_1 = x, T_1 = s) = 0 \text{ if } x > u + cs.$$

Hence

$$\varphi(u) = \int_0^\infty \int_0^{u+cs} \Pr(U(t) \geq 0 \, \forall t \geq 0 \mid X_1 = x, T_1 = s)f_X(x)dx\lambda e^{-\lambda s}\,ds.$$

When $x \leq u + cs$, we know that ruin cannot occur before the time s of the first claim, so that, in order to check that ruin does not occur, we only need to check the surplus process after the first claim. The assumptions of the classical risk model imply that the probabilistic behaviour of the surplus process after the first claim is the same as that of the original surplus process, except that it starts with capital $u + cs - x$. Thus we have, for $x \leq u + cs$,

$$\Pr(U(t) \geq 0 \text{ for all } t \geq 0 \mid X_1 = x, T_1 = s) = \varphi(u + cs - x),$$

and so we obtain

$$\varphi(u) = \int_0^\infty \int_0^{u+cs} \varphi(u + cs - x)f_X(x)\lambda e^{-\lambda s}\,dx\,ds.$$

In the outer integral with respect to s, we make the substitution to a new variable $z = u + cs$:

$$\varphi(u) = \int_u^\infty \frac{\lambda}{c} e^{-\lambda(z-u)/c} \int_0^z \varphi(z-x) f_X(x) dx\, dz$$
$$= \frac{\lambda}{c} e^{\lambda u/c} \int_u^\infty e^{-\lambda z/c} \int_0^z \varphi(z-x) f_X(x) dx\, dz. \tag{6.25}$$

The right-hand side of (6.25) is the product of two functions of u,

$$\varphi(u) = h_1(u) h_2(u),$$

where

$$h_1(u) = \frac{\lambda}{c} e^{\lambda u/c} \quad \text{and} \quad h_2(u) = \int_u^\infty r(z) dz,$$

where $r(z) = e^{-\lambda z/c} \int_0^z \varphi(z-x) f_X(x) dx$. It is easy to check that $r(z) \geq 0$ and

$$\int_0^\infty r(z) dx \leq \int_0^\infty e^{-\lambda z/c} dz < \infty.$$

Using the continuity of φ in Lemma 6.7, it can be shown that $r(z)$ is continuous. Then, by elementary calculus (see, for example, th. 5.1 in Apostol (1967)), we have

$$h_2'(u) = -r(u).$$

We also have $h_1'(u) = (\lambda/c) h_1(u)$. Hence $\varphi(u)$ is differentiable with

$$\varphi'(u) = \frac{\lambda}{c} h_1(u) h_2(u) - h_1(u) r(u)$$
$$= \frac{\lambda}{c} \varphi(u) - \frac{\lambda}{c} e^{\lambda u/c} e^{-\lambda u/c} \int_0^u \varphi(u-x) f_X(x) dx$$
$$= \frac{\lambda}{c} \varphi(u) - \frac{\lambda}{c} \int_0^u \varphi(u-x) f_X(x) dx,$$

and we have proved the integro-differential equation (6.23) for φ.

To obtain the corresponding equation for $\psi(u)$, we replace $\varphi(u)$ in (6.23) by $1 - \psi(u)$ to get

$$-\psi'(u) = \frac{\lambda}{c}(1 - \psi(u)) - \frac{\lambda}{c} \int_0^u (1 - \psi(u-x)) f_X(x) dx$$
$$= \frac{\lambda}{c}(1 - \psi(u)) - \frac{\lambda}{c} F_X(u) + \frac{\lambda}{c} \int_0^u \psi(u-x) f_X(x) dx.$$

This means that

$$\psi'(u) = \frac{\lambda}{c}\psi(u) - \frac{\lambda}{c}(1 - F_X(u)) - \frac{\lambda}{c} \int_0^u \psi(u-x) f_X(x) dx,$$

which is the integro-differential equation (6.24) for ψ. $\qquad\square$

In Lemma 6.9, we use the integro-differential equation for $\varphi(u)$ in order to obtain an integral equation for $\varphi(u)$.

Lemma 6.9 *In the classical risk model with Poisson rate λ, premium rate c, claim-size distribution function F_X and positive relative safety loading, the survival probability $\varphi(u)$ satisfies*

$$\varphi(u) = \varphi(0) + \frac{\lambda}{c} \int_0^u \varphi(u - x)(1 - F_X(x))dx. \qquad (6.26)$$

Proof The integro-differential equation (6.23) for φ says that, for all $v > 0$,

$$\varphi'(v) = \frac{\lambda}{c}\varphi(v) - \frac{\lambda}{c} \int_0^v \varphi(v - x)f_X(x)dx,$$

and integrating this from 0 to u gives

$$\int_0^u \varphi'(v)dv = \frac{\lambda}{c} \int_0^u \varphi(v)dv - \frac{\lambda}{c} \int_0^u \int_0^v \varphi(v - x)f_X(x)dx\,dv.$$

The left-hand side is $\varphi(u) - \varphi(0)$, so that

$$\varphi(u) = \varphi(0) + \frac{\lambda}{c} \int_0^u \varphi(v)dv - \frac{\lambda}{c} \int_0^u \int_0^v \varphi(v - x)f_X(x)dx\,dv. \qquad (6.27)$$

We next concentrate on the inner integral in the double integral term. Using $f_X(x) = -\frac{d}{dx}(1 - F_X(x))$, $F_X(0) = 0$ and integration by parts, we get

$$\int_0^v \varphi(v - x)f_X(x)dx$$

$$= [-\varphi(v - x)(1 - F_X(x))]_{x=0}^v - \int_0^v \varphi'(v - x)(1 - F_X(x))dx$$

$$= -\varphi(0)(1 - F_X(v)) + \varphi(v) - \int_0^v \varphi'(v - x)(1 - F_X(x))dx.$$

The double integral in (6.27) is then as follows:

$$\int_0^u \int_0^v \varphi(v - x)f_X(x)dx\,dv = -\varphi(0) \int_0^u (1 - F_X(v))dv + \int_0^u \varphi(v)dv$$

$$- \int_0^u \int_0^v \varphi'(v - x)(1 - F_X(x))dx\,dv.$$

$$(6.28)$$

We now interchange the order of integration in the third term on the right-hand side of (6.28) to see that

$$
\int_0^u \int_0^v \varphi'(v - x)(1 - F_X(x)) \, dx \, dv
$$

$$
= \int_0^u (1 - F_X(x)) \int_x^u \varphi'(v - x) \, dv \, dx
$$

$$
= \int_0^u (1 - F_X(x))(\varphi(u - x) - \varphi(0)) \, dx
$$

$$
= \int_0^u \varphi(u - x)(1 - F_X(x)) \, dx - \varphi(0) \int_0^u (1 - F_X(x)) \, dx.
$$

Insert this into (6.28) to see that

$$
\int_0^u \int_0^v \varphi(v - x) f_X(x) \, dx \, dv
$$

$$
= -\varphi(0) \int_0^u (1 - F_X(x)) \, dx + \int_0^u \varphi(v) \, dv
$$

$$
- \int_0^u \varphi(u - x)(1 - F_X(x)) \, dx + \varphi(0) \int_0^u (1 - F_X(x)) \, dx
$$

$$
= \int_0^u \varphi(v) \, dv - \int_0^u \varphi(u - x)(1 - F_X(x)) \, dx.
$$

By substituting this into (6.27), it is easy to check that

$$
\varphi(u) = \varphi(0) + \frac{\lambda}{c} \int_0^u \varphi(u - x)(1 - F_X(x)) \, dx,
$$

and the lemma is proved. $\qquad\square$

The above result gives rise to simple expressions for $\varphi(0)$ and $\psi(0)$ in the following lemma. These simple expressions show that $\varphi(0)$ and $\psi(0)$ depend on the claim-size distribution only through its mean.

Lemma 6.10 *In the classical risk model with Poisson rate λ, premium rate c, mean claim size μ and positive relative safety loading θ, the survival and ruin probabilities with zero initial capital are, respectively, given by*

$$
\varphi(0) = \frac{\theta}{1 + \theta} = 1 - \frac{\lambda\mu}{c} \quad and \quad \psi(0) = \frac{1}{1 + \theta} = \frac{\lambda\mu}{c}.
$$

Proof We aim to let $u \to \infty$ in the integral equation (6.26) for φ, so we first consider $\varphi(u)$ as $u \to \infty$. We know that $\varphi(u)$ is bounded (because it is a probability), that it is non-decreasing in u (because $\psi(u)$ is non-increasing in u; see (6.7)), and hence $\lim_{u \to \infty} \varphi(u)$ exists and is finite.

Intuitively, we expect $\lim_{u \to \infty} \varphi(u)$ to be 1, and we now indicate how this may be shown. From the definition of $\varphi(u)$ in (6.8), we have

$$\varphi(u) = \Pr(u + ct - S(t) \geq 0 \text{ for all } t \geq 0)$$
$$= \Pr(S(t) - ct \leq u \text{ for all } t \geq 0)$$
$$= \Pr(L \leq u),$$

where $L = \sup_{t \geq 0} (S(t) - ct)$ is the *maximum aggregate loss*.

We now show that L is a *proper* random variable, that is we now show that L is finite with probability 1. To do this, we consider the sample paths of the process $\{S(t) - ct : t \geq 0\}$, and we want to show that the supremum of this process, L, is finite with probability 1. We have $\mathbb{E}[S(t) - ct] = (\lambda\mu - c)t$, and we know that this is negative because of positive safety loading, so we might expect that eventually the sample paths of $\{S(t) - ct\}$ are indeed negative with probability 1. This is indeed so, as can be shown using deeper results, for example using ths. I.2.3 and II.5.1 in Gut (1988). The sample paths of $\{S(t) - ct : t \geq 0\}$ decrease linearly between claims and have upward jumps at claim times, so that, with probability 1, $L = \sup_{t \geq 0} (S(t) - ct)$ is finite, which means that L is a proper random variable. Therefore we have

$$\varphi(u) = \Pr(L \leq u) \to 1 \text{ as } u \to \infty.$$

We now let u tend to infinity in (6.26), which, using the Monotone Convergence Theorem (see, for example, th. 1.26 in Rudin (1986)) leads to

$$1 = \varphi(0) + \frac{\lambda}{c} \int_0^\infty (1 - F_X(x))dx. \tag{6.29}$$

We can write

$$\int_0^\infty (1 - F_X(x))dx = \int_0^\infty \int_{(x,\infty)} F_X(dt)dx,$$

and, interchanging the order of integration, we find that this is

$$\int_{(0,\infty)} \int_0^t dx F_X(dt) = \int_{(0,\infty)} t F_X(dt) = \mu.$$

So we have shown that

$$\int_0^\infty (1 - F_X(x))dx = \mu, \tag{6.30}$$

a very useful result, and one that we will use again in this chapter. Hence (6.29) becomes

$$1 = \varphi(0) + \frac{\lambda}{c}\mu,$$

and we obtain

$$\varphi(0) = 1 - \frac{\lambda\mu}{c} = \frac{\theta}{1+\theta}.$$

Using $\psi(u) = 1 - \varphi(u)$, we see that $\psi(0) = (1+\theta)^{-1}$, as given in the lemma.   □

Lemmas 6.9 and 6.10 are used to obtain an integral equation for $\psi(u)$ in the following lemma.

Lemma 6.11   *In the classical risk model with Poisson rate λ, premium rate c, claim-size distribution function F_X and positive relative safety loading, the ruin probability $\psi(u)$ satisfies*

$$\psi(u) = \frac{\lambda}{c}\int_u^\infty (1 - F_X(x))dx + \frac{\lambda}{c}\int_0^u \psi(u-x)(1 - F_X(x))dx. \qquad (6.31)$$

Proof   Starting with the integral equation (6.26) for φ, we put $\varphi(u) = 1 - \psi(u)$ and get

$$1 - \psi(u) = 1 - \psi(0) + \frac{\lambda}{c}\int_0^u (1 - \psi(u-x))(1 - F_X(x))dx.$$

By the expression for $\psi(0)$ in Lemma 6.10, and rearranging, this becomes

$$\psi(u) = \frac{\lambda\mu}{c} - \frac{\lambda}{c}\int_0^u (1 - F_X(x))dx + \frac{\lambda}{c}\int_0^u \psi(u-x)(1 - F_X(x))dx$$

$$= \frac{\lambda}{c}\int_u^\infty (1 - F_X(x))dx + \frac{\lambda}{c}\int_0^u \psi(u-x)(1 - F_X(x))dx,$$

where we have used again the fact that $\mu = \int_0^\infty (1 - F_X(x))dx$ (see (6.30)), and the lemma is proved.   □

The equations in this section form the starting point for explicit and asymptotic results for $\varphi(u)$ and $\psi(u)$ in §6.4 and §6.5. They also may be used to give explicit solutions for $\varphi(u)$ and $\psi(u)$ in special cases, and the next example illustrates the use of the integro-differential equations to obtain $\psi(u)$ when the claims are exponentially distributed.

Example 6.12   In a classical risk model, suppose the claims are exponentially distributed with mean μ. Then (6.24) becomes

$$\psi'(u) = \frac{\lambda}{c}\psi(u) - \frac{\lambda}{c}e^{-u/\mu} - \frac{\lambda}{c\mu}\int_0^u \psi(u-x)e^{-x/\mu}\,dx.$$

We make the substitution $z = u - x$ in the integral to get

$$\psi'(u) = \frac{\lambda}{c}\psi(u) - \frac{\lambda}{c}e^{-u/\mu} - \frac{\lambda}{c\mu}e^{-u/\mu}\int_0^u \psi(z)e^{z/\mu}\,dz. \qquad (6.32)$$

The right-hand side of (6.32) is a differentiable function of u, and so, using similar techniques as in the proof of Lemma 6.8, we find that

$$\psi''(u) = \frac{\lambda}{c}\psi'(u) + \frac{\lambda}{c\mu}e^{-u/\mu} + \frac{\lambda}{c\mu^2}e^{-u/\mu}\int_0^u \psi(z)e^{z/\mu}\,dz$$
$$- \frac{\lambda}{c\mu}e^{-u/\mu}\psi(u)e^{u/\mu}$$
$$= \frac{\lambda}{c}\psi'(u) - \frac{1}{\mu}\left(\frac{\lambda}{c}\psi(u) - \frac{\lambda}{c}e^{-u/\mu} - \frac{\lambda}{c\mu}e^{-u/\mu}\int_0^u \psi(z)e^{z/\mu}\,dz\right)$$
$$= \left(\frac{\lambda}{c} - \frac{1}{\mu}\right)\psi'(u),$$

where (6.32) has been used for the last line. From Example 6.2 we have

$$\frac{\lambda}{c} - \frac{1}{\mu} = -\frac{\theta}{(1+\theta)\mu} = -R,$$

so that

$$\frac{\psi''(u)}{\psi'(u)} = -R.$$

Integrating, we find that

$$\log(\psi'(u)) = -Ru + c_1,$$

where here and below we use c_1, c_2, c_3, c_4 to denote constants. The preceding expression means that

$$\psi'(u) = c_2 e^{-Ru}.$$

Another integration yields

$$\psi(u) = c_3 e^{-Ru} + c_4.$$

We know that $\psi(0) = 1/(1+\theta)$ and $\lim_{u\to\infty}\psi(u) = 1 - \lim_{u\to\infty}\varphi(u) = 0$ (see Lemma 6.10 and its proof). These conditions give $c_3 = 1/(1+\theta)$ and $c_4 = 0$, and so, for exponentially distributed claims,

$$\psi(u) = \frac{1}{1+\theta}e^{-Ru} = \frac{1}{1+\theta}e^{-\theta u/((1+\theta)\mu)}.$$

Note that, because $\theta > 0$, the above expression for $\psi(u)$ is indeed smaller than the Lundberg upper bound e^{-Ru} (see Theorem 6.1).

6.4 Compound geometric representations for $\psi(u)$ and $\varphi(u)$: the ruin probability and the survival probability

In this section, we obtain representations of the ruin probability and survival probability as the tail and distribution function, respectively, of a particular compound geometric distribution. We give here a derivation of the results for the ruin and survival probabilities via Laplace transforms.

In order to state the compound geometric results, we first define the necessary quantities and notation. Let

$$f_I(x) = \frac{1 - F_X(x)}{\mu}, \quad x \geq 0. \tag{6.33}$$

Check that $f_I(x) \geq 0$ for all x, and that $\int_0^\infty f_I(x)dx = 1$ (use (6.30)), so that f_I is a probability density function. The probability distribution with density f_I is called the *equilibrium distribution* associated with F_X. It has distribution function

$$F_I(x) = \int_0^x \frac{(1 - F_X(y))}{\mu} \, dy. \tag{6.34}$$

Equilibrium distributions are often used in applied probability, and the claim-size equilibrium distribution will be useful in both this section and the next.

We need several Laplace transform results, which we give below. Recall that the Laplace transform of a function $h(x)$, $x \geq 0$, is the function

$$\tilde{h}(s) = \int_0^\infty e^{-sx} h(x)dx,$$

for real s-values (where the integral is defined), see chap. XIII in Feller (1971). In the following, where we write an equation between Laplace transforms, we tacitly understand that the relationship holds for those s values for which both sides of the equation are finite.

(LT1) Let X be a non-negative random variable with probability density function $f(x)$ and distribution function $F(x)$. The Laplace transform of the tail function $1 - F(x)$ is

$$(1 - F)\tilde{}(s) = \frac{(1 - \tilde{f}(s))}{s}, \quad s > 0,$$

(see eq. (2.7) in chap. XIII of Feller (1971)).

(LT2) If $h(x) = \int_0^x f(x - y)g(y)dy$, and f and g are integrable, then

$$\tilde{h}(s) = \tilde{f}(s)\tilde{g}(s), \tag{6.35}$$

(see sect. 2 in chap. XIII of Feller (1971)). We need to use this when g is a positive bounded function and f is a probability density function, and in this case it is easy to verify that the Laplace transforms are finite for $s > 0$ and that (6.35) is satisfied.

(LT3) If $h(x) = \alpha f(x) + \beta g(x)$, where f and g are functions and α and β are constants, then the Laplace transform of h is $\tilde{h}(s) = \alpha \tilde{f}(s) + \beta \tilde{g}(s)$ (immediate from the definition of the Laplace transform).

(LT4) If Y is a random variable with probability density function f_Y and moment generating function M_Y, then $\tilde{f}_Y(s) = \mathbb{E}[e^{-sY}] = M_Y(-s)$ (immediate from the definitions of the Laplace transform and the moment generating function).

(LT5) A continuous function h is uniquely determined by its Laplace transform $\tilde{h}(s)$ in some interval $a < s < \infty$ (see the corollary to th. 1.4 in sect. XIII.1 of Feller (1971)).

We now state the main theorem of this section. The expression for $\psi(u)$ in (6.36) below is called the *Pollaczek–Khintchine* formula. We discuss the interpretation of the theorem in terms of compound geometric distributions after the proof. However, you might like to see whether you can spot a compound geometric in (6.37) below before you look at the proof or discussion.

Theorem 6.13 *In the classical risk model with Poisson rate λ, premium rate c, mean claim size μ and positive relative safety loading, the ruin probability satisfies*

$$\psi(u) = \sum_{n=1}^{\infty} \left(1 - \frac{\lambda\mu}{c}\right)\left(\frac{\lambda\mu}{c}\right)^n (1 - F_I^{\star n}(u)), \qquad (6.36)$$

where F_I is as in (6.34) and $F_I^{\star n}$ denotes the n-fold convolution of F_I (see §3.2.1). The survival probability is given by

$$\varphi(u) = \sum_{n=0}^{\infty} \left(1 - \frac{\lambda\mu}{c}\right)\left(\frac{\lambda\mu}{c}\right)^n F_I^{\star n}(u). \qquad (6.37)$$

Proof We start with the integral equation (6.31) for $\psi(u)$, which may be written

$$\psi(u) = \frac{\lambda\mu}{c}(1 - F_I(u)) + \frac{\lambda\mu}{c}\int_0^u \psi(u - x)f_I(x)dx, \qquad (6.38)$$

where f_I is as in (6.33). Note that $\psi(u)$ is a probability, and hence bounded, so it certainly has a Laplace transform for all $s > 0$. Taking the Laplace transform of (6.38) we obtain, using (LT1), (LT2) and (LT3),

$$\tilde{\psi}(s) = \frac{\lambda\mu}{c}(1 - F_I)\tilde{\ }(s) + \frac{\lambda\mu}{c}\tilde{\psi}(s)\tilde{f}_I(s)$$

$$= \frac{\lambda\mu}{c}\frac{(1 - \tilde{f}_I(s))}{s} + \frac{\lambda\mu}{c}\tilde{\psi}(s)\tilde{f}_I(s).$$

Rearranging the above, we find that

$$\tilde{\psi}(s) = \frac{1}{s}\frac{\lambda\mu}{c}\frac{(1 - \tilde{f}_I(s))}{\left(1 - \frac{\lambda\mu}{c}\tilde{f}_I(s)\right)}. \tag{6.39}$$

We now aim to link this to results in §3.4.3 for the moment generating function of a compound geometric distribution. By (LT4), we know that $\tilde{f}_I(s)$ is $M_I(-s)$, where M_I is the moment generating function of the claim-size equilibrium distribution. From (3.24) we know that

$$\frac{\left(1 - \frac{\lambda\mu}{c}\right)M_I(r)}{\left(1 - \frac{\lambda\mu}{c}M_I(r)\right)} \tag{6.40}$$

is the moment generating function $M_{G_1}(r)$ of a compound geometric distribution function G_1 with an $\widetilde{nb}(1, 1 - \lambda\mu/c)$-distributed counting random variable and a step random variable distributed as the claim-size equilibrium distribution. Recall that if $M \sim \widetilde{nb}(1, p)$ then

$$\Pr(M = n) = (1 - p)^{n-1}p, \quad n = 1, 2, \ldots,$$

so that $\Pr(M = 0) = 0$. Thus, from (3.7), this compound distribution function G_1 has a density g_1, say. By (LT4) and (6.40), the Laplace transform of g_1 is

$$\tilde{g}_1(s) = M_{G_1}(-s) = \frac{\left(1 - \frac{\lambda\mu}{c}\right)M_I(-s)}{\left(1 - \frac{\lambda\mu}{c}M_I(-s)\right)} = \frac{\left(1 - \frac{\lambda\mu}{c}\right)\tilde{f}_I(s)}{\left(1 - \frac{\lambda\mu}{c}\tilde{f}_I(s)\right)},$$

and so

$$1 - \tilde{g}_1(s) = \frac{1 - \tilde{f}_I(s)}{1 - \frac{\lambda\mu}{c}\tilde{f}_I(s)}.$$

This means that, from (6.39), we have

$$\tilde{\psi}(s) = \frac{\lambda\mu}{c}\frac{(1 - \tilde{g}_1(s))}{s}.$$

Using (LT1), (LT3) and (LT5) (using that ψ is continuous, see Lemma 6.7) we deduce that

$$\psi(u) = \frac{\lambda\mu}{c}(1 - G_1(u)). \tag{6.41}$$

As a check, we note that the right-hand side gives the correct values when $u = 0$ and when $u \to \infty$. Substituting the convolution series expression (3.5) for the compound distribution function into (6.41), we get

$$\psi(u) = \frac{\lambda\mu}{c}\left(1 - \sum_{n=1}^{\infty}\left(1 - \frac{\lambda\mu}{c}\right)\left(\frac{\lambda\mu}{c}\right)^{n-1} F_I^{\star n}(u)\right)$$

$$= \frac{\lambda\mu}{c}\left(\sum_{n=1}^{\infty}\left(1 - \frac{\lambda\mu}{c}\right)\left(\frac{\lambda\mu}{c}\right)^{n-1} - \sum_{n=1}^{\infty}\left(1 - \frac{\lambda\mu}{c}\right)\left(\frac{\lambda\mu}{c}\right)^{n-1} F_I^{\star n}(u)\right)$$

$$= \sum_{n=1}^{\infty}\left(1 - \frac{\lambda\mu}{c}\right)\left(\frac{\lambda\mu}{c}\right)^{n}(1 - F_I^{\star n}(u)),$$

and (6.36) is proved.

For $\varphi(u)$ ($= 1 - \psi(u)$), from (6.41) we see that

$$\varphi(u) = 1 - \frac{\lambda\mu}{c}(1 - G_1(u))$$

$$= 1 - \frac{\lambda\mu}{c} + \frac{\lambda\mu}{c}\sum_{n=1}^{\infty}\left(1 - \frac{\lambda\mu}{c}\right)\left(\frac{\lambda\mu}{c}\right)^{n-1} F_I^{\star n}(u)$$

$$= 1 - \frac{\lambda\mu}{c} + \sum_{n=1}^{\infty}\left(1 - \frac{\lambda\mu}{c}\right)\left(\frac{\lambda\mu}{c}\right)^{n} F_I^{\star n}(u)$$

$$= \sum_{n=0}^{\infty}\left(1 - \frac{\lambda\mu}{c}\right)\left(\frac{\lambda\mu}{c}\right)^{n} F_I^{\star n}(u),$$

where we have used $F^{\star 0}(u) = 1$ for all $u \geq 0$ (see Definition 3.6), and (6.37) is proved.   □

By the convolution series expression in (3.5) the right-hand side of (6.37) is the distribution function G of a random sum $S = Z_1 + \cdots + Z_N$, where the counting random variable N has a geometric distribution with

$$\Pr(N = n) = \left(1 - \frac{\lambda\mu}{c}\right)\left(\frac{\lambda\mu}{c}\right)^{n}, \quad n = 0, 1, 2, \ldots, \tag{6.42}$$

so that $N \sim \mathrm{nb}(1, 1 - \lambda\mu/c)$ (see §3.4.3), and the step random variables, that is the Z_i, are distributed as the equilibrium distribution associated with the claim sizes. So (6.37) says that $\varphi(u) = G(u)$, and thus the *survival probability* is the same as the *compound geometric distribution function* G.

Note that (6.36) can be written

$$\psi(u) = \sum_{n=0}^{\infty}\left(1 - \frac{\lambda\mu}{c}\right)\left(\frac{\lambda\mu}{c}\right)^{n}(1 - F_I^{\star n}(u)), \tag{6.43}$$

where we have added the $n = 0$ term. This works because in the $n = 0$ term, the factor $1 - F_I^{\star 0}(u)$ is zero (see the definition of $F^{\star 0}$ for a distribution function F in Definition 3.6). Rearranging the right-hand side of (6.43) gives

$$\psi(u) = \sum_{n=0}^{\infty}\left(1 - \frac{\lambda\mu}{c}\right)\left(\frac{\lambda\mu}{c}\right)^n - \sum_{n=0}^{\infty}\left(1 - \frac{\lambda\mu}{c}\right)\left(\frac{\lambda\mu}{c}\right)^n F_I^{\star n}(u)$$

$$= 1 - \sum_{n=0}^{\infty}\left(1 - \frac{\lambda\mu}{c}\right)\left(\frac{\lambda\mu}{c}\right)^n F_I^{\star n}(u), \tag{6.44}$$

and this is just $\psi(u) = 1 - G(u)$, where G is as above. This is as expected from $\psi(u) = 1 - \varphi(u) = 1 - G(u)$. Thus the *ruin probability* is the *tail of the compound geometric distribution G*. We obtain the following theorem.

Theorem 6.14 *In the classical risk model with Poisson rate λ, premium rate c, mean claim size μ and positive relative safety loading, the survival probability $\varphi(u)$ and the ruin probability $\psi(u)$ are given by*

$$\varphi(u) = G(u) \quad \text{and} \quad \psi(u) = 1 - G(u), \quad u \geq 0,$$

where G is the compound geometric distribution function

$$G(u) = \sum_{n=0}^{\infty}\left(1 - \frac{\lambda\mu}{c}\right)\left(\frac{\lambda\mu}{c}\right)^n F_I^{\star n}(u),$$

and where F_I is the claim-size equilibrium distribution function in (6.34).

The compound geometric representations of $\varphi(u)$ and $\psi(u)$ are useful because compound geometric results and techniques from Chapter 3 are now available for application to survival and ruin probabilities.

In Example 6.15, we illustrate how the compound geometric representation works when the claims are exponentially distributed. Recall that in Example 6.12, we used the integro-differential equation for $\psi(u)$ to find an exact expression for the ruin probability for exponential claims. The compound geometric approach provides an alternative way to find $\psi(u)$ in this case.

Example 6.15 For exponential claims with mean μ, the equilibrium density and distribution functions are

$$f_I(x) = \frac{1}{\mu}e^{-x/\mu} \quad \text{and} \quad F_I(x) = 1 - e^{-x/\mu}, \quad x > 0,$$

so that the equilibrium distribution is also exponentially distributed with mean μ. We note that the distribution of the counting random variable N in (6.42) can be written in terms of the relative safety loading θ (using $\lambda\mu/c = 1/(1+\theta)$) to give

$$\Pr(N = n) = \frac{\theta}{1+\theta}\left(\frac{1}{1+\theta}\right)^n, \quad n = 0, 1, 2, \ldots,$$

and so

$$N \sim \text{nb}\left(1, \frac{\theta}{1+\theta}\right).$$

We obtained an explicit form for a compound distribution function when the counting random variable is nb(1, p) and the steps are exponentially distributed in Example 3.18. Therefore, replacing p and q in Example 3.18 by $\theta/(1 + \theta)$ and $1/(1 + \theta)$, respectively, we find from (3.26) that

$$\varphi(u) = 1 - \frac{1}{1+\theta} e^{-\theta u/((1+\theta)\mu)}, \quad u \geq 0.$$

Thus we obtain

$$\psi(u) = 1 - \varphi(u) = \frac{1}{1+\theta} e^{-\theta u/((1+\theta)\mu)} = \frac{1}{1+\theta} e^{-Ru}, \quad u \geq 0,$$

where R is the adjustment coefficient (see Example 6.2). This expression for $\psi(u)$ is the same as that obtained in Example 6.12.

6.5 Asymptotics for the probability of ruin

We next turn our attention to the behaviour of the ruin probability $\psi(u)$ as u becomes large in a classical risk model with positive relative safety loading. We consider the "small claims" case, so that the adjustment coefficient exists. The main result of this section, Theorem 6.20, is a classical result in ruin theory. The proof requires results from renewal theory, which we will quote without proof (see, for example, chap. XI in Feller (1971) and chap. V in Asmussen (2003)). We summarise here the results that we need.

A *renewal-type equation* for an unknown quantity $Z(u)$, defined for $u \geq 0$, is of the form

$$Z(u) = z(u) + \int_0^u Z(u - x) f(x) dx, \qquad (6.45)$$

where $z(u)$ is a known function and $f(x)$ is a known *proper* probability density function of a positive random variable, where "proper" means that

$$\int_0^\infty f(x) dx = 1.$$

We assume that the functions Z and f are such that the integral is defined. Equations like (6.45) have been much studied, and there are many general results about the solution $Z(u)$.

We note in passing that this sort of equation is similar to the integral equations that we found for φ and ψ in (6.26) and (6.31). These two equations can be put into a form that resembles (6.45), but with $f(x) = (\lambda\mu/c)f_I(x)$ (see Exercise 6.16). This $f(x)$ is not a proper probability density function because

$$\int_0^\infty f(x)dx = \frac{\lambda\mu}{c}\int_0^\infty f_I(x)dx = \frac{\lambda\mu}{c},$$

and $\lambda\mu/c < 1$ because we have positive relative safety loading. This means that (6.26) and (6.31) are *not* strictly renewal-type equations. However, they are known as *defective* renewal-type equations.

Returning now to (proper) renewal-type equations as in (6.45), we quote below (in Theorem 6.19) the main renewal theory result that we need about the solution $Z(u)$. First we define an integrability property that appears in Theorem 6.19.

Definition 6.16 Let z be a non-negative function on $[0, \infty)$. For $h \geq 0$ and k a non-negative integer, let $\overline{m}_k(h)$ and $\underline{m}_k(h)$ be defined by

$$\overline{m}_k(h) = \sup\{z(y) : kh < y \leq (k+1)h\},$$
$$\underline{m}_k(h) = \inf\{z(y) : kh < y \leq (k+1)h\}.$$

Define the upper and lower sums $\overline{\sigma}(h)$ and $\underline{\sigma}(h)$ by

$$\overline{\sigma}(h) = h\sum_{k=0}^\infty \overline{m}_k(h),$$

$$\underline{\sigma}(h) = h\sum_{k=0}^\infty \underline{m}_k(h).$$

Then the function z is *directly Riemann integrable* if

$$\overline{\sigma}(h) < \infty \text{ and } \underline{\sigma}(h) < \infty \text{ for all } h > 0,$$

and

$$\overline{\sigma}(h) - \underline{\sigma}(h) \to 0 \text{ as } h \downarrow 0.$$

Example 6.17 Consider the function $z(x) = ae^{-bx}$ for $x \geq 0$, where a and b are positive constants. We show that z is directly Riemann integrable. First note that z is decreasing, so that

$$\overline{m}_k(h) = ae^{-bkh} \qquad \text{and} \qquad \underline{m}_k(h) = ae^{-b(k+1)h}.$$

The upper and lower sums are

$$\overline{\sigma}(h) = ah\sum_{k=0}^\infty e^{-bkh} = \frac{ah}{1 - e^{-bh}}.$$

and

$$\underline{\sigma}(h) = ah \sum_{k=0}^{\infty} e^{-b(k+1)h} = \frac{ahe^{-bh}}{1 - e^{-bh}},$$

which are both finite for all $h > 0$. Further we have

$$\overline{\sigma}(h) - \underline{\sigma}(h) = \frac{ah}{1 - e^{-bh}} - \frac{ahe^{-bh}}{1 - e^{-bh}}$$

$$= \frac{ah}{1 - e^{-bh}} \left(1 - e^{-bh}\right)$$

$$= ah,$$

and this tends to zero as $h \downarrow 0$. Hence the function $z(x) = ae^{-bx}$ is directly Riemann integrable.

Remark 6.18   From Asmussen (2003), a function z is directly Riemann integrable if it is bounded and continuous and if there exists a directly Riemann integrable function z^* such that $z \leq z^*$.

For more on directly Riemann integrable functions, see sect. 1 of chap. XI in Feller (1971), sect. V.4 in Asmussen (2003) and sect. 6.14 of Rolski *et al.* (1999).

We are now ready to quote the relevant renewal theorem, called the *Key Renewal Theorem* (see th. V.4.3 in Asmussen (2003)).

Theorem 6.19   *Suppose that $Z(u)$ satisfies the renewal-type equation (6.45). If $z \geq 0$ and is directly Riemann integrable, then*

$$Z(u) \rightarrow \frac{\int_0^{\infty} z(x)dx}{\int_0^{\infty} xf(x)dx} \quad \text{as } u \rightarrow \infty,$$

where the right-hand side is interpreted as zero if $\int_0^{\infty} xf(x)dx = \infty$.

Theorem 6.20 presents the Cramér–Lundberg asymptotic result for $\psi(u)$. We consider claim sizes that satisfy the conditions in Lemma 6.5, so that heavy-tailed distributions are ruled out and the adjustment coefficient R exists, satisfying $R > 0$ and

$$M_X(R) - 1 = cR/\lambda. \tag{6.46}$$

Theorem 6.20   *In the classical risk model with positive relative safety loading θ, suppose that the claim-size moment generating function $M_X(r)$ satisfies conditions (i) and (ii) of Lemma 6.5, and let R be the adjustment coefficient. Then*

$$e^{Ru}\psi(u) \rightarrow A \quad \text{as } u \rightarrow \infty, \tag{6.47}$$

where

$$A = \frac{\theta}{R \int_0^\infty x e^{Rx} f_I(x) dx} = \frac{\mu\theta}{M'_X(R) - \mu(1 + \theta)}, \tag{6.48}$$

and where $f_I(x)$ is as in (6.33).

Proof From the integral equation for $\psi(u)$ in Lemma 6.11, it follows that

$$\psi(u) = \frac{\lambda\mu}{c} \int_u^\infty f_I(x) dx + \frac{\lambda\mu}{c} \int_0^u \psi(u - x) f_I(x) dx. \tag{6.49}$$

We aim to use Theorem 6.19 to find an asymptotic result for $e^{Ru}\psi(u)$, and the first step is to find a renewal-type equation of the form (6.45) for $e^{Ru}\psi(u)$. To this end, multiply (6.49) by e^{Ru} to obtain

$$e^{Ru}\psi(u) = \frac{\lambda\mu}{c} e^{Ru} \int_u^\infty f_I(x) dx + \frac{\lambda\mu}{c} \int_0^u e^{R(u-x)}\psi(u - x) e^{Rx} f_I(x) dx.$$

This is of the form in (6.45) with

$$Z(u) = e^{Ru}\psi(u), \quad u \geq 0$$
$$z(u) = \frac{\lambda\mu}{c} e^{Ru} \int_u^\infty f_I(x) dx, \quad u \geq 0$$
$$f(x) = \frac{\lambda\mu}{c} e^{Rx} f_I(x), \quad x \geq 0. \tag{6.50}$$

It is clear that $z \geq 0$, so, in order to apply Theorem 6.19, we need to check two things: (i) that f is a proper probability density function, and (ii) that z is directly Riemann integrable.

For (i), it is immediate that $f(x) \geq 0$, and, using integration by parts, we have

$$\int_0^\infty f(x) dx = \frac{\lambda\mu}{c} \int_0^\infty e^{Rx} \frac{(1 - F_X(x))}{\mu} dx$$
$$= \frac{\lambda}{c} \left(\left[\frac{e^{Rx}}{R} (1 - F_X(x)) \right]_0^\infty + \frac{1}{R} \int_0^\infty e^{Rx} f_X(x) dx \right)$$
$$= \frac{\lambda}{c} \left(\left[\frac{e^{Rx}}{R} (1 - F_X(x)) \right]_0^\infty + \frac{1}{R} M_X(R) \right). \tag{6.51}$$

In order to evaluate the square brackets, we need to consider

$$\lim_{x \to \infty} e^{Rx}(1 - F_X(x)).$$

By the bound on the tail of the claim-size distribution given in (6.19), we can choose r_0 in (R, r_∞), where r_∞ is as in Lemma 6.5, such that

$$1 - F_X(x) \le Ke^{-r_0 x}, \tag{6.52}$$

where $K = M_X(r_0) < \infty$. Hence we have

$$\lim_{x \to \infty} e^{Rx}(1 - F_X(x)) \le \lim_{x \to \infty} Ke^{-(r_0 - R)x} = 0. \tag{6.53}$$

So, from (6.51) and (6.46), we find

$$\int_0^\infty f(x)dx = \frac{\lambda}{cR}(M_X(R) - 1) = 1.$$

Thus $f(x)$ is a proper probability density function.

For (ii), we show that z satisfies the conditions of Remark 6.18. We first note that z is non-negative and continous (because it is the product of two continuous functions). By (6.52), we have, for all $u \ge 0$,

$$
\begin{aligned}
z(u) &= \frac{\lambda\mu}{c}e^{Ru}\int_u^\infty \frac{(1 - F_X(x))}{\mu}\,dx \\
&\le \frac{\lambda K}{c}e^{Ru}\int_u^\infty e^{-r_0 x}\,dx \\
&= \frac{\lambda K}{cr_0}e^{-(r_0 - R)u},
\end{aligned}
$$

where $0 < R < r_0$. This shows that z is bounded above by $\lambda K/(cr_0)$. Let $z^*(u) = ae^{-bu}$, where $a = \lambda K/(cr_0)$ and $b = r_0 - R$. Then by Example 6.17 the function z^* is directly Riemann integrable. Hence by Remark 6.18 the function z is directly Riemann integrable, and (ii) is satisfied.

Thus we may apply Theorem 6.19 to conclude that

$$e^{Ru}\psi(u) \to \frac{\int_0^\infty z(x)dx}{\int_0^\infty xf(x)dx} \quad \text{as } u \to \infty. \tag{6.54}$$

To evaluate the limit, we see that

$$\int_0^\infty z(x)dx = \frac{\lambda\mu}{c}\int_0^\infty e^{Rx}\int_x^\infty f_I(y)dy\,dx.$$

Interchanging the order of integration and using the definition of $f(x)$ in (6.50) gives

$$\int_0^\infty z(x)dx = \frac{\lambda\mu}{c}\int_0^\infty f_I(y)\int_0^y e^{Rx}\,dx\,dy$$

$$= \frac{\lambda\mu}{c}\int_0^\infty f_I(y)\frac{(e^{Ry}-1)}{R}\,dy$$

$$= \frac{1}{R}\int_0^\infty \frac{\lambda\mu}{c}e^{Ry}f_I(y)dy - \frac{\lambda\mu}{cR}$$

$$= \frac{1}{R}\int_0^\infty f(y)dy - \frac{\lambda\mu}{cR}$$

$$= \frac{1}{R}\left(1 - \frac{\lambda\mu}{c}\right). \tag{6.55}$$

Then the limit in (6.54) is given by

$$\frac{\frac{1}{R}\left(1-\frac{\lambda\mu}{c}\right)}{\frac{\lambda\mu}{c}\int_0^\infty xe^{Rx}f_I(x)dx} = \frac{\theta}{R\int_0^\infty xe^{Rx}f_I(x)dx}, \tag{6.56}$$

where the equality follows on recalling that $\lambda\mu/c = 1/(1+\theta)$. This means that (6.47) and the first expression for A in (6.48) are proved.

The final step is to show that the limit in (6.47) is equal to the right-hand expression in (6.48). We observe that, by (6.52),

$$\int_0^\infty xe^{Rx}f_I(x)dx \le \frac{K}{\mu}\int_0^\infty xe^{-(r_0-R)x}\,dx,$$

and this is finite because $R < r_0$. Integrating by parts, we see that the denominator in (6.54) is

$$\int_0^\infty xf(x)dx = \frac{\lambda}{c}\int_0^\infty xe^{Rx}(1-F_X(x))dx$$

$$= \frac{\lambda}{c}\left\{\left[\left(\frac{x}{R}-\frac{1}{R^2}\right)e^{Rx}(1-F_X(x))\right]_0^\infty\right.$$

$$\left. + \int_0^\infty \left(\frac{x}{R}-\frac{1}{R^2}\right)e^{Rx}f_X(x)dx\right\}$$

$$= \frac{\lambda}{c}\left\{\frac{1}{R^2} + \frac{1}{R}\int_0^\infty xe^{Rx}f_X(x)dx\right.$$

$$\left. - \frac{1}{R^2}\int_0^\infty e^{Rx}f_X(x)dx\right\},$$

where we have used (6.53) and the fact that $\lim_{x\to\infty} xe^{Rx}(1-F_X(x)) = 0$, which can be seen by applying (6.52). Thus we have, by (6.18),

$$\int_0^\infty xf(x)dx = \frac{\lambda}{c}\left\{\frac{1}{R^2} + \frac{M_X'(R)}{R} - \frac{M_X(R)}{R^2}\right\}$$

$$= \frac{\lambda}{c}\left\{\frac{M_X'(R)}{R} - \frac{1}{R^2}(M_X(R) - 1)\right\}.$$

We know that $M_X(R) - 1 = cR/\lambda$ by (6.46), so that

$$\int_0^\infty xf(x)dx = \frac{\lambda}{c}\left\{\frac{M_X'(R)}{R} - \frac{c}{\lambda R}\right\} = \frac{\lambda}{cR}\left\{M_X'(R) - \frac{c}{\lambda}\right\}. \tag{6.57}$$

Hence the limit is given by

$$\frac{\int_0^\infty z(x)dx}{\int_0^\infty xf(x)dx} = \frac{\frac{1}{R}\left(1 - \frac{\lambda\mu}{c}\right)}{\frac{\lambda}{cR}\left(M_X'(R) - \frac{c}{\lambda}\right)} = \frac{\mu\theta}{M_X'(R) - (1 + \theta)\mu},$$

where the last step is obtained using $\lambda\mu/c = 1/(1 + \theta)$ and simplifying. □

Theorem 6.20 implies the *Cramér–Lundberg approximation*:

$$\psi(u) \approx Ae^{-Ru} \quad \text{for large } u, \tag{6.58}$$

where A is given in (6.48). Recall the the Lundberg bound $\psi(u) \le e^{-Ru}$ for all $u \ge 0$. The Cramér–Lundberg approximation (6.58) shows that the adjustment coefficient is even more important than we have previously thought because, as well as being the exponential decay rate of an upper bound for the ruin probability in the Lundberg inequality, the adjustment coefficient also gives the *correct* asymptotic decay rate. However, note that Theorem 6.20 does not tell us how large u has to be for the approximation to be any good.

In order to calculate the Cramér–Lundberg approximation, we need to find R and A. Both of these quantities depend on the whole claim-size distribution via its moment generating function. They are not simple functions that depend only on a few moments of the claim-size distribution, and thus it can be difficult to evaluate the approximation explicitly, except in certain special cases. In the following example we find A in the case of exponential claims.

Example 6.21 When claims are exponentially distributed with mean μ, we know that $R = \theta/((1 + \theta)\mu)$ (see Example 6.2). Further, since $M_X(r) = (1 - \mu r)^{-1}$, we have

$$M_X'(r) = \frac{\mu}{(1 - \mu r)^2},$$

so that

$$M_X'(R) = \frac{\mu}{\left(1 - \dfrac{\theta}{1 + \theta}\right)^2} = \mu(1 + \theta)^2.$$

Hence, from the second expression for A in (6.48), we obtain

$$A = \frac{\mu\theta}{\mu(1 + \theta)^2 - \mu(1 + \theta)} = \frac{1}{1 + \theta}.$$

Then Theorem 6.20 gives

$$e^{Ru}\psi(u) \to \frac{1}{1 + \theta} \quad \text{as } u \to \infty,$$

and the right-hand side of (6.58) shows that the Cramér–Lundberg approximation for $\psi(u)$ is

$$Ae^{-Ru} = \frac{1}{1 + \theta}e^{-Ru}.$$

From Examples 6.12 and 6.15 we know that

$$\psi(u) = \frac{1}{1 + \theta}e^{-Ru},$$

so that, for exponential claims, the Cramér–Lundberg approximation is the same as the true $\psi(u)$.

6.6 Numerical methods for ruin quantities

In this section we consider the question of calculating the adjustment coefficient R or the probability of ruin $\psi(u)$ for a classical risk model where we know the claim-arrival rate λ and the claim-size distribution F_X, and with known premium accrual rate c (or, equivalently, with known relative safety loading θ). As we have seen in earlier sections of this chapter, for certain claim-size distributions, there are explicit expressions for R and $\psi(u)$ in terms of λ, c (or θ) and the parameters of the claim-size distribution. But what about claim-size distributions for which there are no such easy expressions? In this section, we consider numerical methods which allow for the calculation of (numerical approximations to) R and $\psi(u)$ in these cases.

6.6.1 Numerical calculation of the adjustment coefficient

Recall from (6.10) that the adjustment coefficient R is the unique positive solution of

$$M_X(r) - 1 - (1 + \theta)\mu r = 0, \tag{6.59}$$

where we assume that the conditions of Lemma 6.5 are satisfied, so that we are sure that there is a unique positive solution of (6.59). As in the proof of Lemma 6.5, let

$$h(r) = M_X(r) - 1 - (1 + \theta)\mu r,$$

so that R is the positive solution to $h(r) = 0$. Numerical search methods may be used to find the positive solution to this equation, and we illustrate this in Example 6.22. For details of such methods, see, for example, Conte and de Boor (1980) or Epperson (2007).

Example 6.22 We demonstrate a numerical approach to finding R for exponentially distributed claims with mean μ and positive relative safety loading θ. For the purposes of illustrating the numerical approach, we pretend that we do not know the true expression for R. Instead we give a numerical procedure that starts with the input parameters of the risk model and from these produces a sequence $r_0, r_1, r_2, \ldots$ such that $r_n \to R$ as $n \to \infty$.

In this example, we have

$$h(r) = \frac{1}{1 - \mu r} - 1 - (1 + \theta)\mu r,$$

and this is defined for $r < r_\infty$, where $r_\infty = 1/\mu$. Here r_∞ is as in Lemma 6.5, and Example 6.6(ii) shows that $r_\infty = 1/\mu$ for the exponential distribution. We want to find the root of $h(r) = 0$ in the interval $(0, 1/\mu)$. The Newton–Raphson algorithm is a numerical procedure defined by

$$r_{n+1} = r_n - \frac{h(r_n)}{h'(r_n)}, \quad n = 1, 2, \ldots,$$

where, once we have selected an appropriate r_0, the subsequent r_n are calculated recursively. One way to find r_0 is to use the upper bound on R given in Lemma 6.4, which in this case is

$$R \leq \frac{2\mu\theta}{\mathbb{E}[X_1^2]} = \frac{\theta}{\mu},$$

so that R is in the interval $(0, \theta/\mu)$. We could then carry out a binary search of this interval to find r_0 close to R, and then use the Newton–Raphson algorithm; we illustrate this below.

For a specific example, suppose that $\mu = 1$ and $\theta = 0.2$. We note in passing that (6.59) does not depend on λ (except through θ) so that we do not need to specify a value for λ. Thus we know that R is in $(0, \theta/\mu) = (0, 0.2)$. For this example, we have

$$h(r) = \frac{1}{1 - r} - 1 - 1.2r.$$

We evaluate h at the midpoint of the interval $(0, 0.2)$, and we recall that the shape of $h(r)$ is given in Figure 6.2. We find (using a computer) that

$$h(0.1) = -0.009, \quad \text{which means that } R \text{ is in } (0.1, 0.2).$$

Evaluating h at the midpoint of the interval $(0.1, 0.2)$, and repeating, we obtain

$$h(0.15) = -0.004, \quad \text{which means that } R \text{ is in } (0.15, 0.2);$$
$$h(0.175) = 0.002, \quad \text{which means that } R \text{ is in } (0.15, 0.175).$$

If we take $r_0 = 0.1625$ (the midpoint of the interval $(0.15, 0.175)$), then use the Newton–Raphson algorithm (on the computer), we find that

$$r_1 = 0.1667983,$$
$$r_2 = 0.1666668,$$
$$r_3 = 0.1666667.$$

We stop the algorithm when two successive r_n are within a specified distance from each other. If this distance is 10^{-6}, say, then the algorithm stops with r_3 and the value 0.1666667 is returned as the numerical approximation for R.

The advantage of using exponentially distributed claims for this illustrative example is that we can see how well this procedure performs because we know the true value of R. From Example 6.2, we have $R = \theta/((1 + \theta)\mu)$, so the true value of R is $1/6$. This means that $R = 0.1666667$ to seven decimal places, so r_2 is correct to six decimal places. The above calculations were calculated using the software package **R** (be careful: **R** is not to be confused with the adjustment coefficient R). In Exercise 6.19 you are asked to write your own **R** function and to investigate other choices of r_0. Note especially that an inappropriate choice of r_0 can give rise to r_n that converge to zero, which is another solution of $h(r) = 0$, and not to the adjustment coefficient R.

6.6.2 Numerical calculation of the probability of ruin

In this subsection, we use numerical methods to evaluate the probability of ruin $\psi(u)$. The approach that follows exploits the representation in (6.44) of $\psi(u)$ as the tail of a compound geometric distribution,

$$\psi(u) = 1 - G(u),$$

where $G(u)$ is a compound geometric distribution function for which the counting random variable N has, for $n = 0, 1, 2, \ldots$,

$$\Pr(N = n) = \left(1 - \frac{\lambda\mu}{c}\right)\left(\frac{\lambda\mu}{c}\right)^n = \left(1 - \frac{1}{1 + \theta}\right)\left(\frac{1}{1 + \theta}\right)^n, \quad (6.60)$$

and where the step random variables are iid with distribution function F_I (recall that F_I is the claim-size equilibrium distribution function, defined in (6.34)). In Example 3.22 we saw that numerical calculation of compound geometric distributions can be carried out using either the Panjer recursion algorithm or the FFT algorithm. Hence one or other of these numerical methods provides the core of a numerical method for calculating $\psi(u)$.

Here we describe how to do this using the FFT algorithm for a classical risk model with positive relative safety loading θ, with claim-size distribution function F_X and claim-size mean μ. We illustrate the method for exponentially distributed claims. In order to obtain the numerical approximation to the distribution function G, the first step is to discretise the step distribution, that is to discretise the equilibrium distribution F_I associated with F_X. We will be working with the FFT algorithm, so we choose a discretisation parameter h (a small, positive real number) and a truncation parameter m (a large positive integer that is a power of 2) as we did in §3.5.2. In general, we define a $(1 \times m)$-dimensional array `equil` in R such that

$$\texttt{equil} = (\texttt{equil[1]}, \ \texttt{equil[2]}, \ \dots \ , \ \texttt{equil[m]})$$

where, for k taking the value $k \geq 1$, the entry `equil[k+1]` contains the value of

$$F_I((k + 0.5)h) - F_I((k - 0.5)h) = \int_{(k-0.5)h}^{(k+0.5)h} \frac{1 - F_X(x)}{\mu}\, dx,$$

and `equil[1]` contains the value of

$$\int_0^{0.5h} \frac{1 - F_X(x)}{\mu}\, dx.$$

For exponentially distributed claims with mean μ, we obtain an explicit formula for `equil[k+1]` as follows. We know that the equilibrium distribution associated with this particular claim-size distribution is also an $\text{Exp}(1/\mu)$ distribution (see Example 6.15), and so, for k taking the value $k \geq 1$, we have that `equil[k+1]` contains the value of $e^{-(k-0.5)h/\mu}\left(1 - e^{-h/\mu}\right)$, while `equil[1]` contains the value of $1 - e^{-h/(2\mu)}$. For other choices of claim-size distribution, it may not be possible to evaluate `equil` explicitly in this way, and then the elements of the $(1 \times m)$-dimensional array `equil` may be evaluated numerically. For another approach, see Pitts (2006).

With the $(1 \times m)$-dimensional array `equil` as a truncated discretised approximation for F_I, we now calculate the tail of the compound geometric distribution with counting random variable as in (6.60) and with this truncated discretised distribution in equil as the step distribution. We use this tail as an approximation to $1 - G(u)$ and hence as an approximation to $\psi(u)$. We give the **R** code for achieving this using the FFT algorithm. Assume that the **R** objects m and h already contain the chosen values of the truncation and discretisation parameters m and h, respectively. In the **R** code, we illustrate the method for exponentially distributed claims with $\mu = 1$ and with $\theta = 0.1$:

```
mu = 1
theta = 0.1
grid = 0:(m-1)
equil = exp(-(grid-0.5 )*h/mu)*(1-exp(-h/mu))
equil[1] = 1 - exp(-h/(2*mu))
q = 1/(1+theta)
gfft = (1-q)*(1-q*fft(equil))^(-1)
g = Ro(fft( gfft, inverse-T)/m)
psi = 1-cumsum(g)
```

(see §3.5.2). Let S be a random variable with a compound geometric distribution given by the true G. Then, in the preceding **R** code, the **R** object g is a $(1 \times m)$-dimensional array, where the value g_{k+1} in g[k+1] is a numerical approximation to $\Pr((k - 0.5)h < S \leq (k + 0.5)h)$. The **R** code gives psi as a $(1 \times m)$-dimensional array, where psi[k+1] contains the value of $1 - \sum_{i=1}^{k+1} g_i$, so that psi[k+1] contains a numerical approximation to $\Pr(S > (k + 0.5)h)$, that is to $1 - G((k + 0.5)h)$. Hence the value in psi[k+1] is a numerical approximation to $\psi((k + 0.5)h)$.

If we want approximations to $\psi(kh)$, then one possibility is to take the average of the approximations to $\psi((k - 0.5)h)$ and $\psi((k + 0.5)h)$, so that we calculate (psi[k] + psi[k+1])/2. Doing this, with $m = 8192$ and $h = 0.01$ (see Example 3.22), and selecting the appropriate elements from the resulting array, we obtain the values in the third column of Table 6.1.

For exponentially distributed claims, Example 6.15 gives an explicit expression for the true $\psi(u)$, and we show these true values in the second column of Table 6.1; this means that we can evaluate the quality of the FFT numerical approximations. The absolute error is given in the fourth column of Table 6.1. We repeated the calculations with $m = 16\,384$ (with the same $h = 0.01$), and the results are given in the final two columns of the table. We see that when $m = 8192$ the FFT approximation is accurate to three decimal places, except for $u = 20$ and $u = 25$, which are accurate to two decimal places. When $m = 16\,384$ the FFT approximations are correct to five decimal places. For an approach using the Panjer recursion algorithm for this example, see Dickson (2005), sect. 7.9.3.

Table 6.1. *Ruin probabilities for exponential claims with mean* 1 *and relative safety loading* 0.1, *with* $h = 0.01$

u	True $\psi(u)$	FFT ($m = 8192$)	Absolute error	FFT ($m = 16\,384$)	Absolute error
5	0.57703311	0.57683781	0.0001952987	0.57703165	1.45×10^{-6}
10	0.36626393	0.36594565	0.0003182753	0.36626237	1.56×10^{-6}
15	0.23248105	0.23208491	0.0003961444	0.23247963	1.43×10^{-6}
20	0.14756419	0.14711874	0.0004454511	0.14756296	1.23×10^{-6}
25	0.09366437	0.09318769	0.0004766719	0.09366334	1.02×10^{-6}
30	0.05945218	0.05895574	0.0004964407	0.05945134	8.47×10^{-7}

6.7 Statistics for ruin quantities

So far in this chapter we have been concerned with ruin quantities such as the adjustment coefficient R and the probability of ruin $\psi(u)$ for a classical risk model, where the claim-arrival rate λ, the premium accrual rate c (or equivalently the relative safety loading θ) and the claim-size distribution F_X are all known. However, as for §3.7, in practice λ and F_X are often unknown and must be statistically estimated from data.

For example, suppose we have a sample $T_1, T_2, \ldots, T_n$ of n inter-claim arrival times in a classical risk model, so that the T_i are iid exponentially distributed random variables with mean $1/\lambda$. As in §3.7 (see also §2.4), the maximum likelihood estimator of λ is

$$\hat{\lambda} = \left(\frac{1}{n} \sum_{i=1}^{n} T_i \right)^{-1}.$$

Suppose that we also have n observations $X_1, X_2, \ldots, X_n$ of the claim sizes, iid with distribution function F_X. One approach is to suppose that the claim-size distribution belongs to a parametric family with density $f(x; \nu)$, say, where ν is a (possibly vector-valued) parameter. We could use the data $X_1, \ldots, X_n$ to construct an estimator, for example the maximum likelihood estimator, $\hat{\nu}$ of ν. The adjustment coefficient would then be estimated by the adjustment coefficient belonging to a classical risk model with Poisson parameter $\hat{\lambda}$ and claim-size density $f(x; \hat{\nu})$ (see Exercise 6.22).

For certain parametric families of claim-size distributions, there are explicit expressions for R and $\psi(u)$ in terms of λ, ν and c (or θ). In Example 6.23 we consider statistical estimation of $\psi(u)$ in one such case when the claims are exponentially distributed.

Example 6.23 Suppose that the claims are iid with an exponential distribution with mean μ, and that μ and λ are unknown. We assume that c is known and that $c > \lambda\mu$. Assume further that the available data consist of a sample $T_1, \ldots, T_n$ of claim inter-arrival times and a sample $X_1, \ldots, X_n$ of claim sizes. This is the same set-up as in §3.7. There we obtained maximum likelihood estimators of λ and μ to be

$$\hat{\lambda} = \bar{T}^{-1} \text{ and } \hat{\mu} = \bar{X},$$

respectively, where $\bar{T} = \sum_{i=1}^n T_i/n$ and $\bar{X} = \sum_{i=1}^n X_i/n$.

From Example 6.15, the true (but unknown) ruin probability is given by

$$\psi(u) = \frac{1}{1+\theta} \exp\left(-\frac{\theta}{(1+\theta)\mu} u\right)$$

$$= \frac{\lambda\mu}{c} \exp\left(-\left(\frac{1}{\mu} - \frac{\lambda}{c}\right) u\right)$$

$$= g(\lambda, \mu),$$

say. The plug-in estimator is the probability of ruin that belongs to a classical risk model with Poisson rate $\hat{\lambda}$, exponentially distributed claims with mean $\hat{\mu}$ and with premium accrual rate c. It is possible that we have samples such that $c < \hat{\lambda}\hat{\mu}$, even though the true model has $c > \lambda\mu$. When we do not have positive relative safety loading, it is intuitively plausible that ruin is certain, and this can be proved (see Asmussen (2000), cor. III.1.4). Hence we define our estimator of $\psi(u)$ to be

$$\hat{\psi}(u) = \begin{cases} \dfrac{\hat{\lambda}\hat{\mu}}{c} \exp\left(-\left(\dfrac{1}{\hat{\mu}} - \dfrac{\hat{\lambda}}{c}\right) u\right) & \text{if } c > \hat{\lambda}\hat{\mu} \\ 1 & \text{otherwise.} \end{cases}$$

As the sample size n increases, the Strong Law of Large Numbers means that, with probability 1, $\hat{\lambda}$ converges to λ and $\hat{\mu}$ converges to μ, so that, with probability 1, eventually we will have $c > \hat{\lambda}\hat{\mu}$, and hence eventually $\hat{\psi}(u)$ is the same as $g(\hat{\lambda}, \hat{\mu})$. We use the delta method (see §3.7) to obtain

$$\sqrt{n}\left(\hat{\psi}(u) - \psi(u)\right) \to_d \mathrm{N}\left(0, \sigma^2(\lambda, \mu)\right),$$

where $\sigma^2(\lambda, \mu)$ is obtained using (3.39). This yields (see Exercise 6.21)

$$\sigma^2(\lambda, \mu) = \psi^2(u)\left(\left(1 + \frac{\lambda u}{c}\right)^2 + \left(1 + \frac{u}{\mu}\right)^2\right).$$

The above asymptotic normality result for $\hat{\psi}(u)$ gives rise to an approximate asymptotic $100\alpha\%$ confidence interval for $\psi(u)$, and this confidence interval has end points

$$\hat{\psi}(u) \pm \frac{z_{\alpha/2}}{\sqrt{n}} \sqrt{\sigma^2(\hat{\lambda}, \hat{\mu})},$$

where $z_{\alpha/2}$ is the upper $100\alpha/2 \%$ standard normal percentage point.

When claims are exponentially distributed, a plug-in estimator for R is defined similarly, and a similar asymptotic normality result holds (see Exercise 6.22, where you will find that the calculations for R are simpler than those for $\psi(u)$).

In parametric cases where there is no explicit formula for R (or $\psi(u)$) in terms of λ and ν, then we can still estimate R (or $\psi(u)$) by the adjustment coefficient (or the probability of ruin) that belongs to a classical risk model with claim-arrival rate $\hat{\lambda}$ and claim-size density $f(x; \hat{\nu})$, but now we use numerical calculation to evaluate the estimators.

In an alternative non-parametric approach, we could estimate the claim-size distribution function by the empirical distribution function $\hat{F}_n$ based on the sample $X_1, \ldots, X_n$, given by

$$\hat{F}_n(x) = \frac{1}{n} \sum_{i=1}^{n} 1(X_i \le x). \tag{6.61}$$

This means that $\hat{F}_n$ is the distribution function of a probability measure that assigns mass n^{-1} to each of $X_1, \ldots, X_n$. The adjustment coefficient is then estimated by the adjustment coefficient belonging to a classical risk model with Poisson parameter $\hat{\lambda}$ and claim-size distribution function $\hat{F}_n$ (see Grandell (1979) and sect. 1.3 in Grandell (1991)). Note that statistical estimation of the adjustment coefficient has also been considered by various authors, for example Christ and Steinebach (1995), Csörgő and Steinebach (1991), Csörgő and Teugels (1990), Embrechts and Mikosch (1991), Herkenrath (1986) and Pitts *et al.* (1996).

Exercises

6.1   In a classical risk model, claims arrive at rate λ per accounting period and premiums arrive at rate £2.1λ per accounting period. The claims have the density function

$$f(x) = \frac{1}{2}e^{-x} + \frac{1}{4}e^{-x/2}, \quad x > 0.$$

Find the Lundberg upper bound on the probability of ruin.

6.2 Consider a classical risk model where the claims $X_1, X_2, \ldots$ are deterministic with $\Pr(X_1 = \mu) = 1$ and with relative safety loading $\theta > 0$.
  (a) Show that the conditions of Lemma 6.5 are satisfied with $r_\infty = \infty$.
  (b) Write down an equation satisfied by the adjustment coefficient R.
  (c) Let $R_{\exp}$ be the adjustment coefficient when claims are exponentially distributed with mean μ and the relative safety loading is $\theta > 0$. Show that $R_{\exp} < R$, and compare the corresponding Lundberg bounds.

6.3 Consider a classical risk model with relative safety loading $\theta > 0$ and with iid claims $X_1, X_2, \ldots$ such that there exists a constant $m > 0$ with $\Pr(X_1 < m) = 1$.
  (a) Show that the conditions of Lemma 6.5 are satisfied with $r_\infty = \infty$.
  (b) Show that the adjustment coefficient R satisfies

$$R > \frac{1}{m} \log(1 + \theta).$$

Hint: For $0 \le x \le m$ and $R > 0$, first show that

$$e^{Rx} \le \frac{x}{m}(e^{Rm} - 1) + 1.$$

6.4 In a classical risk model with positive relative safety loading, define, for $n = 1, 2, \ldots$,

$\psi_n(u) = \Pr(\text{ruin occurs on or before the } n\text{th claim, initial capital } u)$.

Show that $\psi_n(u) \to \psi(u)$ as $n \to \infty$, that is show (6.12).
Hint: Let $A_{n,u}$ be the event that ruin occurs on or before the nth claim when the initial capital is u.

6.5 Show that the adjustment coefficient for a classical risk model with relative safety loading $\theta > 0$ and claims that have a gamma distribution with mean 1 and variance 0.5 is given by

$$R = \frac{3 + 4\theta - \sqrt{9 + 8\theta}}{2(1 + \theta)}.$$

6.6 If the claims in Exercise 6.5 are mistakenly assumed to be exponentially distributed with mean 1, compare the resulting adjustment coefficient $R_{\exp}$ with the true R found in Exercise 6.5. Compare the resulting Lundberg bounds, and comment.

6.7 In a classical risk model, individual claim sizes are iid with density function

$$f(x) = \frac{1 + 3x}{4} e^{-x}, \quad x > 0,$$

and the relative safety loading is $\theta = 1/4$. Find the adjustment coefficient.

6.8   Suppose claim sizes $X_1, X_2, \ldots$ are iid with probability density function

$$f_X(x) = \frac{1}{(1+x)^2}, \quad x > 0,$$

and let $M_X(r)$ be the moment generating function of X_1. Show that, for all positive r and M,

$$M_X(r) > \frac{1}{(1+M)^2} \int_0^M e^{rx}\, dx.$$

Show that the right-hand side converges to infinity as $M \to \infty$, and hence that $M_X(r)$ is not finite for $r > 0$ (see Example 6.6 (i)).

6.9   Suppose that the claim sizes $X_1, X_2, \ldots$ in a classical risk model have a distribution that is the convolution of two exponential distributions, with means 1 and 0.5, respectively, so that the claim-size density is

$$f_X(x) = g * h(x) = \int_0^x g(x-t)h(t)\,dt,$$

where $g(x) = e^{-x}$ and $h(x) = 2e^{-2x}$ for $x > 0$. The relative safety loading is $3/8$. Find the adjustment coefficient R. Compare R to the adjustment coefficient R_g for claims with density g, and to the adjustment coefficient R_h for claims with density h.

Hint: The moment generating function for a convolution is the product of the separate moment generating functions.

6.10   Suppose that the adjustment coefficient exists for a classical risk model with positive relative safety loading. Let $\varphi(u)$ be the survival probability with initial capital $u \geq 0$. Use the Lundberg inequality to show that $\varphi(u) \to 1$ as $u \to \infty$.

6.11   In the set-up of Exercise 6.1, show that the moment generating function of the equilibrium distribution F_I associated with the claim-size distribution is

$$M_I(r) = \frac{3 - 4r}{3(1-r)(1-2r)}.$$

Show that the moment generating function of the compound geometric distribution G with counting distribution as in (6.42) and step distribution F_I can be written as follows:

$$M(r) = \frac{2}{7} + \frac{A}{1-6r} + \frac{B}{6-7r},$$

for some constants A and B which you should find. Hence find $\varphi(u)$ using the representation of φ as the distribution function G (given in Theorem 6.14). Show that

$$\psi(u) = \frac{20}{29}e^{-u/6} + \frac{5}{203}e^{-6u/7}.$$

6.12 In this exercise we present an alternative way to find the ruin probability for Exercise 6.11. With the same set-up as in Exercise 6.1, write down the integro-differential equation (6.23) and show that

$$\varphi'(u) = \frac{10\varphi(u)}{21} - \frac{5e^{-u}}{21}I_1(u) - \frac{5e^{-u/2}}{42}I_2(u),$$

where $I_1(u) = \int_0^u \varphi(z)e^z \, dz$ and $I_2(u) = \int_0^u \varphi(z)e^{z/2} \, dz$. Hence show that

$$\varphi''(u) = -\frac{5}{42}\varphi(u) - \frac{1}{42}\varphi'(u) + \frac{5e^{-u}}{42}I_1(u).$$

Show that

$$\varphi'''(u) + \frac{43}{42}\varphi''(u) + \frac{6}{42}\varphi'(u) = 0.$$

Solve this second order differential equation for $\varphi'(u)$, and hence show that

$$\varphi(u) = C + De^{-u/6} + Ee^{-6u/7},$$

for constants C, D and E. Use equations for $\varphi(0)$, $\varphi(\infty)$ and $\varphi'(0)$ to find C, D and E. Hence find $\psi(u)$, and check that you get the same answer as in Exercise 6.11.

6.13 In the set-up of Exercise 6.1, using the results of Exercises 6.11 and 6.12 for $\psi(u)$ and Exercise 6.1 for the Lundberg upper bound, construct a plot (use a computer) showing $\psi(u)$ and the Lundberg upper bound, and compare these two quantities.

6.14 In the set-up of Exercise 6.5, show that the equilibrium distribution F_I associated with the claim-size distribution has moment generating function

$$M_I(r) = \frac{4 - r}{(2 - r)^2}.$$

Using an approach similar to Exercise 6.11, use the compound geometric representation to find $\varphi(u)$, and hence show that $\psi(u)$ is given by

$$\psi(u) = Ae^{-Ru} + Be^{-\alpha u},$$

where R is the adjustment coefficient; you should specify α, A and B in terms of θ.

6.15 Let $\varphi(u)$ be the survival probability with initial capital u in Exercise 6.14. Use the integro-differential equation method (see Exercise 6.12) to show that

$$(1 + \theta)\varphi'(u) = \varphi(u) - 4e^{-2}I(u),$$

where $I(u) = \int_0^u \varphi(x)e^{2x}\,dx$. Show that

$$(1 + \theta)\varphi'''(u) + (3 + 4\theta)\varphi''(u) + 4\theta\varphi'(u) = 0.$$

Hence find $\varphi(u)$ and the ruin probability, and check that you get the same answer as in Exercise 6.14.

6.16 In the classical risk model with positive relative safety loading, starting with the integral equation (6.31) for $\psi(u)$, show that $\psi(u)$ satisfies a defective renewal-type equation

$$\psi(u) = z(u) + \int_0^u \psi(u - x)f(x)dx,$$

where $f(x) \geq 0$ for all $x > 0$ and $\int_0^\infty f(x)dx < 1$, that is $f(x)$ is a defective probability density function, and where you should specify $z(u)$ in terms of the equilibrium distribution F_I, see (6.34), associated with the claim-size distribution.

6.17 Find the Cramér–Lundberg approximation (6.58) for the probability of ruin for a classical risk model with claim sizes and relative safety loading as in Exercise 6.1. Investigate how good the approximation is by comparing it to the true probability of ruin (found in Exercises 6.11 and 6.12).

6.18 Consider a classical risk model with positive relative safety loading. Suppose that the adjustment coefficient R exists and that

$$\psi(u) = ae^{-cu} + be^{-du}, \quad u \geq 0,$$

where a, b, c and d are constants satisfying $a > 0$, $0 < a + b < 1$ and $0 < c < d < \infty$. Find the adjustment coefficient and the relative safety loading in terms of a, b, c and d, and write down the Cramér–Lundberg approximation for $\psi(u)$.

6.19 For a classical risk model with exponentially distributed claims with mean μ and with positive safety loading θ, write an *R* function to find the adjustment coefficient R using the Newton–Raphson algorithm with starting point r_0.

When $\mu = 1$ and $\theta = 0.2$, as in Example 6.22, try $r_0 = 0.1625$ and check that you get r_1, r_2 and r_3 as in the example.

What happens if you take $r_0 = 0.05$? Can you explain what is going on?

6.20 For a classical risk model with deterministic claims equal to μ with probability 1 and relative safety loading θ, write an **R** function to find the adjustment coefficient R using the Newton–Raphson method. Use your function to find R when $\mu = 1$ and $\theta = 0.1$. Compare your numerical answer to the adjustment coefficient for exponentially distributed claims with mean 1 and $\theta = 0.1$. (See also Exercise 6.2.)

6.21 Consider the estimation of $\psi(u)$ in Example 6.23. Show that the quantity $\sigma^2(\lambda, \mu)$ in the example is given by

$$\sigma^2(\lambda, \mu) = \psi^2(u)\left(\left(1 + \frac{\lambda u}{c}\right)^2 + \left(1 + \frac{u}{\mu}\right)^2\right).$$

6.22 In a classical risk model with known premium income rate c and unknown claim-arrival rate λ, suppose that claims are exponentially distributed with unknown mean μ. Assume that $c > \lambda\mu$. Suppose that a sample $T_1, \ldots, T_n$ of inter-arrival times and an independent sample $X_1, \ldots, X_n$ of claims are available. Construct a plug-in estimator $\hat{R}$ of the adjustment coefficient R. Obtain an asymptotic normality result for $\hat{R}$, and use this to find an approximate asymptotic $100\alpha\%$ confidence interval for R.

7
Case studies

7.1 Case study 1: comparing premium setting principles

We examine the use of different premium setting principles under different models for an aggregate insurance loss S. We want to compare the different approaches – in particular we want to obtain measures of the uncertainty associated with the premiums set so that we can ascertain how "precise" and reliable the premiums set by the different principles are.

We will use theory and simulated data as far as we can, and also turn to the bootstrap resampling technique for additional enlightenment.

We will compare premiums set using some or all of the following principles (as described in §4.1):

(1) *EVP* (expected value principle)
(2) *SDP* (standard deviation principle)
(3) *VP* (variance principle)
(4) *QP* (quantile principle)
(5) *EPP* (exponential premium principle)

under various distributional assumptions for the risk/aggregate loss S.

We study first the case in which we have an assumed model for the distribution of S, and then the case in which we base our premiums solely on the information in an observed sample of values of S (with no assumed model).

7.1.1 Case 1 – in the presence of an assumed model

Let us assume that S has a compound Poisson distribution $S \sim \text{CP}(\lambda, F_X)$, where λ is the claim rate and X is the individual loss variable. Further, let us assume, for illustrative purposes, that $X \sim \text{Exp}(1/\mu)$ (with mean μ). Using various results in Chapters 2 and 3, we have

$$\mathbb{E}[X] = \mu, \ \mathbb{E}[X^2] = 2\mu^2, \ \mathbb{E}[S] = \lambda\mu \text{ and } \mathrm{Var}[S] = 2\lambda\mu^2,$$

and the moment generating function of S is

$$M_S(t) = \exp[\lambda\{(1 - \mu t)^{-1} - 1\}].$$

The *EVP* premium using relative security loading α_1 is given by

$$P_{EVP} = \lambda\mu + \alpha_1 \times \lambda\mu = (1 + \alpha_1)\lambda\mu.$$

The *SDP* premium using relative security loading α_2 is given by

$$P_{SDP} = \lambda\mu + \alpha_2 \times (2\lambda\mu^2)^{1/2}.$$

The *VP* premium using relative security loading α_3 is given by

$$P_{VP} = \lambda\mu + \alpha_3 \times 2\lambda\mu^2.$$

The *QP* premium set at the $100(1 - \alpha_4)$th percentile of the distribution of S is given by P_{QP}, where

$$\Pr(S > P_{QP}) = \alpha_4.$$

The *EPP* premium using the insurer's utility parameter α_5 is given by

$$P_{EPP} = \frac{1}{\alpha_5} \log M_S(\alpha_5) = \frac{\lambda}{\alpha_5}[(1 - \mu\alpha_5)^{-1} - 1].$$

If we assume that the values of λ, μ and α_1 are known, then we can, for example, identify the values of α_2, α_3 and α_5 that produce *SDP*, *VP* and *EPP* premiums which match the *EVP* premium. We can, in fact, force the *SDP*, *VP* and *EPP* premiums to match the *EVP* premium by choosing

$$\alpha_2 = \left(\frac{\lambda}{2}\right)^{1/2}\alpha_1, \quad \alpha_3 = \left(\frac{1}{2\mu}\right)\alpha_1 \quad \text{and} \quad \alpha_5 = \frac{\alpha_1}{\mu(1 + \alpha_1)},$$

respectively.

To include the *QP* premium we need to have information on the (cumulative) distribution function (cdf) of S. One method of getting this distributional information is to use Panjer's iterative approach to evaluating a discretised version of the distribution function (see §3.5.1). We will use this approach in what follows and also gain insight by identifying quantiles of simulated samples.

To illustrate these ideas, let us suppose that $\lambda = 100$ and $\mu = 1$ (we are taking the individual expected loss as our monetary unit for convenience), and that the insurer sets an *EVP* premium with 50% security loading – that is, $\alpha_1 = 0.5$ and $P_{EVP} = 150$. To match this with *SDP* and *VP*, the insurer requires $\alpha_2 = 5/\sqrt{2} = 3.536$ and $\alpha_3 = 0.25$. To match with *EPP*, the insurer should choose the utility function parameter to be $\alpha_5 = 1/3$. Using Panjer's approach, we find that $\Pr(S \leq 150) \approx 0.999$, and so, to match the other premiums, the *QP* premium would have to be set at or near the 99.9th percentile of the distribution of S. A simulation of one million values of S also produced a relative frequency of values less than 150 of 0.999. This very high demand in terms of quantiles reflects the light tail of the exponential and the fact that the standard deviation of S is only $\sqrt{200} = 14.14$ – the value 150 is more than 3.5 standard deviations above the mean of S. It also suggests that, under the given model, we should consider the 50% security loading in the *EVP* premium as being excessively high.

As in §3.6.1, we use a normal approximation to the distribution of the compound Poisson random variable $S \sim CP(100, F_X)$. The aggregate loss S has mean 100 and variance 200, and we get $\Pr(S \leq 150) \approx \Phi(3.536) = 0.9998$, suggesting that the percentile we should use is even higher than the 99.9% one.

A situation as above in which an insurer can be 99.9% sure of meeting all claims using premium income only and without recall to any reserves (which are attracting interest or other gains) is quite unrealistic – another company will offer policies with premiums much lower than 150. So let us consider instead an *EVP* premium with only 10% security loading – that is $\alpha_1 = 0.1$ and $P_{EVP} = 110$. To match this with the *SDP*, the insurer requires $\alpha_2 = 1/\sqrt{2} = 0.7071$ (which is much more consistent with the oft-quoted commercial "rule of thumb", which sets a premium at "mean plus half standard deviation"). The matching *VP* and *EPP* premiums have $\alpha_3 = 0.05$ and $\alpha_5 = 1/11$. In practice, the insurer will use reserves if required to meet any shortfall in aggregate payout – if the insurer sets aside 40 units for this purpose, then there is a total sum available of $110 + 40 = 150$, and we know from the above that there is then a probability of about 0.999 of being able to meet the commitments on this business.

Using a normal approximation to the distribution of S gives $\Pr(S \leq 110) \approx \Phi(0.707) = 0.760$. In addition, a Panjer recursion gave the percentile of the distribution of S corresponding to $S = 110$ as the 75.6th percentile, and a simulation of one million values of S produced a relative frequency of values less than 110 of 0.766. Taking all this into account, we will set *QP* premiums at the 76th percentile of the distribution of S.

In practice, the values of λ and μ used in the preceding expressions for *EVP, SDP, VP* and *EPP* are not known parameters, but are sample estimates based on past data, say $\hat{\lambda}$ and $\hat{\mu}$. The estimates here will be simply the corresponding sample means for relevant data over recent years (assuming stationary behaviour over the years, these are not only the obvious estimates obtained by the method of moments, but are also the maximum likelihood estimates; see §2.4). These estimates have their own sampling distributions, means and standard deviations. In turn, the *EVP, SDP, VP* and *EPP* premiums, with known security loadings, are *statistics*, that is functions of sample data with no unknown parameters and whose values can therefore be calculated when we have the data. The *QP* premium is also a statistic, as it is based on an estimated distribution function.

The (estimates of the) *EVP* and *SDP* premiums, for example, are now

$$P_{EVP} = (1 + \alpha_1)\hat{\lambda}\hat{\mu} \text{ and } P_{SDP} = \hat{\lambda}\hat{\mu} + \sqrt{2}\alpha_2\hat{\lambda}^{1/2}\hat{\mu},$$

respectively. Clearly the premiums are now complicated expressions, and their sampling properties will be hard to establish – consider, for example, the matter of establishing the standard error of P_{SDP} or P_{QP}.

With a model in place for the distributions of X and S, we can assess the distribution of the premiums by repeated simulation. Consider the case in which we assume the model $S \sim CP(\lambda, F_X)$, where $X \sim Exp(1/\mu)$. To represent the estimation of λ and μ from past data over, say, ten years, a simulation of ten values of the number of losses N is performed (in the simulation $N \sim Poi(100)$ was used) and the mean number is calculated, giving the estimate $\hat{\lambda}$. For each simulated value of N, the appropriate number of losses is simulated (in the simulation $X \sim Exp(1)$ with mean $\mu = 1$ was used) and the mean of all the losses over the ten years is calculated, giving $\hat{\mu}$. We then calculate our first simulated value of each of the four premiums P_{EVP}, P_{SDP}, P_{VP} and P_{EPP} (using $\alpha_1 = 0.1$, $\alpha_2 = 1/\sqrt{2}$, $\alpha_3 = 0.05$ and $\alpha_5 = 1/11$). We repeat this process 10 000 times, giving vectors containing that number of values of each premium – these vectors are then summarised, revealing properties of the premiums set by the four different principles.

Selected results (from R output) are shown in Table 7.1. The results show consistency across the four premium setting principles considered – the levels of uncertainty (lack of precision, as measured by the standard deviations and ranges of the simulated premiums) associated with the principles appear to be very similar.

Simulation details See Simulation note 1 at the end of the case study.

Table 7.1. *Simulation results for exponential losses*

	Number	Min.	Median	Mean	Max.	SD	Range
P_{EVP}	10 000	90.41	110.0	110.1	130.3	4.904	39.90
P_{SDP}	10 000	90.89	110.0	110.1	129.7	4.794	38.83
P_{VP}	10 000	89.75	110.0	110.1	131.1	5.132	41.38
P_{EPP}	10 000	89.69	110.0	110.1	131.2	5.155	41.52

Table 7.2. *Simulation results for exponential losses from direct simulation of S*

	Number	Min.	Median	Mean	Max.	SD	Range
P_{EVP}	10 000	108.3	110.0	110.0	111.8	0.4965	3.516
P_{SDP}	10 000	108.1	110.0	110.0	112.1	0.5339	4.051
P_{VP}	10 000	107.3	110.0	110.0	112.7	0.6847	5.374
P_{EPP}	10 000	107.2	110.0	110.1	112.8	0.7164	5.609
P_{QP}	10 000	107.3	109.7	109.7	112.8	0.6581	5.453

A second approach does not involve estimating λ and μ separately from past data each time – it is not designed to include an allowance for parameter uncertainty. In this approach we simply simulate a sample of values of the aggregate loss S (here a sample size of 1000 was used) and use the sample mean and variance as estimates of $\mathbb{E}[S]$ and $\text{Var}[S]$. These estimates are then used in the calculation of the *EVP, SDP* and *VP* premiums. For the *EPP* premium, we do need estimates of λ and μ, and we get these indirectly from the sample mean and variance (using method of moments estimation). For illustrative purposes, we also include the *QP* premium as found by identifying the 76th percentile of the sample values of S. We then repeat this process 10 000 times, giving vectors containing that number of values of each premium – these vectors are then summarised, revealing properties of the premiums set by the five different principles. Selected results are presented in Table 7.2.

The uncertainty (lack of precision) associated with the methods is very much lower than is the case using the first approach, and the results again show reasonable consistency across the five premium setting principles considered (but less markedly so than in the first study). The results for the *QP* premiums are more in line with those for the *VP* and *EPP* principles than with the others.

The results presented in Tables 7.1 and 7.2, when taken together, indicate that the various premium setting principles we have considered perform with reasonably similar levels of precision. We do note, however, that in each study separately the uncertainty associated with the *VP* and *EPP* approaches is higher than that with the *EVP* and *SDP* approaches. In each study, the *EPP* approach has produced the premiums with the highest uncertainty of those approaches included.

Based on the standard deviations, one could tentatively suggest that the approaches fall into three groups – {*EVP, SDP*}, {*QP*} and {*VP, EPP*}. Further work, especially with other models for the individual losses, is required.

Table 7.3. *Simulation results for lognormal losses with* $\mathbb{E}[X] = \text{Var}[X] = 1$

	Number	Min.	Median	Mean	Max.	SD	Range
P_{EVP}	10 000	108.1	110.0	110.0	111.9	0.4903	3.774
P_{SDP}	10 000	107.5	110.0	110.0	112.0	0.5397	4.414
P_{VP}	10 000	106.9	110.0	110.0	112.7	0.6999	5.802
P_{QP}	10 000	107.1	109.6	109.6	112.8	0.6491	4.901

Simulation details See Simulation note 2 at the end of the case study.

We now repeat the second analysis above in the case that the loss variable X has a lognormal(μ, σ) distribution. We will consider *SDP, VP* and *QP* premiums set to match the *EVP* premium with 10% security loading. We again use $\lambda = 100$, and we will consider two sets of lognormal parameters, both of which give $\mathbb{E}[X] = 1$ (and $\mathbb{E}[S] = 100$) as before.

Lognormal (1) $X \sim$ lognormal(μ, σ) with $\mu = -0.5 \log 2$ and $\sigma = (\log 2)^{0.5}$. From earlier results, we have $\mathbb{E}[X] = \text{Var}[X] = 1$, $\mathbb{E}[S] = 100$, $\text{Var}[S] = 200$, $\alpha_1 = 0.1$, $\alpha_2 = 1/\sqrt{2}$ and $\alpha_3 = 0.05$, all as before.

A simulation of one million values of S produced a relative frequency of values less than 110 of 0.769. The normal approximation for $\Pr(X \leq 110)$ is as before (0.760), and we will set *QP* premiums again at the 76th percentile of the distribution of S. Selected results of 10 000 simulations of the premiums are presented in Table 7.3. The results are very similar to those in the case $X \sim$ Exp(1). The *VP* approach has produced the premiums with the highest uncertainty of the approaches included.

Simulation details See Simulation note 3 at the end of the case study.

Lognormal (2) $X \sim$ lognormal(μ, σ) with $\mu = -0.5 \log 5$ and $\sigma = (\log 5)^{0.5}$. In this case we have $\mathbb{E}[X] = 1$, $\text{Var}[X] = 4$, $\mathbb{E}[S] = 100$, $\text{Var}[S] = 500$, $\alpha_1 = 0.1$, $\alpha_2 = 1/\sqrt{5}$ and $\alpha_3 = 0.02$, reflecting the change in the value of $\text{Var}[S]$ from 200 to 500.

A simulation of one million values of S produced a relative frequency of values less than 110 of 0.717. The normal approximation for $\Pr(X \leq 110)$ is now lower, at 0.673. Taking all this into account, we will set *QP* premiums at the 70th percentile of the distribution of S. Selected results of 10 000 simulations of the premiums are presented in Table 7.4.

The results for this case ($X \sim$ lognormal with $\text{Var}[X] = 4$ and $\text{Var}[S] = 500$) are noticeably different from the earlier case ($X \sim$ lognormal with $\text{Var}[X] = 1$ and $\text{Var}[S] = 200$). As a result of the increase in the values of $\text{Var}[X]$ and

Table 7.4. *Simulation results for lognormal losses with* $\mathbb{E}[X] = 1$
and $\mathrm{Var}[X] = 4$

	Number	Min.	Median	Mean	Max.	SD	Range
P_{EVP}	10 000	107.3	110.0	110.0	112.9	0.7810	5.657
P_{SDP}	10 000	106.7	110.0	110.0	117.7	0.9528	10.99
P_{VP}	10 000	106.0	109.9	110.0	129.2	1.298	23.26
P_{QP}	10 000	105.4	108.8	108.8	112.3	0.9745	6.844

$\mathrm{Var}[S]$, the uncertainty associated with the *SDP* principle is considerably higher than before. In the case of the *VP* approach, the increase in uncertainty is even more striking. The increase in uncertainty associated with the *EVP* and *QP* approaches is more modest. This suggests that the *VP* approach is the least robust to increased uncertainty in the individual and aggregate loss distributions – this is consistent with the fact that variance itself is a non-robust measure of spread, being highly susceptible to unusually high observations – in the simulation one sample had an especially high variance, producing a *VP* premium as high as 129.2, considerably higher than that produced by the other approaches.

Simulation details See Simulation note 4 at the end of the case study.

It is left as an exercise for the reader to investigate the effects of using a Pareto distribution instead of a lognormal distribution for the individual losses; $X \sim \mathrm{Pa}(2.5, 1.5)$ and $X \sim \mathrm{Pa}(1.8, 0.8)$ are suggested models (they are used in Case study 2).

7.1.2 Case 2 – without model assumptions, using bootstrap resampling

To assess the precision of an estimator (for example, a sample statistic such as the mean, median, maximum, standard deviation, or, in the case of paired data, the correlation coefficient), we require information on the sampling distribution of the estimator. In some situations we have an exact distribution to work with – for example, if our data comprise a random sample of size n from a $N(\mu, \sigma^2)$ distribution, then we know that $\bar{X}$, the usual estimator of μ, has distribution $\bar{X} \sim N(\mu, \sigma^2/n)$, giving the standard error of estimation as $s.e.(\bar{X}) = \sigma/\sqrt{n}$. In some cases, with large samples, we can appeal to the asymptotic distribution of the estimator (for example, under fairly general conditions, maximum likelihood estimators have well-known and usable large-sample distributions).

In many situations, however, we do not have sufficient (if any) knowledge about the underlying population distribution to justify adopting a particular model for the distribution of our estimator. The question then arises of how to assess the precision of our estimator when we do not have an expression for its standard error and all we do have is a sample of data from the unknown population distribution.

This difficulty faces us when we try to assess the precision of premiums set by the various principles when we do not have a model in place for the distribution of the losses and all we have is a sample of such losses.

Bootstrap estimation is an imaginative technique which essentially replaces distributional assumptions by the use of computing power to perform repeated simulations of samples and consequent calculations. The method is attractively simple and is easily implemented – it can provide answers to questions which defy traditional approaches to statistical analysis. The methodology of the bootstrap was proposed by Efron (see Efron (1979)) – the technique has become widely known and applied since then.

The bootstrap technique is based on using the *empirical (cumulative) distribution function* (ecdf) (see (6.61)) of the sample we do have, in place of the unknown (cumulative) distribution function of the underlying population variable. (We do not have the distribution we need, so we "pick ourselves up by our bootstraps" and use the only thing available – the equivalent sample version – instead.) We now regard the ecdf as a proxy population distribution function and sample repeatedly from it. The samples, each of which is called a *bootstrap sample*, are taken with replacement. For each such sample drawn, we calculate the value of our estimator, and, over a succession of samples, this provides an observed sampling distribution of our estimator. The bootstrap technique is an example of a *resampling technique* (the name coming from the use of repeated samples taken from the ecdf of our actual sample).

Example 7.1 To assist the reader to appreciate the technique, we illustrate the bootstrap technique first with a simple problem: estimate the precision of the sample mean and median claim amounts as estimators of the mean and median claim amounts, respectively, in the underlying population, given the following random sample of 50 claim amounts (in some suitable units, and sorted for convenience):

14	24	39	50	104	111	114	138	181	204
259	379	407	420	438	453	503	550	587	607
632	645	653	666	772	795	821	860	1017	1172
1278	1398	1424	1583	1794	1917	1918	1963	2074	2085
2252	2347	2460	2559	2743	3151	3189	3351	8618	10026

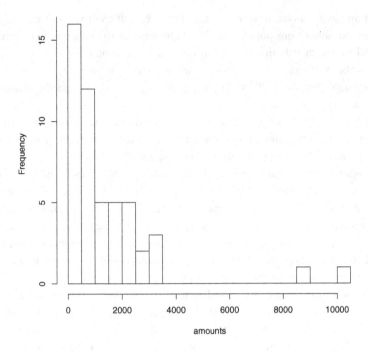

Figure 7.1. Histogram of sample of 50 claim amounts.

A histogram of the claim amounts is presented in Figure 7.1 – the data are strongly positively skewed. The sample mean and standard deviation are $\bar{x} = 1434.9$ and $s = 1885.3$. The sample median is 783.5. Using standard statistical theory we can estimate the standard error of estimation using $\bar{x}$ as $s.e.(\bar{x}) = 1885.3/\sqrt{50} = 266.6$. We cannot find a corresponding approximation to the standard error of the sample median without inappropriate assumptions about, or knowledge of, the probability density function of the underlying population variable, knowledge we do not have.

We now assess the variation of the sample mean and median using the bootstrap technique by taking 1000 samples from the ecdf of our sample. A graph of the ecdf is given in Figure 7.2.

We sample from the ecdf by taking 1000 samples (each of size 50) with replacement, one after another, from the original sample, the set of claim amounts. We save the mean and median of each sample and then summarise the collections of these 1000 *bootstrap means* and *bootstrap medians*. The standard deviations of these collections give us our estimated standard errors of estimation. Figure 7.3 displays histograms of the bootstrap means and medians. The summary results are given in Table 7.5.

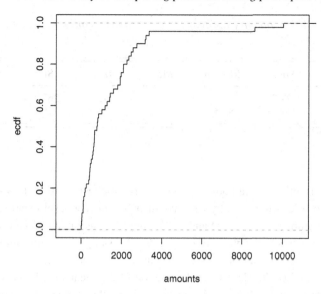

Figure 7.2. Empirical (cumulative) distribution function of sample of 50 claim amounts.

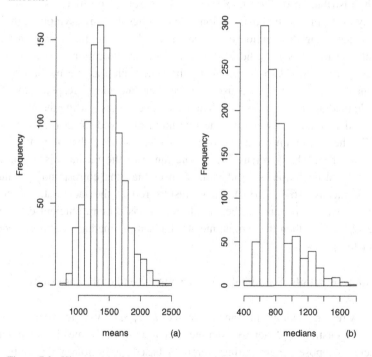

Figure 7.3. Histograms of sample means (a) and medians (b) for 1000 bootstrap samples from the original sample of 50 claim amounts.

Table 7.5. *Summary results for the bootstrap means and bootstrap*
medians in Example 7.1

	Min.	Median	Mean	Max.	SD	Range
Means	727.7	1400	1426	2427	263.4	1699
Medians	438.0	783.5	813.3	1794	215.5	1356

The distribution of the bootstrap means has a modest positive skew – but it will be modelled quite well by a normal distribution. The level/location of the means is summarised as 1400 and 1426 using the median and mean, respectively. The inclusion of very high values in the sample of claim amounts is reflected in the extremes of the set of means, in particular a maximum of 2427 and a range of 1699. The standard deviation of the means is 263.4, which is in good agreement with our earlier estimate of the standard error of estimation $s.e.(\bar{x}) = 266.6$.

The distribution of the sample medians is strongly positively skewed and clearly far from a normal distribution – this implies that any asymptotic theory for the sampling distribution of a median based on sampling from a normal population may not be valid in this case. The level/location of the medians is summarised as 783.5 (fortuitously, the same value as the median of the original amounts) and 813.3 using the median and mean, respectively. The sample medians are less variable than the sample means – while the minimum observed median is lower than the minimum observed mean (438.0 versus 727.7), the maximum observed median is much smaller than the maximum observed mean (1794 versus 2427). The range of the medians is 1356, and the standard deviation is 215.5, much lower than the corresponding values for the means (1699, 263.4). These results reflect the fact that, for positively skewed distributions, the sample median is a more efficient estimator of level/location than the sample mean – the sample median is a more *robust* estimator.

Simulation details See Simulation note 5 at the end of the case study.

We now return to comparing the precision of *EVP, SDP, VP* and *QP* premiums, using a bootstrap approach to resample from a set of aggregate claim amounts.

Let us suppose we are setting premiums based on the following sample of 100 aggregate claim amounts (sorted):

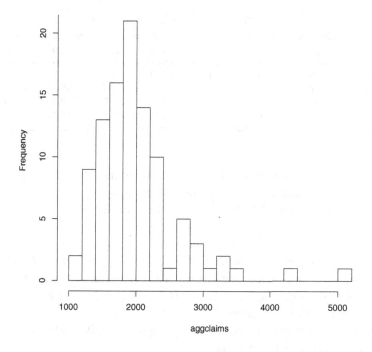

Figure 7.4. Histogram of sample of 100 aggregate claim amounts.

1091	1171	1229	1233	1285	1327	1334	1358	1367	1369
1388	1402	1424	1450	1462	1490	1498	1510	1519	1537
1543	1556	1566	1568	1618	1637	1643	1654	1663	1707
1714	1716	1718	1739	1748	1753	1754	1755	1757	1759
1814	1816	1819	1834	1837	1838	1843	1844	1859	1864
1873	1884	1885	1885	1889	1897	1899	1913	1949	1955
1999	2005	2030	2033	2051	2061	2064	2067	2096	2098
2119	2170	2180	2187	2194	2240	2240	2245	2267	2276
2314	2323	2344	2361	2368	2416	2640	2714	2715	2745
2779	2850	2970	2993	3175	3205	3380	3523	4343	5065

A histogram of the amounts is given in Figure 7.4 – the data are again strongly positively skewed. The sample mean, variance and standard deviation are $\bar{x} = 2002.54$, $s^2 = 395\,605$ and $s = 628.971$, respectively.

We will base our analysis on an *EVP* premium with 10% security loading, which, based on the sample above, is $1.1 \times 2002.54 = 2202.8$. To match this with *SDP* and *VP* premiums, we require $\alpha_2 = 0.3184$ and $\alpha_3 = 0.0005062$. The percentile of the data closest to 2202.8 is the 75th percentile, so we will set *QP* premiums at this percentile.

Table 7.6. *Summary results using bootstrap samples*

	Number	Min.	Median	Mean	Max.	SD	Range
P_{EVP}	10 000	1970	2200	2202	2510	69.22	539.4
P_{SDP}	10 000	1925	2196	2200	2597	85.48	672.0
P_{VP}	10 000	1876	2192	2200	2774	110.4	898.3
P_{QP}	10 000	1949	2206	2204	2714	75.44	765.3

We now take 10 000 bootstrap samples from our data set consisting of the 100 aggregate claim amounts and, for each bootstrap sample, calculate the premiums using each of the four methods. The summary results are given in Table 7.6.

In this study the *EVP* approach gives the most precise (greatest consistency, lowest uncertainty) results, followed in order by the *QP, SDP* and *VP* approaches (as measured by the standard deviation). The *VP* approach – perhaps not surprisingly – again reveals itself as performing poorly in this regard relative to the other approaches.

As a result of this study, one can perhaps suggest that *EVP, SDP* and *QP* fall into one group, with *VP* on its own.

Simulation details See Simulation note 6 at the end of the case study.

Simulation notes

Simulation note 1 The simulation was carried out by executing a function, here called `simprems` and previously stored as a text file, as given below. The output from the function is an **R** object – a list containing four vectors called `premslist.list`. The command

```
premsim1 = simprems(10000,100,1)
```

executes the function once, carrying out 10 000 simulations of the numbers of claims and corresponding aggregate claims over ten years, the calculation of estimates of λ and μ each time, and the calculation of the premiums as given by each of the four setting principles being considered. The results (the list of four vectors) are held in the object `premsim1` and can be retrieved using, for example,

```
summary(premsim1[[2]])
```

to obtain a summary of the 10 000 simulated values of P_{SDP}. The text defining
the function is

```
simprems = function(n,lambda,mu){
pevp = psdp = pvp = pepp = (1:n)*0
for(i in 1:n){
nc = rpois(10,lambda)
nctot = sum(nc)
lambdahat= nctot/10
muhat = sum(rexp(nctot,1/mu))/nctot
pevp[i] = 1.1*lambdahat*muhat
psdp[i] = lambdahat*muhat + muhat*lambdahat^(0.5)
pvp[i] = lambdahat*muhat + 0.1*lambdahat*muhat^2
pepp[i] = 11*lambdahat*((1 - muhat/11)^(-1) - 1) }
premslist.list = list(pevp,psdp,pvp,pepp) }
```

Simulation note 2   The simulation was carried out by executing a function,
here called `simprems2` and previously stored as a text file, as given below.
The output from the function is an *R* object – a list containing five vectors
called `premslist.list`. The command

```
premsim2 = simprems2(10000,100,1)
```

executes the function once, carrying out 10 000 simulations of the values of
$\mathbb{E}[S]$ and $\mathrm{Var}[S]$ (using the sample mean and variance of a sample of 1000
observations of S each time) and the calculation of the premiums as given
by each of the five setting principles being considered. The results (the list of
five vectors) are held in the object `premsim2` and can be retrieved using, for
example,

```
summary(premsim2[[4]])
```

to obtain a summary of the 10 000 simulated values of P_{EPP}. The text defining
the function is

```
simprems2 = function(n,lambda,mu){
pevp = psdp = pvp = pepp = pqp = (1:n)*0
for (j in 1:n){
nc = rpois(1000,lambda)
aggclaims = (1:1000)*0
for (i in 1:1000){
aggclaims[i] = sum(rexp(nc[i],1/mu))}
sbar = mean(aggclaims)
```

```
ssd = sd(aggclaims)
lambdaest = 2*sbar^2/ssd^2
muest = ssd^2/(2*sbar)
pevp[j] = 1.1*sbar
psdp[j] = sbar + (1/2^(0.5))*ssd
pvp[j]  = sbar + 0.05*ssd^2
pepp[j] = 11*lambdaest*((1-muest/11)^(-1) - 1)
pqp[j] =  quantile(aggclaims,0.76) }
premslist.list=list(pevp,psdp,pvp,pepp,pqp) }
```

Simulation note 3   The simulation was carried out by executing a function, here called `simprems3` and previously stored as a text file, as given below. The output from the function is an *R* object – a list containing four vectors called `premslist.list`. The command

```
premsim3 = simprems3(10000,-0.5*log(2),(log(2))^(0.5))
```

executes the function once, carrying out 10 000 simulations of samples of values of S of size 1000, and using the sample mean and variance as estimates of $\mathbb{E}[S]$ and $\mathrm{Var}[S]$ in the calculation of the premiums each time. The text defining the function is

```
simprems3 = function(n,lambda,m,s){
pevp = psdp = pvp = pqp = (1:n)*0
for (j in 1:n){
nc = rpois(1000,lambda)
aggclaims = (1:1000)*0
for (i in 1:1000){
aggclaims[i] = sum(rlnorm(nc[i],m,s))}
sbar = mean(aggclaims)
ssd = sd(aggclaims)
pevp[j] = 1.1*sbar
psdp[j] = sbar + (1/2^(0.5))*ssd
pvp[j]  = sbar + 0.05*ssd^2
pqp[j] =  quantile(aggclaims,0.76) }
premslist.list=list(pevp,psdp,pvp,pqp) }
```

Simulation note 4   This is as in simulation note 3, except that the lines defining `psd[j]`, `pvp[j]` and `pqp[j]` in `simprems3` are changed to

```
psdp[j] = sbar + (1/5^(0.5))*ssd
pvp[j]  = sbar + 0.02*ssd^2
pqp[j] =  quantile(aggclaims,0.7)
```

This gives the function `simprems4`, which is executed as

```
premsim4 = simprems4(10000,-0.5*log(5),(log(5))^(0.5))
```

Simulation note 5   We define a function `boot1` as follows:

```
boot1 = function(data,n){
sampmean = sampmedian=(1:n)*0
for(i in 1:n){
bootsamp=sample(data,replace=T)
sampmean[i] = mean(bootsamp)
sampmedian[i] = median(bootsamp) }
outlist.list = list(sampmean,sampmedian) }
```

and execute it using

```
bootsim1 = boot1(amounts,1000)
```

where `amounts` is the vector containing the sample of 50 claim amounts. The object `bootsim1` contains two vectors, the first of which contains 1000 *bootstrap means* – each component is the mean of a random sample of size 50 selected with replacement from the vector `amounts`. Similarly, the second vector contains 1000 *bootstrap medians* – the components are the medians of the bootstrap samples whose means comprise the first vector.

Simulation note 6   We define a function `boot2` as follows:

```
boot2 = function(data,n){
pevp = psdp = pvp = pqp = (1:n)*0
for(i in 1:n){
bootsamp = sample(data,replace=T)
m1 = mean(bootsamp)
s1 = sd(bootsamp)
v1 = var(bootsamp)
pevp[i] = 1.1*m1
psdp[i] = m1 + 0.3184*s1
pvp[i] = m1 + 0005062*v1
pqp[i] = quantile(bootsamp,0.75) }
outlist.list = list(pevp,psdp,pvp,pqp) }
```

and execute it using

```
bootsim2 = boot2(aggclaims,10000)
```

where `aggclaims` is the vector containing the sample of 100 aggregate claim amounts. The object `bootsim2` contains four vectors, the first of which contains 10 000 *bootstrap EVP premiums* – similarly the other vectors contain the *SDP, VP* and *QP* premiums (based on the same bootstrap samples).

7.2 Case study 2: shared liabilities – who pays what?

We examine individual and aggregate insurance losses in the situation in which there is both a deductible and an excess of loss reinsurance arrangement in place for each loss, as described in Chapter 5. We want to investigate how the losses break down into payments by the three parties involved, namely the policyholder, the direct insurer and the reinsurer.

Let us suppose that an insurer writes policies with deductible (excess) D per individual loss, so the insured party bears the first D units of any loss and submits a claim if the loss exceeds D. The insured loss is $X - D \mid (X > D)$. In addition, we suppose the insurer has entered into an excess of loss reinsurance contract with another company whereby the direct insurer pays a maximum of M units in respect of each individual claim, with the remaining amount being borne by the reinsurer. The policyholder is involved in all losses, the direct insurer is involved in losses which exceed D, and the reinsurer is involved in losses which exceed $D + M$. While in practice the insurers are only interested in insured losses, it is of interest to study the basic partition of a general loss into the three basic component parts, averaging over all losses.

We will model a general individual loss (gross – before taking account of the deductible and reinsurance) as a random variable X, and let C, Y and Z denote the components attributable to the policyholder (the customer), the direct insurer and the reinsurer, respectively. So we have $X = C + Y + Z$ in all cases.

We can express C, Y and Z as follows:

$$C = \begin{cases} X & \text{if } X \le D \\ D & \text{if } X > D, \end{cases}$$

$$Y = \begin{cases} 0 & \text{if } X \le D \\ X - D & \text{if } D < X \le D + M \\ M & \text{if } X > D + M, \end{cases}$$

$$Z = \begin{cases} 0 & \text{if } X \le D + M \\ X - (D + M) & \text{if } X > D + M. \end{cases}$$

We will derive some general theoretical expressions under three models for the loss X: (1) an exponential distribution (for illustration), (2) a two-parameter

Pareto distribution, and (3) a lognormal distribution. We will also compare our theoretical results with those of simulations. We will also study the corresponding results for aggregate losses, examining the expected payouts by the three parties involved.

7.2.1 Case 1 – exponential losses

Suppose, for illustration purposes, that the individual loss variable X has an exponential distribution with mean 1 (we are again taking the expected loss as our monetary unit for convenience). We note that in this case $\Pr(X > k) = e^{-k}$, for $k \geq 0$.

We can find expressions for $\mathbb{E}[C]$, $\mathbb{E}[Y]$ and $\mathbb{E}[Z]$ in turn quite easily by evaluating the integrals which define these expectations. But it is easier to find $\mathbb{E}[Z]$ first and then deduce the others. We have

$$\mathbb{E}[Z] = \int_{D+M}^{\infty} \{x - (D+M)\}e^{-x}\,dx = \int_0^{\infty} ue^{-(u+D+M)}\,du$$

$$= e^{-(D+M)} \int_0^{\infty} ue^{-u} = e^{-(D+M)}.$$

Alternatively, we can introduce the reinsurance claim variable Z^*, where

$$Z^* \equiv Z \mid (X > D+M) \equiv X - (D+M) \mid (X > D+M).$$

Then in this exponential case we have (by appealing to the lack of memory property (2.17)) that Z^* has the same distribution as X, with $\mathbb{E}[Z^*] = 1$. Using the relationship

$$\mathbb{E}[Z] = \mathbb{E}[Z^*]\Pr(X > D+M),$$

we then have again $\mathbb{E}[Z] = e^{-(D+M)}$.

Now consider the variable $Y + Z$ ($\equiv X - C$), which is the amount attributable to the direct insurer and reinsurer together and is easily defined as follows:

$$Y + Z = \begin{cases} 0 & \text{if } X \leq D \\ X - D & \text{if } X > D. \end{cases}$$

The variable $Y + Z$ has the same structure as Z, with $D + M$ replaced by D. Since $\mathbb{E}[Z] = e^{-(D+M)}$, it follows immediately that $\mathbb{E}[Y + Z] = e^{-D}$, and hence that $\mathbb{E}[Y] = e^{-D} - e^{-(D+M)}$.

Finally, since $1 = \mathbb{E}[X] = \mathbb{E}[C] + \mathbb{E}[Y] + \mathbb{E}[Z]$, we find $\mathbb{E}[C] = 1 - e^{-D}$.

To sum up:

$$\mathbb{E}[C] = 1 - e^{-D}, \quad \mathbb{E}[Y] = e^{-D} - e^{-(D+M)} \text{ and } \mathbb{E}[Z] = e^{-(D+M)}.$$

Note The fact that the expressions for $\mathbb{E}[C], \mathbb{E}[Y]$ and $\mathbb{E}[Z]$ are the same as those for $\Pr(X \leq D), \Pr(D < X \leq D + M)$ and $\Pr(X > D + M)$, respectively, is a feature of this particular case, in which $X \sim \text{Exp}(1)$.

We now find the expected sizes of losses conditional on the loss having a value less than D, between D and $D + M$, and greater than $D + M$. We know that $\mathbb{E}[Z^*] = \mathbb{E}[X - (D + M) \mid (X > D + M)] = 1$, and so we have

$$\mathbb{E}[X \mid (X > D + M)] = D + M + 1.$$

Similarly, we have $\mathbb{E}[X - D \mid (X > D)] = 1$, and so $\mathbb{E}[X \mid (X > D)] = D + 1$. Now, from

$$\mathbb{E}[X] = \mathbb{E}[X \mid (X \leq D)] \Pr(X \leq D) + \mathbb{E}[X \mid (X > D)] \Pr(X > D),$$

we have $1 = \mathbb{E}[X \mid (X \leq D)](1 - e^{-D}) + (D + 1)e^{-D}$, and so

$$\mathbb{E}[X \mid (X \leq D)] = 1 - \frac{De^{-D}}{1 - e^{-D}}.$$

Finally, from

$$\mathbb{E}[X] = \mathbb{E}[X \mid (X \leq D)] \Pr(X \leq D)$$
$$+ \mathbb{E}[X \mid (D < X \leq D + M)] \Pr(D < X \leq D + M)$$
$$+ \mathbb{E}[X \mid (X > D + M)] \Pr(X > D + M),$$

we have

$$1 = \left\{ 1 - \frac{De^{-D}}{1 - e^{-D}} \right\} (1 - e^{-D})$$
$$+ \mathbb{E}[X \mid (D < X \leq D + M)](e^{-D} - e^{-(D+M)})$$
$$+ (D + M + 1)e^{-(D+M)},$$

from which we get

$$\mathbb{E}[X \mid (D < X \leq D + M)] = 1 + D - \frac{Me^{-M}}{1 - e^{-M}}.$$

To sum up:

$$\mathbb{E}[X \mid (X \leq D)] = 1 - \frac{De^{-D}}{1 - e^{-D}},$$
$$\mathbb{E}[X \mid (D < X \leq D + M)] = 1 + D - \frac{Me^{-M}}{1 - e^{-M}},$$
$$\mathbb{E}[X \mid (X > D + M)] = D + M + 1.$$

Some numerical values are presented in Table 7.7. We note the obvious fact that $\Pr(X \leq D)$ increases as the deductible D increases. For fixed retention M, both

Table 7.7. *Probabilities for exponential losses*

	$M = 3$ $D = 0.1$	$M = 3$ $D = 0.3$	$M = 5$ $D = 0.1$	$M = 5$ $D = 0.3$
$\Pr(X \leq D)$	0.0952	0.2592	0.0952	0.2592
$\Pr(D < X \leq D + M)$	0.8598	0.7039	0.8987	0.7358
$\Pr(X > D + M)$	0.0450	0.0369	0.0061	0.0050

Table 7.8. *Expected values for exponential losses*

	$M = 3$ $D = 0.1$	$M = 3$ $D = 0.3$	$M = 5$ $D = 0.1$	$M = 5$ $D = 0.3$
$\mathbb{E}[C]$	0.0952	0.2592	0.0952	0.2592
$\mathbb{E}[Y]$	0.8598	0.7039	0.8987	0.7358
$\mathbb{E}[Z]$	0.0450	0.0369	0.0061	0.0050
$\mathbb{E}[XLO]$	0.0492	0.1425	0.0492	0.1425
$\mathbb{E}[XMED]$	0.9428	1.1428	1.0661	1.2661
$\mathbb{E}[XHI]$	4.1	4.3	6.1	6.3

Table 7.9. *Expected values (£) when one monetary unit is £5000*

	$M = 15\,000$ $D = 500$	$M = 15\,000$ $D = 1500$	$M = 25\,000$ $D = 500$	$M = 25\,000$ $D = 1500$
$\mathbb{E}[C]$	476	1296	476	1296
$\mathbb{E}[Y]$	4299	3520	4494	3679
$\mathbb{E}[Z]$	225	184	30	25
$\mathbb{E}[XLO]$	246	713	246	713
$\mathbb{E}[XMED]$	4714	5714	5330	6330
$\mathbb{E}[XHI]$	20\,500	21\,500	30\,500	31\,500

$\Pr(D < X \leq D + M)$ and $\Pr(X > D + M)$ decrease as the deductible increases, while, for fixed deductible, $\Pr(D < X \leq D + M)$ increases and $\Pr(X > D + M)$ decreases as the retention increases.

Let $\mathbb{E}[XLO]$, $\mathbb{E}[XMED]$ and $\mathbb{E}[XHI]$ represent $\mathbb{E}[X \mid (X \leq D)]$, $\mathbb{E}[X \mid (D < X \leq D + M)]$ and $\mathbb{E}[X \mid X > D + M]$, respectively; see Table 7.8. In the case that our unit ($\mathbb{E}[X]$) represents £5000, we have Table 7.9 (entries in £).

We note that, as we would expect, both $\mathbb{E}[C]$ and $\mathbb{E}[X \mid (X \leq D)]$ increase as the deductible D increases. For fixed retention M, both $\mathbb{E}[Y]$ and $\mathbb{E}[Z]$ decrease as the deductible increases, while for fixed deductible $\mathbb{E}[Y]$ increases and $\mathbb{E}[Z]$ decreases as the retention increases.

Table 7.10. *Theoretical results for exponential losses*

	Number	Expected number	Mean
X	10 000	–	1
C	10 000	–	0.2592
Y	10 000	–	0.7039
Z	10 000	–	0.0369
Z^*	–	369	1
$X:\ 0 < X \leq 0.3$	–	2592	0.1425
$X:\ 0.3 < X \leq 3.3$	–	7039	1.1428
$X:\ X > 3.3$	–	369	4.3

Table 7.11. *Simulation results for exponential losses*

	Number	Min.	Median	Mean	Max.
x	10 000	0.000	0.699	1.013	9.336
c	10 000	0.000	0.300	0.259	0.300
y	10 000	0.000	0.399	0.713	3.000
z	10 000	0.000	0.000	0.040	6.036
zstar	394	0.004	0.662	1.023	6.036
x: 0 < x <= 0.3	2606	0.000	0.142	0.144	0.300
x: 0.3 < x <= 3.3	7000	0.300	0.948	1.150	3.299
x: x > 3.3	394	3.304	3.962	4.323	9.336

To illustrate these results, a simulation of 10 000 losses was carried out (using **R**), using $X \sim \mathrm{Exp}(1)$, $D = 0.3$ and $M = 3$. The loss vector x was manipulated to produce vectors c, y, z and zstar containing the values indicated by these vectors' names, and the data were then summarised. Vectors containing the losses less than 0.3, between 0.3 and 3.3, and greater than 3.3 were also constructed and summarised.

The corresponding theoretical results are given in Table 7.10. The simulation results are given in Table 7.11.

Note   The simple **R** code used to produce the required vectors of data was as follows (xlo, xmed and xhi are the parts of the x-vector such that $0 < x \leq 0.3$, $0.3 < x \leq 3.3$ and $x > 3.3$, respectively):

```
d = 0.3
m = 3
x = rexp(10000,1)
c = pmin(x,d)       #'parallel min' c[i]=min(x[i],d[i])
z = pmax(0,x-d-m)   #'parallel max'
```

```
y = x - c - z
xlo = x[x <= d]
x2 = x[x > d]
xmed = x2[x2 <= d+m]
xhi = x[x > d+m]
zstar = xhi - (d+m)
```

We extend the situation now to one in which we study an aggregate loss variable S, where S has a compound Poisson distribution with claim rate λ and individual claim size variable $X \sim \text{Exp}(1)$. Each loss arises on a policy with deductible D in force. In addition, an excess of loss reinsurance arrangement is in place whereby the direct insurer pays a maximum of M on each loss.

Let S_C, S_I and S_R represent the aggregate amounts of the losses paid by the policyholder (the customer), the direct insurer and the reinsurer, respectively. Then we know the following:

$$S \sim \text{CP}(\lambda, F_X) \text{ and } \mathbb{E}[S] = \lambda\mathbb{E}[X] = \lambda,$$
$$S_C \sim \text{CP}(\lambda, F_C) \text{ and } \mathbb{E}[S_C] = \lambda\mathbb{E}[C] = \lambda(1 - e^{-D}),$$
$$S_I \sim \text{CP}(\lambda, F_Y) \text{ and } \mathbb{E}[S_I] = \lambda\mathbb{E}[Y] = \lambda\{e^{-D} - e^{-(D+M)}\},$$
$$S_R \sim \text{CP}(\lambda, F_Z) \text{ and } \mathbb{E}[S_R] = \lambda\mathbb{E}[Z] = \lambda e^{-(D+M)}.$$

To illustrate these results, a simulation of 10 000 aggregate losses was carried out (using R), using $X \sim \text{Exp}(1)$, $\lambda = 100$, $D = 0.3$ and $M = 3$. In this case the expected payouts are simply 100 times what they are for individual losses, that is $\mathbb{E}[S] = 100 \times 1 = 100$, $\mathbb{E}[S_C] = 100(1 - e^{-0.3}) = 25.92$, $\mathbb{E}[S_I] = 100(e^{-0.3} - e^{-3.3}) = 70.39$, $\mathbb{E}[S_R] = 100e^{-3.3} = 3.688$. The simulation results are given in Table 7.12. We note that the number of losses ($N \sim \text{Poi}(100)$) ranges from 63 to 139, and that the means of the simulated data are in close agreement with the theoretical results.

Note   The simulation was carried out by executing a function, here called cpsimrel and previously stored as a text file, as given below. The output

Table 7.12. *Further simulation results for exponential losses*

	Number	Min.	Median	Mean	Max.
N	10 000	63	100.0	99.81	139
S	10 000	49.65	99.39	99.73	156.5
S_C	10 000	15.66	25.80	25.86	36.97
S_I	10 000	30.47	69.84	70.18	115.1
S_R	10 000	0.000	3.189	3.700	17.89

from the function is an **R** object – a list containing five vectors – called
payout.list. The command

```
case1sim = cpsimrel(10000,100,1,0.3,3)
```

executes the function once, carrying out 10 000 simulations of the number of
claims and resulting aggregate payout variables in the case $\lambda = 100$, $\mu = 1$,
$D = 0.3$ and $M = 3$. The results (the list of five vectors) are held in the object
case1sim and can be retrieved using, for example,

```
summary(case1sim[[3]])
```

to obtain a summary of the values of S_C. The text defining the function is

```
cpsimrel = function(n,lam,mu,d,m){
xagg = cagg = yagg = zagg = numcl = (1:n)*0
for (i in 1:n){nc = rpois(1,lam)
if(nc == 0){
numcl[i] = xagg[i] = cagg[i] = yagg[i] = zagg[i] = 0}
else{x = rexp(nc,1/mu)
c = pmin(x,d)
z = pmax(x - d - m,0)
y = x - c - z
numcl[i] = nc
xagg[i] = sum(x)
cagg[i] = sum(c)
yagg[i] = sum(y)
zagg[i] = sum(z) } }
payout.list = list(numcl,xagg,cagg,yagg,zagg) }
```

7.2.2 Case 2 – Pareto losses

Suppose now that the individual loss variable X has a Pareto distribution, $X \sim$
Pa(α, λ). This distribution is fat-tailed, whereas the exponential is thin-tailed
(see Chapter 2). For the distribution to have finite mean we require $\alpha > 1$,
and for it to have finite variance we require $\alpha > 2$; it should be noted that, in
many applications of the Pareto distribution as a model for insurance losses,
the value of α estimated from data is less than 2.

Before proceeding, the reader should verify that, for $X \sim$ Pa(α, λ) with $\alpha > 1$
and for $b > 0$,

$$\int_b^\infty xf(x)dx = \int_b^\infty x\frac{\alpha\lambda^\alpha}{(\lambda + x)^{\alpha+1}} dx = \left(\frac{\lambda}{\lambda + b}\right)^\alpha \left(\frac{\lambda + b\alpha}{\alpha - 1}\right).$$

As in Case 1 we will take the mean of the distribution of X to be our monetary unit, that is $\mathbb{E}[X] = \lambda/(\alpha - 1) = 1$, and so we will take $\alpha = \lambda + 1$. We then have

$$\Pr(X > k) = \left(\frac{\lambda}{\lambda + k}\right)^{\lambda+1}, \text{ for } k \geq 0.$$

As in Case 1, we can find expressions for $\mathbb{E}[C], \mathbb{E}[Y]$ and $\mathbb{E}[Z]$ in turn quite easily by evaluating the integrals which define these expectations. But again it is easier to find $\mathbb{E}[Z]$ first and then deduce the others:

$$\begin{aligned}
\mathbb{E}[Z] &= \int_{D+M}^{\infty} \{x - (D + M)\} f(x) dx \\
&= \int_{D+M}^{\infty} x f(x) dx - (D + M) \Pr(X > D + M) \\
&= \left(\frac{\lambda}{\lambda + D + M}\right)^{\lambda+1} \left(\frac{\lambda + D + M}{\lambda} + D + M\right) \\
&\quad - (D + M)\left(\frac{\lambda}{\lambda + D + M}\right)^{\lambda+1} \\
&= \left(\frac{\lambda}{\lambda + D + M}\right)^{\lambda}.
\end{aligned}$$

It follows (see Case 1), on replacing $D + M$ by D in this result, that

$$\mathbb{E}[Y + Z] = \left(\frac{\lambda}{\lambda + D}\right)^{\lambda},$$

and hence

$$\mathbb{E}[Y] = \left(\frac{\lambda}{\lambda + D}\right)^{\lambda} - \left(\frac{\lambda}{\lambda + D + M}\right)^{\lambda}.$$

Finally, we have

$$\mathbb{E}[C] = 1 - \left(\frac{\lambda}{\lambda + D}\right)^{\lambda}.$$

We give some numerical values follow for the case $\alpha = 12$ ($X \sim \text{Pa}(12, 11)$, with mean 1 and variance 1.2 – which gives a model with variance fairly close to that of the exponential model used in Case 1), and using the same values for D and M, in Table 7.13.

These values are quite close to those obtained using $X \sim \text{Exp}(1)$ in Case 1. An explanation may be found in the choice of the parameter values: $\alpha = 12$ corresponds to a much less fat-tailed distribution than is the case for lower values of α, and is very much higher than would normally be appropriate for modelling insurance losses. Tables 7.14 and 7.15 follow for the more realistic cases $\alpha = 2.5$ ($X \sim \text{Pa}(2.5, 1.5)$, with mean 1 and variance 5) and $\alpha = 1.8$ ($X \sim \text{Pa}(1.8, 0.8)$, with mean 1 but without a finite variance).

Table 7.13. *Probabilities when X ~ Pa(12, 11)*

	M = 3 D = 0.1	M = 3 D = 0.3	M = 5 D = 0.1	M = 5 D = 0.3
Pr(X ≤ D)	0.1029	0.2759	0.1029	0.2759
Pr(D < X ≤ D + M)	0.8463	0.6811	0.8867	0.7151
Pr(X > D + M)	0.0508	0.0429	0.0103	0.0089

Table 7.14. *Probabilities when X ~ Pa(2.5, 1.5)*

	M = 3 D = 0.1	M = 3 D = 0.3	M = 5 D = 0.1	M = 5 D = 0.3
Pr(X ≤ D)	0.1490	0.3661	0.1490	0.3661
Pr(D < X ≤ D + M)	0.7903	0.5793	0.8264	0.6111
Pr(X > D + M)	0.0607	0.0546	0.0246	0.0229

Table 7.15. *Probabilities when X ~ Pa(1.8, 0.8)*

	M = 3 D = 0.1	M = 3 D = 0.3	M = 5 D = 0.1	M = 5 D = 0.3
Pr(X ≤ D)	0.1910	0.4363	0.1910	0.4363
Pr(D < X ≤ D + M)	0.7512	0.5109	0.7815	0.5379
Pr(X > D + M)	0.0578	0.0528	0.0274	0.0258

Within each table we can note the same behaviour as we found in Case 1 – but drawing general conclusions from a comparison of the tables must be approached with care, bearing in mind that having fixed the mean loss at 1, the loss distribution is constrained to be of the form $Pa(\lambda + 1, \lambda)$. For example, under this constraint,

$$\Pr(X > D + M) = \left(\frac{\lambda}{\lambda + D + M}\right)^{\lambda+1},$$

and for fixed $D+M$ this probability is not monotonic (increasing or decreasing) in λ.

Table 7.16 of tail probabilities shows the values of $\Pr(X > 3.1)$ again, alongside columns for smaller tails, namely $\Pr(X > 6)$, $\Pr(X > 8)$ and $\Pr(X > 10)$ – in the last three columns the probabilities are increasing as α decreases. The expected sizes of the components of the loss under $X \sim Pa(1.8, 0.8)$ are given in Table 7.17. In the case that our unit ($\mathbb{E}[X]$) represents £5000, we have Table 7.18 (entries in £).

Table 7.16. *Tail probabilities when* $X \sim Pa(\alpha, \alpha - 1)$

	Pr($X > 3.1$)	Pr($X > 6$)	Pr($X > 8$)	Pr($X > 10$)
$\alpha = 12$	0.0508	0.0054	0.0014	0.0004
$\alpha = 2.5$	0.0607	0.0179	0.0099	0.0061
$\alpha = 1.8$	0.0578	0.0212	0.0134	0.0092

Table 7.17. *Expected values when* $X \sim Pa(1.8, 0.8)$

	$M = 3$ $D = 0.1$	$M = 3$ $D = 0.3$	$M = 5$ $D = 0.1$	$M = 5$ $D = 0.3$
$\mathbb{E}[C]$	0.0899	0.2249	0.0899	0.2249
$\mathbb{E}[Y]$	0.6285	0.5046	0.7079	0.5782
$\mathbb{E}[Z]$	0.2816	0.2705	0.2022	0.1969

Table 7.18. *Expected values (£) when one monetary unit is £5000*

	$M = 15\,000$ $D = 500$	$M = 15\,000$ $D = 1500$	$M = 25\,000$ $D = 500$	$M = 25\,000$ $D = 1500$
$\mathbb{E}[C]$	450	1124	450	1124
$\mathbb{E}[Y]$	3142	2523	3539	2891
$\mathbb{E}[Z]$	1408	1353	1011	984

When we compare these results with those for Case 1 we see that, over all losses, the policyholder has a similar expected payout (to within £200), while that for the direct insurer is considerably reduced and that for the reinsurer is considerably increased (hugely increased in relative terms).

To illustrate these results, a simulation of 10 000 losses was carried out (using *R*), using X with a Pareto distribution with $\alpha = 1.8$, mean = 1, $D = 0.3$ and $M = 3$. As in Case 1, the loss vector x was manipulated to produce vectors c, y, z and zstar, and the data were then summarised. Vectors containing the losses less than 0.3, between 0.3 and 3.3, and greater than 3.3 were again constructed and summarised. The simulation results are given in Table 7.19.

The fat-tailed nature of the Pa(1.8, 0.8) distribution used here as our model for the losses results in a maximum observed loss of 194.5 (compare this with $\mathbb{E}[X] = 1$): in repeated simulations, we find that the maximum observed loss and maximum payout by the reinsurer exhibit high variation.

Table 7.19. *Simulation results when* $X \sim \text{Pa}(1.8, 0.8)$

	Number	Min.	Median	Mean	Max.
x	10 000	0.000	0.377	1.009	194.5
c	10 000	0.000	0.300	0.225	0.300
y	10 000	0.000	0.077	0.511	3.000
z	10 000	0.000	0.000	0.273	191.2
zstar	561	0.001	1.932	4.859	191.2
x: 0 < x <= 0.3	4344	0.000	0.115	0.127	0.300
x: 0.3 < x <= 3.3	5095	0.300	0.732	0.974	3.300
x: x > 3.3	561	3.301	5.232	8.159	194.5

Note The simple *R* code used to produce the required vectors of data is the same as that used in Case 1, except for the third line, in which the command

```
x = rexp(10000,1)
```

is replaced by

```
x = 0.8*(runif(10000)^(-1/1.8) -1)
```

As in Case 1 we extend the situation to one in which S has a compound Poisson distribution with claim rate now given the symbol v and individual claim size variable $X \sim \text{Pa}(\lambda + 1, \lambda)$, with mean 1. Each loss arises on a policy with deductible D in force. In addition, an excess of loss reinsurance arrangement is in place whereby the direct insurer pays a maximum of M on each loss. Let S_C, S_I and S_R be as before. Then we know the following:

$$S \sim \text{CP}(v, F_X) \text{ and } \mathbb{E}[S] = v\mathbb{E}[X] = v,$$
$$S_C \sim \text{CP}(v, F_C) \text{ and } \mathbb{E}[S_C] = v\mathbb{E}[C] = v\left\{1 - \left(\frac{\lambda}{\lambda+D}\right)^\lambda\right\},$$
$$S_I \sim \text{CP}(v, F_Y) \text{ and } \mathbb{E}[S_I] = v\mathbb{E}[Y] = v\left\{\left(\frac{\lambda}{\lambda+D}\right)^\lambda - \left(\frac{\lambda}{\lambda+D+M}\right)^\lambda\right\},$$
$$S_R \sim \text{CP}(v, F_Z) \text{ and } \mathbb{E}[S_R] = v\mathbb{E}[Z] = v\left(\frac{\lambda}{\lambda+D+M}\right)^\lambda.$$

To illustrate these results, a simulation of 10 000 aggregate losses was carried out using $X \sim \text{Pa}(1.8, 0.8)$, $v = 100$, $D = 0.3$ and $M = 3$. The expected payouts are simply 100 times what they are for individual losses calculated over all losses, that is $\mathbb{E}[S] = 100$, $\mathbb{E}[S_C] = 100 \times 0.2249 = 22.49$, $\mathbb{E}[S_I] = 100 \times 0.5046 = 50.46$ and $\mathbb{E}[S_R] = 100 \times 0.2705 = 27.05$.

The simulation results are given in Table 7.20. We note that the number of losses ($N \sim \text{Poi}(100)$) ranges from 66 to 141, and that the means of the simulated data are in close agreement with the theoretical results.

Table 7.20. *Further simulation results when* $X \sim \text{Pa}(1.8, 0.8)$

	Number	Min.	Median	Mean	Max.
N	10000	66	100.0	99.86	141
S	10000	36.05	92.68	100.6	2225
S_C	10000	14.01	22.39	22.45	31.74
S_I	10000	18.92	50.10	50.45	94.37
S_R	10000	0.000	18.08	27.69	2144

The fat-tailed nature of the $\text{Pa}(1.8, 0.8)$ distribution used here as our model for the individual losses results in a maximum observed aggregate loss of 2225 (compare this with $\mathbb{E}[X] = 100$): in repeated simulations, we find that the maximum observed aggregate loss and maximum aggregate payout by the reinsurer exhibit high variation.

Note The simulation was carried out by executing a function as in Case 1 – the function, called 'cpsimre2, is the same as cpsimre1, except that the function is defined as

```
cpsimre2 = function(n,rate,lambda,d,m)
```

and, in line 3, the statement

```
nc = rpois(1,lam)
```

is replaced by

```
nc = rpois(1,rate)
```

and, in line 5, the statement

```
x = rexp(nc,1/mu)
```

is replaced by

```
x = lambda*(runif(nc)^(-1/(lambda + 1)) -1)
```

The command

```
case2sim = cpsimre2(10000,100,0.8,0.3,3)
```

executes the function once, carrying out 10000 simulations of the number of claims and resulting aggregate payout variables in the case $v = 100$, $\lambda = 0.8$, $D = 0.3$ and $M = 3$. The results (the list of five vectors) are held in the object case2sim and can be retrieved using, for example,

```
summary(case2sim[[4]])
```

to obtain a summary of the values S_I.

7.2.3 Case 3 – lognormal losses

Suppose now that the individual loss variable X has a lognormal(μ, σ) distribution (so $\log X \sim N(\mu, \sigma^2)$). Then $\Pr(X > k) = 1 - \Phi((\log k - \mu)/\sigma)$.

For convenience, let us use the symbol g for

$$g(\mu, \sigma) = e^{\mu + \sigma^2/2} = \mathbb{E}[X],$$

$h_1(x)$ for

$$h_1(x, \mu, \sigma) = \Phi\left(\frac{\log x - \mu - \sigma^2}{\sigma}\right)$$

and $h_2(x)$ for

$$h_2(x, \mu, \sigma) = \Phi\left(\frac{\log x - \mu}{\sigma}\right).$$

Before proceeding, the reader should verify that, for $X \sim \text{lognormal}(\mu, \sigma)$,

$$\int_0^a x f(x) dx = g\, h_1(a)$$

(see Exercise 5.7).

The following results are then easily obtained:

$$\mathbb{E}[C] = g\, h_1(D) + D\{1 - h_2(D)\},$$
$$\mathbb{E}[Y] = g\,\{h_1(D + M) - h_1(D)\} - D\{h_2(D + M) - h_2(D)\}$$
$$+ M\{1 - h_2(D + M)\},$$
$$\mathbb{E}[Z] = g\,\{1 - h_1(D + M)\} - (D + M)\{1 - h_2(D + M)\}.$$

As before we will take the mean of the loss distribution to be $\mathbb{E}[X] = 1$, so setting $\mu + \sigma^2/2 = 0$ we have $\sigma^2 = -2\mu$, $g = 1$,

$$h_1(x) = \Phi\left(\frac{\log x + \mu}{\sqrt{-2\mu}}\right), \text{ and } h_2(x) = \Phi\left(\frac{\log x - \mu}{\sqrt{-2\mu}}\right).$$

Some numerical values follow for the two lognormal distributions below:

(1) $\mu = -0.5 \log 2 (= -0.34657)$, $\sigma = (\log 2)^{0.5} (= 0.83255)$, for which $\mathbb{E}[X] = 1$ and $\text{Var}[X] = 1$, and
(2) $\mu = -0.5 \log 5 (= -0.80472)$, $\sigma = (\log 5)^{0.5} (= 1.26864)$, for which $\mathbb{E}[X] = 1$, and $\text{Var}[X] = 4$.

Probabilities of the form $\Pr(X \leq k)$ can be found easily using **R** as in, for example,

Table 7.21. *Probabilities and expected values for lognormal losses with* $\mathbb{E}[X] = \text{Var}[X] = 1$

	M = 3, D = 0.1	M = 3, D = 0.3	M = 5, D = 0.1	M = 5, D = 0.3
Pr($X \le D$)	0.0094	0.1515	0.0094	0.1515
Pr($D < X \le D + M$)	0.9527	0.8163	0.9818	0.8407
Pr($X > D + M$)	0.0379	0.0321	0.0088	0.0078
$\mathbb{E}[C]$	0.0998	0.2858	0.0998	0.2858
$\mathbb{E}[Y]$	0.8449	0.6658	0.8835	0.6991
$\mathbb{E}[Z]$	0.0553	0.0484	0.0167	0.0151

Table 7.22. *Probabilities and expected values for lognormal losses with* $\mathbb{E}[X] = 1$ *and* $\text{Var}[X] = 4$

	M = 3, D = 0.1	M = 3, D = 0.3	M = 5, D = 0.1	M = 5, D = 0.3
Pr($X \le D$)	0.1189	0.3765	0.1189	0.3765
Pr($D < X \le D + M$)	0.8177	0.5659	0.8536	0.5979
Pr($X > D + M$)	0.0635	0.0576	0.0275	0.0257
$\mathbb{E}[C]$	0.0953	0.2437	0.0953	0.2437
$\mathbb{E}[Y]$	0.7031	0.5668	0.7872	0.6441
$\mathbb{E}[Z]$	0.2016	0.1895	0.1175	0.1122

```
a = -0.5*log(2)
b = sqrt(log(2))
d = c(0.1,0.3)
plnorm(d,a,b)
```

which return the values 0.0094016 and 0.1515422 for Pr($X \le 0.1$) and Pr($X \le 0.3$), respectively, in the case $X \sim$ lognormal($-0.5 \log 2, (\log 2)^{0.5}$).

(1) $X \sim$ lognormal with $\mathbb{E}[X] = 1$ and $\text{Var}[X] = 1$ (see Table 7.21);
(2) $X \sim$ lognormal with $\mathbb{E}[X] = 1$ and $\text{Var}[X] = 4$ (see Table 7.22).

Here, with a unit ($\mathbb{E}[X]$) representing £5000, we have the corresponding values (entries in £) in Table 7.23. When we compare these results with those for Case 1 ($X \sim$ Exp with mean £5000) and Case 2 ($X \sim$ Pa(1.8, 0.8) in units of £5000), we see that the policyholder's position is similar in all three cases. Over all losses, the direct insurer's expected payout in this last (lognormal) case lies between those in the exponential and Pareto cases. Again, over all losses, the reinsurer's expected payout in the lognormal case is about 70% of

Table 7.23. *Expected values (£) when one monetary unit is £5000*

	M = 15 000 D = 500	M = 15 000 D = 1500	M = 25 000 D = 500	M = 25 000 D = 1500
$\mathbb{E}[C]$	476	1219	476	1219
$\mathbb{E}[Y]$	3516	2834	3936	3220
$\mathbb{E}[Z]$	1008	947	588	561

Table 7.24. *Simulation results for lognormal losses with $\mathbb{E}[X] = 1$ and $\mathrm{Var}[X] = 4$*

	Number	Min.	Median	Mean	Max.
x	10 000	0.005	0.454	1.016	112.6
c	10 000	0.005	0.300	0.244	0.300
y	10 000	0.000	0.154	0.570	3.000
z	10 000	0.000	0.000	0.202	109.3
zstar	576	0.002	1.757	3.503	109.3
x: 0 < x <= 0.3	3792	0.005	0.146	0.152	0.300
x: 0.3 < x <= 3.3	5632	0.300	0.780	1.006	3.294
x: x > 3.3	576	3.302	5.057	6.803	112.6

that in the Pareto case when the lower retention level (M) is in place, and is a bit less than 60% of it when the higher retention level is in place. In this limited comparison, the direct insurer's liability is highest under the exponential loss model, while the reinsurer's liability is highest under the Pareto loss model.

To illustrate the results, a simulation of 10 000 losses was carried out (using *R*), using $X \sim$ lognormal with $\mu = -0.5 \log 5$, $\sigma = (\log 5)^{0.5}$, $\mathbb{E}[X] = 1$, $\mathrm{Var}[X] = 4$, $D = 0.3$ and $M = 3$. The simulation results are given in Table 7.24. The fat-tailed nature of the lognormal distribution used here as our model for the losses results in a maximum observed loss of 112.6 (compare this with $\mathbb{E}[X] = 1$): in repeated simulations, the maximum observed loss and maximum payout by the reinsurer again exhibit high variation.

Note The simple *R* code used to produce the required vectors of data is the same as that used in Case 1, except for the third line, in which the command

```
x = rexp(10000,1)
```

is replaced by

```
x = rlnorm(10000, -0.5 * log (5), log (5)^(0.5))
```

Table 7.25. *Further simulation results for lognormal losses with*
$\mathbb{E}[X] = 1$ *and* $\text{Var}[X] = 4$

	Number	Min.	Median	Mean	Max.
N	10 000	58	100.0	100.2	146
S	10 000	43.38	98.30	101.0	333.2
S_C	10 000	14.64	24.38	22.44	35.76
S_I	10 000	22.46	56.67	57.03	101.1
S_R	10 000	0.000	15.89	19.53	236.3

As in the earlier cases, we now extend the situation to one in which we study an aggregate loss variable S, where S has a compound Poisson distribution with claim rate λ and individual claim size variable $X \sim$ lognormal(μ, σ). Each loss arises on a policy with deductible D in force. In addition, an excess of loss reinsurance arrangement is in place whereby the direct insurer pays a maximum of M on each loss. Let S_C, S_I and S_R be as before. The expected payouts are simply λ times what they are for individual losses when averaged over all losses.

To illustrate these results, a simulation of 10 000 aggregate losses was carried out (using $\boldsymbol{R}$), using $X \sim$ lognormal with $\mu = -0.5 \log 5$, $\sigma = (\log 5)^{0.5}$, $\lambda = 100$, $D = 0.3$ and $M = 3$. In this case the expected payouts are $\mathbb{E}[S] = 100 \times 1 = 100$, $\mathbb{E}[S_C] = 100 \times 0.2437 = 24.37$, $\mathbb{E}[S_I] = 100 \times 0.5668 = 56.68$ and $\mathbb{E}[S_R] = 100 \times 0.1895 = 18.95$.

The simulation results are as in Table 7.25. We note that the number of losses ($N \sim \text{Poi}(100)$) ranges from 58 to 146, and that the means of the simulated data are in close agreement with the theoretical results.

The fat-tailed nature of the lognormal distribution used here as our model for the individual losses results in a maximum observed aggregate loss in this particular simulation of 333.2 (compare this with $\mathbb{E}[X] = 100$); in repeated simulations, we again find that the maximum observed aggregate loss and maximum aggregate payout by the reinsurer exhibit high variation.

Note The simulation was carried out by executing a function as in Case 1 – the function, called cpsimre3, is the same as cpsimre1, except that the function is defined as

```
cpsimre3 = function(n,lam,mu,sigma,d,m)
```

and, in line 5, the statement

```
x = rexp(nc,1/mu)
```

is replaced by

```
x = rlnorm(nc,mu,sigma)
```

The commands

```
mu = -0.5* log(5)
sigma = log (5)^(0.5)
case3sim = cpsimre3(10000, 100, mu, sigma, 0.3, 3)
```

executed the function once, carrying out 10 000 simulations of the number of claims and resulting aggregate payout variables in the case $\lambda = 100$, $\mu = -0.5 \log 5$, $\sigma = (\log 5)^{0.5}$, $D = 0.3$, and $M = 3$. The results (the list of five vectors) are held in the object case3sim and can be retrieved using, for example,

```
summary(case3sim[[5]])
```

to obtain a summary of the values S_R.

7.3 Case study 3: reinsurance and ruin

7.3.1 Introduction

In this case study we bring together ideas and techniques from Chapters 5 and 6 in order to investigate the effect of reinsurance on various ruin quantities. In particular, as in Chapter 6, we consider the classical risk model in which claims $X_1, X_2, \ldots$ are iid positive random variables with distribution function F_X, moment generating function $M_X(r)$ and finite mean μ_X, arriving in a Poisson process with rate λ, and where the X_i are independent of the claim-arrivals process. The premium income rate is $c = (1 + \theta)\lambda\mu_X$, where θ is the relative safety loading, assumed positive. The initial capital is $u \geq 0$, so that the surplus at time $t \geq 0$ is given by

$$U(t) = u + ct - \sum_{i=1}^{N(t)} X_i, \qquad (7.1)$$

where $N(t)$ is the number of claims arriving in $(0, t]$.

With this model, we consider two ruin quantities. The first is the adjustment coefficient R (if it exists, see §6.2.3), and the second is the ruin probability $\psi(u)$. We hope to have small ruin probabilities. The Lundberg inequality in Theorem 6.1 states that $\psi(u) \leq e^{-Ru}$ for all $u \geq 0$, so that in general we prefer to have larger R so that the upper bound on the ruin probability is smaller. One way to approach reinsurance is to look at its effect on either R or $\psi(u)$, and that is what we do here.

Suppose now that the direct insurer takes out a reinsurance contract such that, on a claim X_i, the direct insurer pays Y_i and the reinsurer pays Z_i, where

$$Y_i = g(X_i) \quad \text{and} \quad Z_i = X_i - g(X_i).$$

Thus the reinsurance is "per claim". Proportional reinsurance with retained proportion β and excess of loss reinsurance with retention limit M both fit into this scheme, with $g(x) = \beta x$ and $g(x) = \min(x, M)$, respectively, and there are other possibilities for $g(\cdot)$, one of which is considered in §7.3.4. We require that the function g takes the non-negative real numbers into the non-negative real numbers, and that

(i) $g(0) = 0$,
(ii) $0 \le g(x) \le x$, and
(iii) $g(x)$ and $x - g(x)$ are non-decreasing.

These are all common sense requirements for $g(\cdot)$: (i) means that, if the original claim X_i is zero, then both the direct insurer's and the reinsurer's payouts are zero; (ii) means that the direct insurer's and the reinsurer's payouts are both non-negative and never larger than the original claim size; and (iii) means that neither the direct insurer's payout nor the reinsurer's payout decreases if the original claim size increases.

The expected payouts for the direct insurer and reinsurer are μ_Y and μ_Z, respectively, where

$$\mu_Y = \mathbb{E}[Y_i] = \int g(x) F_X(dx) \quad \text{and} \quad \mu_Z = \mu_X - \mu_Y, \qquad (7.2)$$

although, as in Chapter 5 and §7.2, in practice it can sometimes be easier to find μ_Z by direct integration and then use $\mu_Y = \mu_X - \mu_Z$ to find μ_Y. By (ii) above, we see that

$$\mu_Y = \int g(x) F_X(dx) \le \int x F_X(dx) = \mu_X \, (<\infty).$$

The claim payouts for the direct insurer are $Y_1, Y_2, \ldots$. These are iid because the original X_i are iid and each Y_i is $g(X_i)$. We write F_Y for the distribution function of Y_i. The Y_i arrive at the direct insurer at the instants when the original claim arrives, that is the Y_i arrive in a Poisson process with rate λ. We further note that this Poisson process is independent of the Y_i because of the independence assumptions in the original risk model without reinsurance.

The reinsurer charges a premium for taking on the reinsurance risk, and we suppose that this is calculated using the expected value principle (see Chapter 4) with security loading ζ, so that the premium rate for the reinsurer is $(1 + \zeta)\lambda\mu_Z$ per unit time. As in §5.4, we require $\zeta > \theta$ in order to avoid the possibility that the direct insurer can make a profit while bearing none of the risk.

The direct insurer's surplus at time t, taking account of reinsurance (that is, net of reinsurance), is $U_I(t)$, where

$$U_I(t) = u + (1 + \theta)\lambda\mu_X t - (1 + \zeta)\lambda\mu_Z t - \sum_{i=1}^{N(t)} Y_i.$$

This is of the same form as (7.1), with c replaced by

$$c_I = (1 + \theta)\lambda\mu_X - (1 + \zeta)\lambda\mu_Z \tag{7.3}$$

and X_i replaced by Y_i. Thus we can model the direct insurer's surplus, net of reinsurance, as that of a classical risk model with Poisson rate λ, initial capital u, claim-size distribution function F_Y and premium rate c_I.

For the net profit condition to be satisfied in the direct insurer's model, net of reinsurance, we require that $c_I > \lambda\mu_Y$, and, as explained in §5.4, this means that $\zeta < \mu_X \theta / \mu_Z$. Putting this together with the requirement that $\zeta > \theta$, we find (as in §5.4) that

$$\theta < \zeta < \frac{\mu_X}{\mu_Z}\theta = \left(1 + \frac{\mu_Y}{\mu_Z}\right)\theta. \tag{7.4}$$

The direct insurer's relative safety loading, taking account of reinsurance, is

$$\begin{aligned} \theta_I &= \frac{c_I - \lambda\mu_Y}{\lambda\mu_Y} = \frac{(1 + \theta)\lambda\mu_X - (1 + \zeta)\lambda\mu_Z - \lambda\mu_Y}{\lambda\mu_Y} \\ &= \frac{\theta\mu_X - \zeta\mu_Z}{\mu_Y} = \theta - (\zeta - \theta)\frac{\mu_Z}{\mu_Y}, \end{aligned} \tag{7.5}$$

where we have used $\mu_X - \mu_Y - \mu_Z = 0$ (see (7.2)).

Similarly, the reinsurer's surplus is also modelled as a classical risk model, with surplus at time t given by

$$U_{Re}(t) = u_{Re} + (1 + \zeta)\lambda\mu_Z t - \sum_{i=1}^{N(t)} Z_i,$$

where u_{Re} is the reinsurer's initial capital, and we note that u_{Re} is not necessarily the same as the direct insurer's initial surplus u. Further, some of the Z_i may be zero, as happens, for example, in the case of excess of loss reinsurance. In this case, we can equivalently model the reinsurer's surplus by only counting the non-zero claims, which arrive according to a Poisson process with rate

$\lambda^* = \lambda \Pr(Z_i > 0)$. To see this, let Z_i^* be as in §5.1.1, so that Z_i^* is distributed as $Z_i \mid Z_i > 0$. Let $N^*(t)$ be the number of non-zero reinsurance claims in $(0, t]$, so that $N^*(t) \sim \text{Poi}(\lambda^* t)$ (see §5.1.3). The process $\{N^*(t) : t \geq 0\}$ inherits the property of stationary independent increments from the corresponding property of the Poisson process $\{N(t)\}$, and so $\{N^*(t)\}$ is a Poisson process with rate λ^*. We also have that

$$\sum_{i=1}^{N(t)} Z_i = \sum_{i=1}^{N^*(t)} Z_i*,$$

since we have only omitted the zero Z_i in the left-hand sum in order to obtain the right-hand sum. Hence the reinsurer's surplus at time t is given by

$$U_{\text{Re}}(t) = u_{\text{Re}} + c^* t - \sum_{i=1}^{N^*(t)} Z_i^*,$$

where $c^* = (1 + \zeta)\lambda^* \mathbb{E}[Z_i^*]$.

In this case study we will be concerned with ruin quantities for the direct insurer, taking account of reinsurance, and for these we will need the moment generating function of the Y_i, given by

$$M_Y(r) = \mathbb{E}[e^{rY_1}] = \mathbb{E}[e^{rg(X_1)}] = \int e^{rg(x)} F_X(dx). \tag{7.6}$$

Because $0 \leq g(x) \leq x$, we have immediately from (7.6) that, for $r \geq 0$, $M_Y(r) \leq M_X(r)$, so that if $M_X(r)$ is finite for some $r > 0$, then $M_Y(r)$ is also finite. We also know that $M_Y(r) < \infty$ for all $r \leq 0$ because Y_i is a non-negative random variable.

In the following sections, we consider specific choices for g and F_X, and we look at the effects of the reinsurance on the adjustment coefficient and/or the probability of ruin.

7.3.2 Proportional reinsurance

In this section, we consider the case of proportional reinsurance (see §5.2) with retained proportion β, $0 < \beta < 1$, where $g(x) = \beta x$ for $x \geq 0$, so that

$$Y_i = \beta X_i \qquad \text{and} \qquad Z_i = (1 - \beta)X_i.$$

Note that the function $g(x) = \beta x$ satisfies (i), (ii) and (iii) in §7.3.1. We have

$$\mu_Y = \beta \mu_X, \quad \mu_Z = (1 - \beta)\mu_X,$$

and, from the expression for c_I in (7.3),

$$
\begin{aligned}
c_I &= (1 + \theta)\lambda\mu_X - (1 + \zeta)\lambda\mu_Z \\
&= (1 + \theta)\lambda\mu_X - (1 + \zeta)\lambda(1 - \beta)\mu_X \\
&= ((1 + \theta) - (1 + \zeta)(1 - \beta))\,\lambda\mu_X \\
&= (\theta - \zeta + (1 + \zeta)\beta)\,\lambda\mu_X.
\end{aligned}
$$

For positive relative safety loading net of reinsurance, we require $c_I > \lambda\mu_Y$, and, for given θ and ζ, this means that

$$(\theta - \zeta + (1 + \zeta)\beta)\,\lambda\mu_X > \beta\lambda\mu_X,$$

or equivalently

$$\theta - \zeta + (1 + \zeta)\beta > \beta.$$

Hence we require

$$\beta > 1 - \frac{\theta}{\zeta}. \tag{7.7}$$

From the expression for θ_I in (7.5), the relative safety loading net of reinsurance is

$$
\begin{aligned}
\theta_I &= \frac{\theta\mu_X - \zeta\mu_Z}{\mu_Y} = \frac{\theta\mu_X - \zeta(1 - \beta)\mu_X}{\beta\mu_X} \\
&= \frac{\theta - \zeta + \beta\zeta}{\beta} = \zeta - \frac{(\zeta - \theta)}{\beta}.
\end{aligned} \tag{7.8}
$$

For the adjustment coefficient, assume that the moment generating function $M_X(r)$ of the X_i satisfy conditions (i) and (ii) of Lemma 6.5, that is assume that there exists $r_{X,\infty}$, $0 < r_{X,\infty} \le \infty$, such that $M_X(r) < \infty$ for all $r < r_{X,\infty}$, and $M_X(r) \to \infty$ as $r \to r_{X,\infty}^-$. By Lemma 6.5, the adjustment coefficient R exists for the original direct insurer's model without reinsurance.

For the direct insurer's model with reinsurance, the moment generating function of Y_i (see (7.6)) is

$$M_Y(r) = \int e^{r\beta x} F_X(dx) = M_X(\beta r).$$

Thus we have $M_Y(r) < \infty$ for all $r < r_{X,\infty}/\beta$, and $M_Y(r) \to \infty$ as $r \to (r_{X,\infty}/\beta)^-$. This means that the conditions of Lemma 6.5 are satisfied by $M_Y(r)$, with r_∞ in Lemma 6.5 replaced by $r_{Y,\infty} = r_{X,\infty}/\beta$, and so the adjustment coefficient R_I exists for the direct insurer taking account of reinsurance. From (6.10) the equation for R_I is

$$M_Y(r) - 1 = (1 + \theta_I)\mu_Y r,$$

and, for proportional reinsurance, this is

$$M_X(\beta r) - 1 = (1 + \theta_I)\beta\mu_X r.$$

7.3.3 Proportional reinsurance with exponential claim sizes

We now suppose that the claims sizes $X_1, X_2, \ldots$ are independent exponentially distributed random variables with mean μ_X. For the direct insurer without reinsurance, we have

$$M_X(r) = \frac{1}{1 - \mu_X r} \text{ for } r < r_{X,\infty} = \frac{1}{\mu_X},$$

and from Example 6.2 the adjustment coefficient is

$$R = \frac{\theta}{(1 + \theta)\mu_X}.$$

Consider proportional reinsurance with retained proportion β, where, from (7.7), we have $\beta \in (1 - \theta/\zeta, 1)$. Then $Y_i = \beta X_i$, so that

$$\Pr(Y_i > y) = \Pr\left(X_i > \frac{y}{\beta}\right) = \exp\left(-\frac{y}{\beta\mu_X}\right),$$

and hence $Y_1, Y_2, \ldots$ are independent exponentially distributed random variables with mean $\beta\mu_X$. Thus the direct insurer's risk model including reinsurance is a classical risk model with exponentially distributed claims; that is, it is of exactly the same type as the original model for the direct insurer without reinsurance, but with a new mean claim size $\beta\mu_X$ and a new premium income rate c_I. This is very good news, as it means that we already know that such a model has an adjustment coefficient, and, moreover, Example 6.2 shows that, when reinsurance is included, the direct insurer has an adjustment coefficient given by

$$
\begin{aligned}
R_I &= \frac{\theta_I}{(1 + \theta_I)\mu_Y} \\
&= \frac{\theta_I}{(1 + \theta_I)\beta\mu_X} \\
&= \frac{\theta - \zeta + \beta\zeta}{(\beta + \theta - \zeta + \beta\zeta)\beta\mu_X},
\end{aligned}
\tag{7.9}
$$

where we have used (7.8) for θ_I. For a concrete example, consider the case where $\mu_X = 1$, $\theta = 0.3$ and $\zeta = 0.4$. Without reinsurance, the direct insurer has adjustment coefficient

$$R = \frac{\theta}{(1 + \theta)\mu_X} = \frac{0.3}{1.3} = 0.23077. \tag{7.10}$$

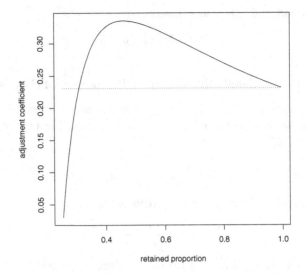

Figure 7.5. The adjustment coefficient R_I with proportional reinsurance plotted against the retained proportion β for exponentially distributed claims with mean 1, $\theta = 0.2$ and $\zeta = 0.4$. The dotted line shows the adjustment coefficient R without reinsurance.

When the direct insurer takes out proportional reinsurance with retained proportion β, by (7.7), we need β to satisfy $\beta > 1 - \theta/\zeta$, that is $\beta > 0.25$. Then we find that, with reinsurance, the direct insurer has adjustment coefficient

$$R_I = \frac{0.4\beta - 0.1}{(1.4\beta - 0.1)\beta} = \frac{4\beta - 1}{14\beta^2 - \beta}.$$

Figure 7.5 shows R_I plotted against β for $\beta > 0.25$ and also shows the adjustment coefficient R without reinsurance. Recall that in general we prefer larger values of R_I over smaller values. We see that when $\beta = 1$ (when the direct insurer retains the whole of every claim), then, as expected, R_I is the same as the adjustment coefficient R for the original model without reinsurance. Further, we note from the figure that for smaller values of β, taking out the reinsurance leads to an adjustment coefficient R_I which is smaller than the original one, whereas for moderate and larger values of β, the reinsurance leads to an increase in the adjustment coefficient. From the graph, we see that there is an optimum retained proportion β_{opt} that maximises the adjustment coefficient. We can find the value of β_{opt} via differentiation as follows. We find (after some simplification) that

$$\frac{dR_I}{d\beta} = \frac{-56\beta^2 + 28\beta - 1}{(14\beta^2 - \beta)^2}.$$

The quadratic in the numerator is zero when $\beta = 0.0387$ and $\beta = 0.461$. Only the second of these is greater than 0.25, and so $\beta_{\text{opt}} = 0.461$ (and from Figure 7.5 this is a maximum). The maximum value of the adjustment coefficient is

$$R_{\text{opt}} = \frac{4\beta_{\text{opt}} - 1}{14\beta_{\text{opt}}^2 - \beta_{\text{opt}}} = 0.3357.$$

We next consider the effect of proportional reinsurance on the probability of ruin in the case of exponential claims. For the classical risk model with exponential claims, we have an exact expression for the probability of ruin, given in Example 6.12. Using this, we find that the probability of ruin with initial capital u for the direct insurer without reinsurance is

$$\psi(u) = \frac{1}{1 + \theta} e^{-Ru}.$$

Recalling that the direct insurer's model with reinsurance is also a classical risk model with exponentially distributed claims, we know that, with reinsurance, the direct insurer's probability of ruin is

$$\psi_I(u) = \frac{1}{1 + \theta_I} e^{-R_I u},$$

where θ_I and R_I are the relative safety loading and the adjustment coefficient for retained proportion β, as given in (7.8) and (7.9), respectively. With $\mu_X = 1$, $\theta = 0.3$ and $\zeta = 0.4$ as above, it is interesting to consider the probability of ruin for the particular choice $\beta = \beta_{\text{opt}}$. Figure 7.6 shows this, together with the probability of ruin without reinsurance. The retained proportion β_{opt} produces the largest adjustment coefficient and hence the smallest upper bound (via Lundberg's inequality) on the probability of ruin. Nevertheless, we note that the resulting probability of ruin is not smaller than the original ruin probability for every value of u. From Figure 7.6, we see that for small values of u the probability of ruin without reinsurance is smaller than the probability of ruin with reinsurance with retained proportion β_{opt}.

For a particular value of u, we might also investigate the effect of other choices of β. Table 7.26 shows $\psi_I(u)$ for $u = 5$ and $u = 10$ for various β-values. The value $\beta = 1$ corresponds to the original model without reinsurance. From the table, we see that, when $u = 5$, the choice $\beta = 0.5$ has a smaller ruin probability $\psi_I(5)$ than that obtained with β_{opt}. Thus, for a particular u, it is possible to have other β-values with a smaller ruin probability than that achieved by β_{opt}. In addition, we see from Table 7.26 that, with some choices of u and β, it is possible for the reinsurance to lead to a situation where the probability of ruin $\psi_I(u)$ with reinsurance is larger than the probability of ruin

Table 7.26. *Ruin probabilities $\psi_I(5)$ and $\psi_I(10)$ with proportional reinsurance with retained proportion β for exponential claims with mean 1, $\theta = 0.3$, $\zeta = 0.4$*

β	0.30	0.40	$\beta_{opt} = 0.46$	0.50	0.60	1
$\psi_I(5)$	0.3308	0.1703	0.1578	0.1574	0.1611	0.2243
$\psi_I(10)$	0.1167	0.03335	0.02945	0.002973	0.03163	0.07653

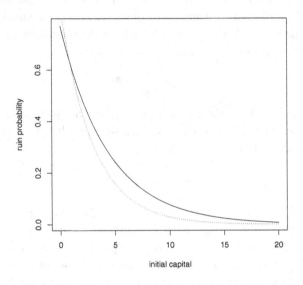

Figure 7.6. The probability of ruin without reinsurance (solid line) and the probability of ruin with proportional reinsurance with retained proportion β_{opt} for exponentially distributed claims with mean 1, $\theta = 0.2$ and $\zeta = 0.4$.

$\psi(u)$ without reinsurance. For example, if $u = 10$, then $\psi(10) = 0.07653$, but with proportional reinsurance with $\beta = 0.3$ has $\psi_I(10) = 0.1167$. We saw a similar effect of low β-values on the direct insurer's profitability in §5.4.

7.3.4 Excess of loss reinsurance in a layer

We now consider a different type of reinsurance where, on a claim X_i, the direct insurer and the reinsurer pay Y_i and Z_i, respectively, where

$$Y_i = \begin{cases} X_i & \text{if } X_i \le M \\ M & \text{if } M < X_i \le M + A \\ X_i - A & \text{if } X_i > M + A \end{cases} \qquad (7.11)$$

and

$$
Z_i = \begin{cases} 0 & \text{if } X_i \leq M \\ X_i - M & \text{if } M < X_i \leq M + A \\ A & \text{if } X_i > M + A, \end{cases}
$$

where M and A are positive real numbers. In this set-up, the reinsurer's payout is the excess of the original claim X_i over the retention limit M up to a maximum reinsurance payout of A, that is this is excess of loss reinsurance with retention M in a layer of size A. It is easy to check that $Y_i + Z_i = X_i$. To avoid trivialities, throughout this section we assume that

$$
\Pr(X_i \leq M) > 0, \Pr(M < X_i \leq M + A) > 0, \Pr(X_i > M + A) > 0. \qquad (7.12)
$$

This reinsurance scheme fits into that of §7.3.1 with, for $x \geq 0$,

$$
g(x) = \begin{cases} x & \text{if } x \leq M \\ M & \text{if } M < x \leq M + A \\ x - A & \text{if } x > M + A, \end{cases}
$$

which may be written

$$
g(x) = \min(x, M) + \max(x - M - A, 0).
$$

Plots of $g(x)$ and $x - g(x)$ against x are shown in Figure 7.7. Note that, in each plot, the scale on the vertical axis is different from the scale on the horizontal axis. From Figure 7.7, we see that $g(0) = 0$, $g(x)$ and $x - g(x)$ are non-decreasing, and $0 \leq g(x) \leq x$, so that the conditions (i), (ii) and (iii) in §7.3.1 are satisfied.

The reinsurer's expected payout on a claim X_i is

$$
\mu_Z = \int_{(M,M+A]} (x - M) F_X(dx) + A \Pr(X_i > M + A) \qquad (7.13)
$$

$$
= \int_{(M,M+A]} x F_X(dx) - M \Pr(M < X_i \leq M + A)
$$
$$
+ A \Pr(X_i > M + A),
$$

and the direct insurer's expected payout per claim net of reinsurance is

$$
\mu_Y = \mu_X - \mu_Z = \int x F_X(dx) - \mu_Z
$$

$$
= \int_{(0,M]} x F_X(dx) + \int_{(M+A,\infty)} x F_X(dx)
$$
$$
+ M \Pr(M < X_i \leq M + A) - A \Pr(X_i > M + A). \qquad (7.14)
$$

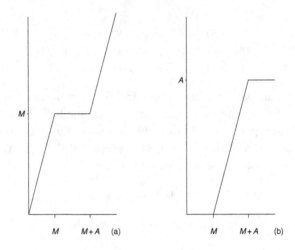

Figure 7.7. The functions $g(x)$ (a) and $x - g(x)$ (b) for excess of loss reinsurance with retention M in a layer size A.

The moment generating function of a typical direct insurer's payout Y_i net of reinsurance is given by (7.6) as

$$
\begin{aligned}
M_Y(r) &= \int e^{rg(x)} F_X(dx) \\
&= \int_{(0,M]} e^{rx} F_X(dx) + e^{rM} \Pr(M < X_i \le M + A) \\
&\quad + \int_{(M+A,\infty)} e^{r(x-A)} F_X(dx).
\end{aligned}
\tag{7.15}
$$

As in §7.3.2, we assume there exists $r_{X,\infty}$, $0 < r_{X,\infty} \le \infty$, such that conditions (i) and (ii) in Lemma 6.5 are satisfied by $M_X(r)$; that is, $M_X(r) < \infty$ for all $r < r_{X,\infty}$ and $M_X(r) \to \infty$ as $r \to r_{X,\infty}^-$, so that the adjustment coefficient R exists for the direct insurer without reinsurance. We now ask whether $M_Y(r)$ also satisfies the same conditions for some $r_{Y,\infty}$, $0 < r_{Y,\infty} \le \infty$, because if these conditions are satisfied, then Lemma 6.5 implies that the adjustment coefficient exists for the direct insurer when the above reinsurance is in place. We show below that the two conditions are indeed satisfied for this reinsurance contract with $r_{Y,\infty} = r_{X,\infty}$.

From the argument below (7.6), we know that $M_Y(r) < \infty$ for all $r < r_{X,\infty}$, so $M_Y(r)$ satisfies (i) in Lemma 6.5 with r_∞ replaced by $r_{X,\infty}$. For (ii), we consider $\lim_{r \to r_{X,\infty}^-} M_Y(r)$. From (7.15), we have

$$M_Y(r) \geq \int_{(M+A,\infty)} e^{r(x-A)} F_X(dx)$$

$$\geq e^{rM} \Pr(X_1 > M + A). \tag{7.16}$$

When $r_{X,\infty} = \infty$, the right-hand side of (7.16) converges to infinity as $r \to r_{X,\infty}^-$ (recall that we assume $\Pr(X_i > M + A) > 0$ in (7.12)), and hence in this case $M_Y(r)$ converges to infinity as $r \to r_{X,\infty}^-$.

The above argument does not work when $r_{X,\infty}$ is finite, because in this case the right-hand side of (7.16) has a finite limit as $r \to r_{X,\infty}^-$. Taking a different approach, we note from (7.15) that, for $0 < r < r_{X,\infty}$,

$$M_Y(r) \geq \int_{(M+A,\infty)} e^{r(x-A)} F_X(dx)$$

$$= e^{-rA} \left(M_X(r) - \int_{(0,M+A]} e^{rx} F_X(dx) \right)$$

$$\geq e^{-rA} \left(M_X(r) - e^{r(M+A)} \Pr(X_i \leq M + A) \right)$$

$$= e^{-rA} M_X(r) - e^{rM} \Pr(X_i \leq M + A). \tag{7.17}$$

When $r_{X,\infty} < \infty$, the right-hand side of (7.17) converges to

$$e^{-r_{X,\infty} A} \lim_{r \to r_{X,\infty}} M_X(r) - e^{r_{X,\infty} M} \Pr(X_i \leq M + A). \tag{7.18}$$

In the finite $r_{X,\infty}$ case, we know that $e^{-r_{X,\infty} A}$ and $e^{r_{X,\infty} M}$ are finite. By assumption, we have $\lim_{r \to r_{X,\infty}^-} M_X(r) = \infty$, and, putting all these things together, we see that (7.18) is ∞. Hence we must have $M_Y(r) \to \infty$ as $r \to r_{X,\infty}^-$. Thus $M_Y(r)$ satisfies conditions (i) and (ii) of Lemma 6.5 with r_∞ in the lemma replaced by $r_{X,\infty}$. From the statement of Lemma 6.5, this means that the adjustment coefficient R_I exists for the direct insurer when reinsurance is included. This adjustment coefficient R_I is the unique positive solution of

$$M_Y(r) - 1 = (1 + \theta_I)\mu_Y r,$$

where M_Y is given by (7.15), μ_Y is given by (7.14), and, from (7.5), we have

$$\theta_I = \frac{\theta\mu_X - \zeta\mu_Z}{\mu_Y}$$

(recall that θ and ζ are the loading factors for the direct insurer's premium without reinsurance and for the reinsurer's premium, respectively).

7.3.5 Excess of loss reinsurance in a layer with exponential claim sizes

In this section we consider excess of loss reinsurance with retention M in a layer of size A as described in §7.3.4, when the claim sizes $X_1, X_2, \ldots$ are independent exponentially distributed random variables with mean μ_X and distribution function $F_X(x) = 1 - e^{-x/\mu_X}$, $x \geq 0$. As in §7.3.3, for the direct insurer without reinsurance, the adjustment coefficient exists and is given by

$$R = \frac{\theta}{(1 + \theta)\mu_X}.$$

With reinsurance in place, the direct insurer's claim sizes $Y_1, Y_2, \ldots$ are iid with Y_i given by (7.11). This means that the distribution of Y_i includes a discrete atom at M with probability $\Pr(Y_i = M) = \Pr(M < X_i \leq M + A)$. For exponentially distributed X_i, this is

$$\Pr(Y_i = M) = e^{-M/\mu_X}\left(1 - e^{-A/\mu_X}\right),$$

so that the distribution function of Y_i has a jump of this size at M. The complete distribution function of Y_i is

$$F_Y(y) = \Pr(Y_i \leq y) = \begin{cases} \Pr(X_i \leq y) & \text{if } 0 \leq y < M \\ \Pr(X_i \leq M + A) & \text{if } y = M \\ \Pr(X_i - A \leq y) & \text{if } y > M, \end{cases}$$

$$= \begin{cases} F_X(y) & \text{if } 0 \leq y < M \\ F_X(M + A) & \text{if } y = M \\ F_X(y + A) & \text{if } y > M, \end{cases}$$

$$= \begin{cases} 1 - e^{-y/\mu_X} & \text{if } 0 \leq y < M \\ 1 - e^{-(M+A)/\mu_X} & \text{if } y = M \\ 1 - e^{-(y+A)/\mu_X} & \text{if } y > M. \end{cases}$$

The distribution function F_Y is shown in Figure 7.8, together with the distribution function F_X. We do not recognise F_Y as one of the standard named distributions, which means that this case is more difficult than the proportional reinsurance case in §7.3.3, where the Y_i were found to be exponentially distributed. In the excess of loss reinsurance in a layer, we must work out quantities such as μ_Z, μ_Y and $M_Y(r)$ from scratch, which we now do.

From (7.13), the expected payout μ_Z on a claim X_i for the reinsurer is

$$\mu_Z = \int_M^{M+A} (x - M)\frac{1}{\mu_X}e^{-x/\mu_X}\, dx + Ae^{-(M+A)/\mu_X}.$$

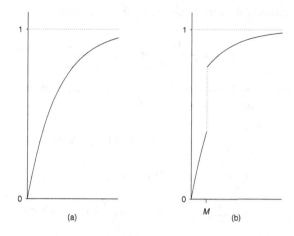

Figure 7.8 The distribution functions F_X (a) and F_Y (b) for excess of loss reinsurance with retention M in a layer size A. The function F_Y has a jump of size $F_X(M + A) - F_X(M)$ at M.

Substituting $w = x - M$, we find that

$$
\begin{aligned}
\mu_Z &= e^{-M/\mu_X} \int_0^A w \frac{1}{\mu_X} e^{-w/\mu_X} \, dw + A e^{-(M+A)/\mu_X} \\
&= e^{-M/\mu_X} \left(\mu_X - \mu_X e^{-A/\mu_X} - A e^{-A/\mu_X} \right) + A e^{-(M+A)/\mu_X} \\
&= \mu_X e^{-M/\mu_X} \left(1 - e^{-A/\mu_X} \right),
\end{aligned}
\tag{7.19}
$$

where we have used integration by parts to obtain the second line in (7.19). Then the expected payout per claim for the direct insurer net of reinsurance is

$$
\mu_Y = \mu_X - \mu_Z = \mu_X \left(1 - e^{-M/\mu_X} + e^{-(M+A)/\mu_X} \right).
\tag{7.20}
$$

From (7.3), the direct insurer's premium rate with reinsurance is

$$
\begin{aligned}
c_I &= (1 + \theta) \lambda \mu_X - (1 + \zeta) \lambda \mu_Z \\
&= (1 + \theta) \lambda \mu_X - (1 + \zeta) \lambda \mu_X e^{-M/\mu_X} \left(1 - e^{-A/\mu_X} \right),
\end{aligned}
$$

and, from (7.5), the direct insurer's relative safety loading inlcuding reinsurance is

$$
\theta_I = \frac{\theta \mu_X - \zeta \mu_Z}{\mu_Y} = \frac{\theta - \zeta e^{-M/\mu_X} \left(1 - e^{-A/\mu_X} \right)}{1 - e^{-M/\mu_X} \left(1 - e^{-A/\mu_X} \right)}.
\tag{7.21}
$$

Condition (7.4) becomes

$$
\theta < \zeta < \frac{\mu_X}{\mu_Z} \theta = \frac{e^{M/\mu_X}}{1 - e^{-A/\mu_X}} \theta.
$$

Without reinsurance, the direct insurer's claim-size moment generating function is

$$M_X(r) = \frac{1}{1 - \mu_X r} \text{ for } r < r_{X,\infty} = \frac{1}{\mu_X}.$$

With reinsurance, the direct insurer's claim-size moment generating function is given by (7.15); that is, for $r < r_{X,\infty} = 1/\mu_X$,

$$M_Y(r) = \int_{(0,M]} e^{rx} F_X(dx) + e^{rM} \Pr(M < X_i \le M + A)$$

$$+ \int_{(M+A,\infty)} e^{r(x-A)} F_X(dx)$$

$$= \int_0^M \frac{1}{\mu_X} e^{(r-1/\mu_X)x} \, dx + e^{rM} \left(e^{-M/\mu_X} - e^{-(M+A)/\mu_X} \right)$$

$$+ e^{-rA} \int_{M+A}^\infty \frac{1}{\mu_X} e^{(r-1/\mu_X)x} \, dx.$$

After a few steps of calculation, this can be shown to be as follows:

$$M_Y(r) = \frac{1}{1 - \mu_X r} + \exp\left(\left(r - \frac{1}{\mu_X} \right) M \right)$$

$$\times \left(1 - e^{-A/\mu_X} \right) \left(1 - \frac{1}{1 - \mu_X r} \right)$$

$$= \frac{1}{1 - \mu_X r} - \exp\left(\left(r - \frac{1}{\mu_X} \right) M \right)$$

$$\times \left(1 - e^{-A/\mu_X} \right) \left(\frac{\mu_X r}{1 - \mu_X r} \right). \tag{7.22}$$

We saw in §7.3.4 that $M_Y(r)$ satisfies the conditions of Lemma 6.5 with r_∞ replaced by $r_{X,\infty}$ and $r_{X,\infty} = 1/\mu_X$ for exponentially distributed claims. Thus the adjustment coefficient R_I for the direct insurer with reinsurance exists, with $R_I \in (0, 1/\mu_X)$. The equation satisfied by R_I is

$$M_Y(r) - 1 = (1 + \theta_I)\mu_Y r,$$

where, for exponentially distributed claims, $M_Y(r)$, θ_I and μ_Y are given by (7.22), (7.21) and (7.20), respectively. This equation for R_I does not have an easy closed-form solution, and so R_I must be found numerically, using the methods of §6.6.1.

To do this, we set

$$h(r) = M_Y(r) - 1 - (1 + \theta_I)\mu_Y r$$

and then use the Newton–Raphson method to find successive approximations $r_1, r_2, \ldots$ to R_I, where

$$r_{n+1} = r_n - \frac{h(r_n)}{h'(r_n)}, \quad n = 1, 2, \ldots,$$

and where we must choose an appropriate initial value r_0. The function $M_Y(r)$ is known explicitly, and its derivative $M'_Y(r)$ may also be found; it is given by

$$M'_Y(r) = \frac{\mu_X}{(1 - \mu_X r)^2}$$
$$+ M \exp\left(\left(r - \frac{1}{\mu_X}\right)M\right)\left(1 - e^{-A/\mu_X}\right)\left(1 - \frac{1}{1 - \mu_X r}\right)$$
$$- \exp\left(\left(r - \frac{1}{\mu_X}\right)M\right)\left(1 - e^{-A/\mu_X}\right)\frac{\mu_X}{(1 - \mu_X r)^2}.$$

To write the functions $M_Y(r)$ and $M'_Y(r)$ neatly, we define

$$h_1(r) = \exp\left(\left(r - \frac{1}{\mu_X}\right)M\right)\left(1 - e^{-A/\mu_X}\right),$$

and then we have

$$M_Y(r) = \frac{1 - h_1(r)\mu_X r}{1 - \mu_X r}$$

and

$$M'_Y(r) = \frac{\mu_X}{(1 - \mu_X r)^2}(1 - h_1(r)) - Mh_1(r)\frac{\mu_X r}{1 - \mu_X r}.$$

As a preliminary to carrying out the Newton–Raphson calculation, we first give **R** functions for calculating $M_Y(r)$ and $M'_Y(r)$ for given M, A and μ_X, which we assume are already in **R** objects M, A and muX. For $M_Y(r)$ we have

```
mgfY = function(muX,M,A,r){
temp = exp((r - (1/muX))*M)*(1 - exp(-A/muX))
mgf = (1-temp*muX*r)/(1-muX*r)
mgf
}
```

and an **R** function for $M'_Y(r)$ is

```
mgfYderiv = function(muX,M,A,r){
temp = exp((r - (1/muX))*M)*(1 - exp(-A/muX))
mgfderiv = muX*(1-muX*r)^(-2)*(1-temp) -M*temp*muX*r/
(1-muX*r)
mgfderiv
}
```

Next, we give an **R** function to find R_I using the Newton–Raphson method.
Look back at §6.6.1 to see the basic structure of such a function. In particular,
n is the number of iterations to be carried out. The **R** function is

```
layerRexp = function(muX,theta,zeta,M,A,n,rzero){
muZ = muX*exp(-M/muX)*(1-exp(-A/muX))
muY = muX-muZ
thetaI = (theta*muX - zeta*muZ)/muY
i=1
s=(1:n)*0
s[1] = rzero
while(i<n){
i=i+1
temp1 = mfgY(muX,M,A,s[i-1]) -1 -(1+thetaI)*muY*s[i-1]
temp2 = mfgYderiv(muX,M,A,s[i-1]) - (1+thetaI)*muY
s[i] = s[i-1]-temp1/temp2
}
s
}
```

We illustrate this when $\mu_X = 1$, $\theta = 0.3$ and $\zeta = 0.4$ (which are the same
values as in §7.3.3) and with $M = 2$ and $A = 1$. From (7.19) and (7.20) we
easily find that

$$\mu_Z = 0.08555 \text{ and } \mu_Y = 0.91445.$$

We also see (using (7.21)) that

$$\theta_I = 0.2906.$$

From (7.10), we know that the adjustment coefficient for the direct insurer
without reinsurance is $R = 0.23077$, and one possibility is to set the initial
value in the Newton–Raphson method to be $r_0 = R$. If we do this, and run
`layerRexp` as follows:

```
RI = layerRexp(1,0.3,0.4,2,1,10,0.23077)
```

then we find that the **R** returns the value 0.27947 for the adjustment coefficient
R_I. We note that $R_I > R$.

The **R** function `layerRexp` can be used to find R_I as a function of M for various A-values. Figure 7.9 shows R_I against M for $A = 0.6$, $A = 1.0$ and $A = 1.4$,
together with the adjustment coefficient R without reinsurance. For the range
of M-values shown, the adjustment coefficient increases as A increases. For

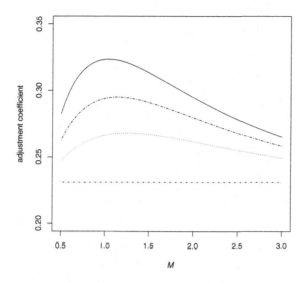

Figure 7.9. The adjustment coefficient R_I net of reinsurance plotted against the retention limit M for exponentially distributed claims with mean 1, for $A = 1.4$ (solid), $A = 1.0$ (dot-dashed) and $A = 0.6$ (dotted line). The adjustment coefficient R without reinsurance is shown as a horizontal line.

each of the three A-values, the adjustment coefficient R_I increases to a maximum for particular maximising M-values (different for different A-values), and then decreases. For the M- and A-values shown, the direct insurer's adjustment coefficient with reinsurance is greater than the original adjustment coefficient without reinsurance.

Figure 7.10 shows R_I against A for three different M-values, and also shows R. Here the situation is more complicated, and we see that the graphs cross over. However, for the M- and A-values shown, R_I is greater than the original R.

Going one stage further, a two-dimensional plot is given in Figure 7.11, which shows R_I as a function of both M and A. The features exhibited are the same as those described above.

If we let $A = \infty$ in §7.3.4, then this corresponds to excess of loss reinsurance with retention M, and results in this case can be compared to those in Dickson and Waters (1996).

A possible direction for extension for this case study is to explore the effect on the adjustment coefficient when the claim sizes are not exponentially distributed, but have another distribution (although remember that for an adjustment coefficient to exist in the original model without reinsurance, this distribution should not be heavy-tailed). You could try this for both proportional reinsurance and for excess of loss reinsurance in a layer.

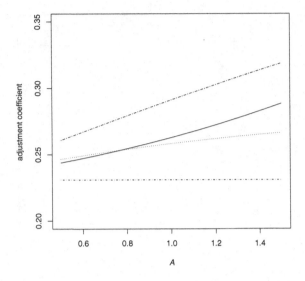

Figure 7.10. The adjustment coefficient R_I net of reinsurance plotted against the layer size A for exponentially distributed claims with mean 1, for $M = 0.5$ (solid), $M = 1.5$ (dot-dashed) and $M = 3.0$ (dotted line). The adjustment coefficient R without reinsurance is shown as a horizontal line.

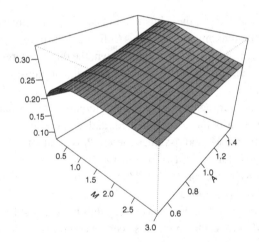

Figure 7.11. The adjustment coefficient R_I net of reinsurance plotted against the retention limit M and the layer size A for exponentially distributed claims with mean 1.

You might like to try a more challenging extension, which is to find the direct insurer's probability of ruin $\psi_I(u)$, net of reinsurance, for excess of loss reinsurance in a layer, with exponentially distributed claims. There is no easy explicit formula for $\psi_I(u)$, so numerical methods would be needed.

For more on the effect of reinsurance on ruin quantities, see Dickson and Waters (1996) and Schmidli (2008), and the references given therein.

Appendix A
Utility theory

For our purposes, we restrict our consideration of *utility theory* to that which is concerned with the concept of the *utility of money*. The *utility* of a sum of money is not its actual monetary value, but is the value placed by an individual (for example an investor) or an institution on gaining or losing that sum of money, in the context of the existing assets of the individual/institution. The following situations illustrate the idea.

- A gift of £1000 will be received with different reactions by two people of very different circumstances. A rich person who has a healthy bank balance, has everything they want, and is content, will probably think little about having an extra £1000 – it will not change their behaviour or contentment in a major way. A poorer person, struggling to pay some accumulated household bills on top of feeding a family, will find the extra £1000 of enormous value – they can perhaps clear their debts and make a fresh start.
- A similar observation can be made about a single individual – an individual with current wealth £2000 will regard a gift of £1000 as being more useful than they would if their current wealth were £100 000.
- Consider a gambler faced with deciding whether or not to enter into a bet in which £1000 will be won or lost, each outcome having probability 0.5. The gambler's expected gain on the bet is zero. For many people, the prospect of not losing the £1000 is more attractive than the prospect of gaining the same sum – a person with such a viewpoint will not accept the bet. Putting it another way, how much would you be prepared to pay for a bet in which you win £1000 or nothing, each outcome having probability 0.5? Your expected gain is £500, but you may not be willing to pay £500 for the bet – if you are willing, you will definitely have to pay out £500 (the cost of the bet) and you may end up with a net gain of £500, but the odds are not stacked in your favour. For an entry fee of £500 and "50:50" odds, you may demand a return,

if you win, of *more than* £1000 – you may want a return of, say, £1200, so that your expected net gain is positive (in this case £100). A person taking this approach is said to be "risk averse" and will prefer a certain return of *less than £k* to a gamble with expected return of £k.

- You may be prepared to pay a premium of £600 to insure a risk (for example, your house) where the expected loss (the average loss per risk over the whole portfolio of comparable risks) is only £400. The value to you of the "extra" £200 is the protection it offers you against a catastrophic loss (for example, the cost of rebuilding or replacing your house if it is destroyed) – the insurer is charging is a "risk premium" of £200, which you are prepared to pay.
- An investor's decisions will be influenced by their attitude to risk – they may decide to choose one investment over another on grounds other than a simple comparison of the expected monetary gains of the two investments.

These illustrations show that monetary values alone do not provide an adequate quantitative description of the value we put on sums of money in differing circumstances – the following section does, however, provide such a description.

A.1 Utility functions

A *utility function* is a mathematical way of representing the value an individual (or institution) places on having specified levels of wealth.

We will denote a general utility function by $u(x)$, which represents a measure of the value the individual places on having wealth $x > 0$. We require $u(x)$ to be a "well-behaved" function of x; specifically, in the light of the preceding examples, we require it to have the following two properties.

(i) We assume that a rational person will value greater wealth at least as highly as lower wealth, and so we require that $u(x)$ is non-decreasing:

$$y > z \Rightarrow u(y) \geq u(z), \text{ that is } u'(x) \geq 0 \text{ for all } x > 0.$$

(ii) We assume that a rational person with given wealth will value additional wealth of a fixed amount at least as highly as they would if they had greater wealth to start with, and so we require that $u'(x)$ is non-increasing:

$$y > z \Rightarrow u'(y) \leq u'(z), \text{ that is } u''(x) \leq 0 \text{ for all } x > 0.$$

A *utility function* is thus *concave*.

Note that utility is a *relative*, not an *absolute*, measure – the utility functions $u(x)$ and $au(x) + b$ (with $a > 0$) are equivalent, and, in their definitions of utility functions, some authors require the strict inequalities $u'(x) > 0$ and $u''(x) < 0$.

A.1.1 Examples of utility functions

Here are some examples of utility functions:

(i) linear $u(x) = x$;
(ii) exponential $u(x) = -e^{-\alpha x}$, $\alpha > 0$;
(iii) quadratic $u(x) = x - \alpha x^2$,   for $0 < x \leq 1/2\alpha$, $\alpha > 0$;
(iv) piecewise

$$u(x) = \begin{cases} x - \alpha x^2 & \text{for } 0 < x \leq 1/2\alpha \\ 1/4\alpha & \text{for } x \geq 1/2\alpha \end{cases} \quad \alpha > 0;$$

(v) log $u(x) = \log x$;
(vi) fractional power $u(x) = x^\gamma$, $0 < \gamma < 1$.

The first of these, the linear function, is included for comparison only – it is not used as a utility function as it simply measures the actual monetary value of wealth. Indeed, the use of a linear function in utility arguments can lead to difficulties of interpretation.

It is sometimes more convenient to use the exponential utility function (ii) in the form $u(x) = 1 - e^{-\alpha x}$.

The restriction on the range of x in the quadratic utility function (iii) is required to satisfy the properties of the first and second derivatives of a utility function.

Another version of the fractional power utility function above is

$$u(x) = \frac{x^\gamma - 1}{\gamma}, \quad \text{for } x \geq 1, \quad 0 < \gamma < 1. \tag{A.1}$$

This is a generalisation of the log utility function (v) for $x \geq 1$: the limit of the function (A.1) as $\gamma \to 0$ gives the log utility function.

Figure A.1 shows the relative shapes of linear, exponential, quadratic and piecewise utility functions (all with $u(0) = 0$ and scaled appropriately). Figure A.2 shows a log utility function and fractional power utility functions (version (A.1)) with $\gamma = 0.75$ and $\gamma = 0.25$ (for $x > 1$ and scaled appropriately).

The property $u''(x) \leq 0$ summarises the concept of *decreasing* (or *diminishing*) *marginal utility*, which recognises that the additional value of an increase in wealth of a fixed amount decreases (strictly speaking, does not increase) as the wealth to which it is added increases. Equivalently, one can express

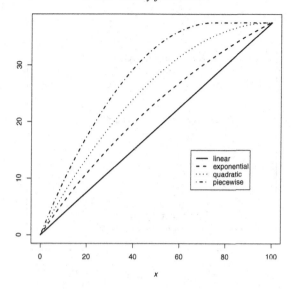

Figure A.1. Comparable linear, exponential, quadratic and piecewise utility functions.

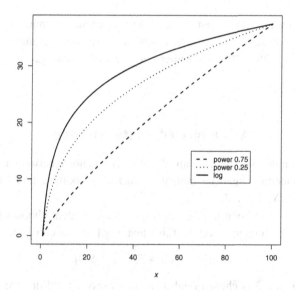

Figure A.2. Comparable log and fractional power utility functions.

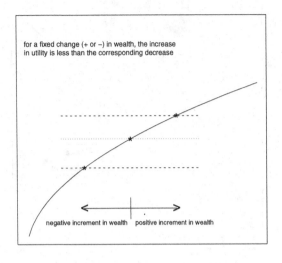

for a fixed change (+ or –) in wealth, the increase
in utility is less than the corresponding decrease

negative increment in wealth   positive increment in wealth

x

Figure A.3. Decreasing marginal utility.

this by saying that the added value of an incremental increase in wealth is
less (strictly speaking, is not more) than the reduction in value of an equal
decrease in wealth – "I would rather not lose £1000 than gain £1000" – see
Figure A.3.

A.2 Expected utility criterion

Consider an individual with wealth W who faces a choice between two actions
that have uncertain financial outcomes, such that taking action i leads to a
financial gain S_i, $i = 1, 2$.

The *expected utility criterion* states that the individual chooses the action
that leads to the higher expected utility, that is action 1 is chosen if

$$\mathbb{E}[u(W + S_1)] > \mathbb{E}[u(W + S_2)];$$

otherwise action 2 is chosen (and if the two expected utilities are equal, the
individual is indifferent to which action is chosen).

With an exponential utility function $u(x) = -e^{-\alpha x}$, the initial wealth does not
affect the decision, since

$$\mathbb{E}[u(W + S_1)] = -\mathbb{E}[\exp(-\alpha(W + S_1))] = -\exp(-\alpha W)\mathbb{E}[\exp(-\alpha S_1)],$$

and the individual therefore chooses action 1 if

$$-\exp(-\alpha W)\mathbb{E}[\exp(-\alpha S_1)] > -\exp(-\alpha W)\mathbb{E}[\exp(-\alpha S_2)],$$

that is, if

$$\mathbb{E}[\exp(-\alpha S_1)] < \mathbb{E}[\exp(-\alpha S_2)],$$

assuming the expectations exist.

This conclusion can be expressed in the notation of moment generating functions: the individual chooses action 1 if $M_{S_1}(-\alpha) < M_{S_2}(-\alpha)$.

Example A.1 Consider an individual who adopts an exponential utility function $u(x) = -e^{-\alpha x}$ and who faces the choice of two actions as above, with the gains $S_1 \sim N(250, 200)$ and $S_2 \sim N(300, 1200)$. Using the standard result for the moment generating function of a normal random variable, we have

$$M_{S_1}(-\alpha) = \exp(-250\alpha + 200\alpha^2/2) = \exp(-250\alpha + 100\alpha^2)$$

and

$$M_{S_2}(-\alpha) = \exp(-300\alpha + 1200\alpha^2/2) = \exp(-300\alpha + 600\alpha^2).$$

Using the expected utility criterion, the individual chooses action 1 if

$$\exp(-250\alpha + 100\alpha^2) < \exp(-300\alpha + 600\alpha^2),$$

that is if $-250\alpha + 100\alpha^2 < -300\alpha + 600\alpha^2$, that is if $\alpha > 0.1$. So the individual chooses the action with the lower expected gain if the parameter of the exponential utility function is high enough. This feature, in which the individual chooses to avoid the much greater uncertainty of the second outcome, illustrates the concept of risk-aversion and the role of the parameter α (see §A.3).

Example A.2 Consider a TV quiz show in which a contestant has the opportunity to answer a series of questions; for each successive correct answer, the contestant's winnings double (or approximately double). Suppose the contestant has reached the £64 000 level of winnings and is faced with the next question – a multiple-choice question with four possible answers. If the contestant answers correctly, the winnings will move up to the £125 000 level; if the contestant answers incorrectly, the winnings will drop back to the £32 000 level.

Suppose a contestant, Susan, sees the next question and the four possible answers, but has no idea about which of the available answers is the correct one. Should she answer the question, or walk away with the £64 000?

We compare two situations in which Susan may find herself, and with three different utility settings.

Situation (1)   Susan has no access to other facilities of the quiz such as "ask the audience" or "50:50" (in which two wrong answers from the four possible answers are removed before the contestant selects the answer from the two remaining), and she simply guesses the answer from the four available answers, so

$$\text{Pr(Susan gives the correct answer)} = 1/4.$$

Situation (2)   Susan has the "50:50" facility still available to her, uses it, and then guesses from the two remaining available answers, so

$$\text{Pr(Susan gives the correct answer)} = 1/2.$$

We will assume that Susan makes her decisions on the basis of the expected utility criterion applied to her winnings after this round of the quiz, and we will work with a monetary unit of £1000. We consider these two situations in each of three utility settings.

- Utility setting A – using monetary values only (a linear utility function).
  Situation (1): Susan's expected winnings after giving her answer are

$$\frac{1}{4} \times 125 + \frac{3}{4} \times 32 = 55.25,$$

  which is less than her guaranteed 64, so she should walk away.
  Situation (2): Susan's expected winnings after giving her answer are

$$\frac{1}{2} \times 125 + \frac{1}{2} \times 32 = 78.5,$$

  which is greater than her guaranteed 64, so she should answer the question.
- Utility setting B – using an exponential utility function $u(x) = -e^{-0.01x}$;
  Susan's utility of winnings if she walks away is $-e^{-0.64} = -0.527$.
  Situation (1): Susan's expected utility of winnings after giving her answer is

$$-\frac{1}{4} \times e^{-1.25} - \frac{3}{4} \times e^{-0.32} = -0.616,$$

  which is less than -0.527, so she should walk away.
  Situation (2): Susan's expected utility of winnings after giving her answer is

$$-\frac{1}{2} \times e^{-1.25} - \frac{1}{2} \times e^{-0.32} = -0.506,$$

  which is greater than -0.527, so she should answer the question.
  The decisions for both settings are the same as those based on monetary value alone.

- Utility setting C – using an exponential utility function with a higher value of α, namely $u(x) = -e^{-0.02x}$; Susan's utility of winnings if she walks away is $-e^{-1.28} = -0.278$.

  Situation (1): Susan's expected utility of winnings after giving her answer is

$$-\frac{1}{4} \times e^{-2.5} - \frac{3}{4} \times e^{-0.64} = -0.416,$$

which is less than −0.278, so she should walk away.

  Situation (2): Susan's expected utility of winnings after giving her answer is

$$-\frac{1}{2} \times e^{-2.5} - \frac{1}{2} \times e^{-0.64} = -0.305,$$

which is less than −0.278, so she should walk away.

  The decision in situation (2) is not the same as that based on monetary value alone, or that using the utility function with the lower value of α. With $\alpha = 0.01$ Susan chooses to answer the question (she chooses to gamble), whereas with $\alpha = 0.02$ she is not willing to gamble and instead walks away with the existing guaranteed winnings.

A.3 Risk aversion

An individual who adopts as his utility function a function $u(x)$ which is strictly concave, that is with $u'(x) > 0$ and $u''(x) < 0$, is described as being *risk-averse*.

There is a generally accepted measure which quantifies the degree of risk aversion, and which we motivate by considering an individual with current wealth x faced with the following gamble:

win an amount h with probability $1/2 + p$,
lose an amount h with probability $1/2 - p$,

where $h > 0$ and $p > 0$.

Using monetary value only, the individual's expected wealth after accepting the gamble is $x + 2ph$. The gamble favours the individual, and more so as p (or h) increases.

With utility considerations, the individual's expected utility after accepting the gamble is given by

$$\text{expected utility} = \left(\frac{1}{2} + p\right) u(x + h) + \left(\frac{1}{2} - p\right) u(x - h).$$

Using Taylor expansions, we get

$$\text{expected utility} = \left(\frac{1}{2} + p\right)\left\{u(x) + hu'(x) + \frac{1}{2}h^2u''(x) + \cdots\right\}$$

$$+ \left(\frac{1}{2} - p\right)\left\{u(x) - hu'(x) + \frac{1}{2}h^2u''(x) - \cdots\right\}$$

$$= u(x) + 2phu'(x) + \frac{1}{2}h^2u''(x) + \cdots$$

$$= u(x) + \frac{1}{2}h^2\left\{4\frac{p}{h}u'(x) + u''(x)\right\} + \cdots.$$

The "indifference" position occurs in the case where the expected utility after accepting the gamble equals the utility without accepting the gamble, namely $u(x)$. For small h, this occurs (approximately) when

$$p = -\frac{1}{4}h\frac{u''(x)}{u'(x)}.$$

An individual who is more risk-averse than another will require p to be higher to persuade them to accept the gamble. This justifies the accepted measure of risk aversion for an individual with utility function $u(\cdot)$ and current wealth x. It is the *coefficient of risk aversion*, denoted $r(x)$, and defined by

$$r(x) = -\frac{u''(x)}{u'(x)}. \tag{A.2}$$

Increasing values of $r(x)$ correspond to increasing risk aversion.

An individual whose risk aversion does not depend on current wealth has $r(x)$ constant ($= \alpha$ say, where $\alpha > 0$) and so this time the utility function satisfies

$$\frac{u''(x)}{u'(x)} = -\alpha, \text{ that is } u''(x) + \alpha u'(x) = 0.$$

A solution of this linear differential equation is the exponential utility function $u(x) = -e^{-\alpha x}$. The risk aversion is independent of wealth, but increases as α increases.

In Example A.1, the level of risk aversion dictates the individual's choice of action. In Example A.2, utility setting C ($\alpha = 0.02$) corresponds to the contestant being more risk-averse than with setting B ($\alpha = 0.01$); the consequence is that, even after using the "50:50" facility, Susan refuses the gamble and walks away; she prefers the guaranteed £64 000 to the uncertainty of the gamble.

For the log utility function $u(x) = \log x$, the risk aversion (A.2) decreases with wealth (see Exercise 4.3).

Example A.3 Consider an investor, Roger, faced with the choice of two actions:

action 1 – invest in a product which returns a guaranteed £5000;
action 2 – invest in a product which returns £1000, £7000 or £12 000, these
outcomes having probabilities 0.4, 0.4 and 0.2, respectively.

Roger decides which action to take using the criterion of maximising the expected utility of his wealth. He works in the context of a monetary unit of £1000 and has initial wealth (in this unit) of W. Letting X represent the return on the investment, Roger's wealth becomes $W + X$.

Case 1 Suppose Roger adopts the quadratic utility function

$$u(x) = x - 0.02x^2, \ 0 < x < 25.$$

If he takes action 1, then his utility is

$$(W + 5) - 0.02(W + 5)^2 = -0.02W^2 + 0.8W + 4.5.$$

If he takes action 2, then his expected utility is

$$
\begin{aligned}
\mathbb{E}[u(W + X)] &= \sum u(W + x) \Pr(X = x) \\
&= \sum [(W + x) - 0.02(W + x)^2] \Pr(X = x) \\
&= -0.02W^2 + (1 - 0.04\mathbb{E}[X])W + \mathbb{E}[X] - 0.02\mathbb{E}[X^2]
\end{aligned}
$$

Here, $\mathbb{E}[X] = 5.6$ and $\mathbb{E}[X^2] = 48.8$, so the expected utility is

$$-0.02W^2 + 0.776W + 4.624.$$

Roger therefore chooses action 1 if

$$-0.02W^2 + 0.8W + 4.5 > -0.02W^2 + 0.776W + 4.624,$$

that is if $W > 5.17$. So, if Roger's initial wealth is greater than a stated amount (£5170), he chooses the investment with the certain return rather than the investment with an uncertain return (albeit the gamble has a higher expected return (£5600) than the fixed return (£5000)).

Noting that $u'(x) = 1 - 0.04x$ and $u''(x) = -0.04$, the coefficient of risk aversion is given by $r(x) = 0.04(1 - 0.04x)^{-1}$, and hence $r(x)$ is an increasing function of x. The higher Roger's initial wealth, the more risk-averse he is – the result reflects this fact.

In this illustration the utility function is almost linear, and for all values of W for which the utility function is sensible, the expected utility for action 2 is very close to the utility for action 1.

Case 2   Suppose Roger adopts the exponential utility function $u(x) = -e^{-\alpha x}$. The initial wealth can be ignored. Let $q = e^{-\alpha}$. If Roger takes action 1, then his utility is $-e^{-5\alpha} = -q^5$. If he takes action 2, then his expected utility is

$$\mathbb{E}[u(X)] = \mathbb{E}[-e^{-\alpha X}] = 0.4 \times (-q) + 0.4 \times (-q^7) + 0.2 \times (-q^{12}).$$

Roger therefore chooses action 1 if

$$0.4q - q^5 + 0.4q^7 + 0.2q^{12} > 0.$$

The reader can verify (by computer search, or some other method) that this inequality is satisfied for $q < 0.9316$, that is for $\alpha > 0.071$. Higher α corresponds to greater risk aversion.

Case 3   Suppose Roger adopts the log utility function $u(x) = \log x$. If he takes action 1, then his utility is $\log(W + 5)$. If he takes action 2, then his expected utility is

$$\mathbb{E}[u(X)] = 0.4 \log(W + 1) + 0.4 \log(W + 7) + 0.2 \log(W + 12)$$
$$= \log\left[(W + 1)^{0.4}(W + 7)^{0.4}(W + 12)^{0.2}\right].$$

Roger therefore chooses action 1 if

$$W + 5 > (W + 1)^{0.4}(W + 7)^{0.4}(W + 12)^{0.2}.$$

The reader can verify (again by computer search, or some other method) that this inequality is satisfied for $W < 9.03$.

The coefficient of risk aversion is given by $r(x) = 1/x$ and is a decreasing function of x. Higher initial wealth corresponds to lower risk aversion – with W greater than a specified amount (£9030), Roger chooses the gamble.

A.4 Jensen's inequality

The following inequality is one of a number of illustrations of a general result published by the Danish telephone engineer and mathematician Johan Jensen in the early part of the twentieth century. The inequality states the following:

Let Y be a random variable and let $u(\cdot)$ be a (strictly) concave function, so $u''(x) < 0$. Then

$$\mathbb{E}[u(Y)] \leq u(\mathbb{E}[Y]), \tag{A.3}$$

with equality if and only if Y is constant.

Proof   We use a Taylor expansion (with remainder)

$$u(y) = u(a) + u'(a)(y - a) + u''(z)(y - z)^2/2,$$

and so

$$u(y) \leq u(a) + u'(a)(y - a),$$

since $u''(z) \leq 0$.

Considering y as an observation of random variable Y, and letting $a = \mathbb{E}[Y]$, we have

$$u(Y) \leq u(\mathbb{E}[Y]) + u'(\mathbb{E}[Y])(Y - \mathbb{E}[Y]),$$

and, taking expected values through this inequality, gives the result. □

Note A graphical justification is available and is left to the reader (show from a graph that $u(y)$ satisfies

$$u(y) \leq u(\mathbb{E}[Y]) + u'(\mathbb{E}[Y])\{y - \mathbb{E}[Y]\},$$

and then proceed as above).

Example A.4 Consider an investor who adopts an exponential utility function

$$u(x) = 1 - e^{-\alpha x},$$

and who makes an investment with financial gain S, which is a random variable with mean μ.

The expected utility of the investor's gain on the investment is

$$\mathbb{E}[u(S)] = \mathbb{E}[1 - e^{-\alpha S}] \leq 1 - e^{-\alpha \mu},$$

by Jensen's inequality (A.3). We note that the upper bound is an increasing function of α.

Consider, for example, the case $\mu = 100$ with two investors, Susan and Roger, who use utility parameters $\alpha = 0.01$ and $\alpha = 0.02$, respectively. Their expected utilities of gain have upper bounds of $1 - e^{-1} = 0.632$ and $1 - e^{-2} = 0.865$. The fact that the second investor, Roger, is more risk-averse than the first, Susan, is balanced by the fact that there is at least the possibility of Roger achieving a higher utility if he *does* invest.

If we have $S \sim N(100, 50^2)$ then the expected utilities of the investors' gains are in fact given by

$$\mathbb{E}[1 - e^{-0.01S}] = 1 - \exp(-100 \times 0.01 + 50^2 \times 0.01^2/2)$$
$$= 1 - e^{-0.875} = 0.583;$$
$$\mathbb{E}[1 - e^{-0.02S}] = 1 - \exp(-100 \times 0.02 + 50^2 \times 0.02^2/2)$$
$$= 1 - e^{-1.5} = 0.777.$$

We note that these expected utilities are, of course, lower than the bounds obtained above using Jensen's inequality.

Appendix B

Answers to exercises

Chapter 2

2.1 (b) Modes at $\lambda - 1$ and λ.

2.3 (a) First note $\Pr(N \geq n) = \Pr(\text{no successes in first } n \text{ trials}) = q^n$; the geometric model possesses a "lack of memory" property.

(b) $\Pr(N \geq 10 \mid N \geq 8) = \Pr(N \geq 2) = 1 - \Pr(N \leq 1) = 0.01$.

2.4 29.

2.6 (b) $f(x) = n\lambda e^{-\lambda x}(1 - e^{-\lambda x})^{n-1}$, $x > 0$.

2.7 (a) 2, 9.

2.8 (a) $e^{-3} = 0.0498$, (b) $Pr(\chi^2_6 > 6) = 0.4232$.

2.9 (b) $\dfrac{2(\beta - \alpha)}{\{\alpha\beta(\alpha + \beta)\}^{1/2}}$, $\dfrac{6(\alpha^2 - \alpha\beta + \beta^2)}{\alpha\beta(\alpha + \beta)} + 3$.

2.10 0.1108.

2.12 (a) 0.2725; (b) 837.29, 1991.2; (c) 0.0319; (d) (763.9, 910.7); (e) 941.1.

2.13 (c) (1) $\dfrac{\lambda + M}{\alpha - 1}$, (2) $\dfrac{2(\lambda + M)^2}{\{(\alpha - 1)(\alpha - 2)\}}$;

(d) coefficient of skewness decreases to 2 as α increases.

2.14 $\alpha = 4.7844$, $\lambda = 926.02$.

2.15 (a) (i) 0.00611, (ii) 180, 116 600; (b) 0.00410;

(c) ignoring heterogeneity under-estimates tail probability.

2.17 (c) Failure rate is (1) decreasing, (2) increasing, (3) constant $(= c)$; $X \sim$ Exp(c), as in part(a).

(d) $q(x) = \dfrac{\alpha \tau x^{\tau - 1}}{\lambda + x^\tau}$; $\tau = 1$ is Pa(α, λ), as in part (b).

2.22 $\dfrac{n}{\gamma} + \sum \log x_i - \dfrac{n \sum x_i^{\widehat{\gamma}} \log x_i}{\sum x_i^{\widehat{\gamma}}} = 0$.

2.23 (a) $\tilde{\alpha} = 1 + 2/\bar{x}$, $\widehat{\alpha} = n / \sum \log(1 + x_i/2)$;

(b) $\tilde{c} = (2/\bar{x})^{1/2}$, $\widehat{c} = n / \sum (x_i^{1/2})$.

2.24 (a) $\tilde{\alpha} = (\bar{x}/s)^2$, $\tilde{\lambda} = \bar{x}/s^2$;

(b) $\tilde{\alpha} = \dfrac{2s^2}{s^2 - \bar{x}^2}$, $\tilde{\lambda} = \dfrac{\bar{x}(s^2 + \bar{x}^2)}{(s^2 - \bar{x}^2)}$.

2.25 (b) 22.52, 58.64, 134.5;

(c) $\chi^2 = 3.6$ on 2 df, *P-value* = 0.17, fit adequate.

2.26 (a) (1) $\tilde{\alpha} = 0.6105$, $\tilde{\lambda} = 0.0008091$; (2) $\tilde{\alpha} = 5.135$, $\tilde{\lambda} = 3120$;

(3) $\tilde{c} = 0.1996$.

(b) (1) $\widehat{\lambda} = 0.001325$; (2) $\widehat{\mu} = 5.914$, $\widehat{\sigma} = 1.423$; (3) $\widehat{\alpha} = 17.28$,

$\widehat{\lambda} = 2.922$.

(c) (1) 0.0786, 0.0809, 0.0706, 0.1179; (2) 2471, 3381, 2260, 3845.

Chapter 3

3.1 $\sum_{n=0}^{\infty} a^n \Pr(N = n) = G_N(a)$.

3.3 $F_S = \Pr(N = 0)1_{[0,\infty)} + \Pr(N \geq 1)\widetilde{F}_S$, so the mixing proportions are $\Pr(N = 0)$ and $\Pr(N \geq 1)$.

$\widetilde{F}_S$ is a compound distribution with step random variables distributed as X_1 and with counting random variable $\widetilde{N}$ satisfying $\Pr(\widetilde{N} = n) = \Pr(N = n)/\Pr(N \geq 1)$ for $n = 1, 2, \ldots$.

3.5 Let $\kappa_{S,j}$, $\kappa_{X,j}$ and $\kappa_{N,j}$ be the jth cumulants of S, X_1 and N, respectively,

$\kappa_{S,1} = \kappa_{X,1}\kappa_{N,1}$;

$\kappa_{S,2} = \kappa_{N,2}\kappa_{X,1}^2 + \kappa_{N,1}\kappa_{X,2}$;

$\kappa_{S,3} = \kappa_{N,3}\kappa_{X,1}^3 + 3\kappa_{N,2}\kappa_{X,1}\kappa_{X,2} + \kappa_{N,1}\kappa_{X,3}$.

3.6 (a) The jth cumulant of S is $\kappa_{S,j} = \lambda\mathbb{E}[X_1^j]$.

The skewness of S is positive.

(b) $\kappa_{S,1} = \frac{kq}{p}\mathbb{E}[X_1]$;

$\kappa_{S,2} = \frac{kq}{p}\mathbb{E}[X_1^2] + \frac{kq^2}{p^2}(\mathbb{E}[X_1])^2$;

$\kappa_{S,3} = \frac{kq}{p}\mathbb{E}[X_1^3] + \frac{3kq^2}{p^2}\mathbb{E}[X_1^2]\mathbb{E}[X_1] + \frac{2kq^3}{p^3}(\mathbb{E}[X_1])^3$.

The skewness of S is positive whether or not β_X is positive.

3.7 $\mathbb{E}[X_1]\text{Var}[N]$.

3.8 (a) Let m_i be $\int x^i F(dx)$:

$\mathbb{E}[S] = nm_1\mathbb{E}[\lambda]$;

$\text{Var}[S] = nm_2\mathbb{E}[\lambda] + nm_1^2\text{Var}[\lambda]$.

(b) $\mathbb{E}[S] = nm_1\mathbb{E}[\lambda]$;

$\text{Var}[S] = nm_2\mathbb{E}[\lambda] + n^2m_1^2\text{Var}[\lambda]$.

3.9 $N \sim \text{nb}\left(2, \frac{2}{2+\mu}\right)$.

3.10 The distribution of T is a mixture of an atom at zero (with mixing proportion p) and an $\text{nb}\left(1, \frac{p\tilde{p}}{1-p(1-\tilde{p})}\right)$ (with mixing proportion $1 - p$).

3.14 The distribution of S is a mixture of an atom at zero (with mixing proportion p^2) and (with mixing proportion $1 - p^2$) a distribution with density

$$f_S(x) = \frac{3p^2\sqrt{1-p}\lambda}{4}\left(e^{-\lambda_2 x} - e^{-\lambda_1 x}\right) + \frac{p^2\sqrt{1-p}\lambda}{4}\left(xe^{-\lambda_2 x} + xe^{-\lambda_1 x}\right),$$

where $\lambda_1 = \lambda(1 + \sqrt{1-p})$ and $\lambda_2 = \lambda(1 - \sqrt{1-p})$.

3.20 $\mathbb{E}[V] = k + \alpha/v$;

$\mathrm{Var}[V] = \alpha/v^2$;

$$\beta_V = \frac{\mathbb{E}[(V - \mathbb{E}[V])^3]}{\left(\mathrm{Var}[V]\right)^{3/2}} = \frac{2}{\sqrt{\alpha}};$$

$$k = \mathbb{E}[S] - \frac{2\sqrt{\mathrm{Var}[S]}}{|\beta_S|}, \quad \alpha = \frac{4}{\beta_S^2}, \quad v = \frac{2}{|\beta_S|\sqrt{\mathrm{Var}[S]}}.$$

3.23 The asymptotic approximation is $\mathrm{Pr}(S > x) \sim \dfrac{p\sqrt{1-p}\,e^{-\left(1-\sqrt{1-p}\right)\lambda x}}{2\left(1 - \sqrt{1-p}\right)}.$

3.24 (b) 884 policies;

(c) 9721 policies.

3.25 (a) $\hat{\mu}_L = \dfrac{2n\hat{\mu}}{\chi^2_{2n}(\alpha/2)}$, $\hat{\mu}_U = \dfrac{2n\hat{\mu}}{\chi^2_{2n}(1-\alpha/2)}$, where $\chi^2_n(\alpha)$ denotes the upper $100\alpha\%$ point of a χ^2_n distribution.

(b) $\hat{\lambda}_L = \dfrac{\chi^2_{2n}(1-\alpha/2)\lambda}{2n}$, $\hat{\lambda}_U = \dfrac{\chi^2_{2n}(\alpha/2)\lambda}{2n}$.

3.28 $\mathbb{E}[T] = \sum_{i=1}^{n} q_i b_i$, the same as in Example 3.26;

$\mathrm{Var}[T] = \sum_{i=1}^{n} q_i(2 - q_i)b_i^2$, greater than or equal to the variance in Example 3.26.

3.29 For example, if $\mu_i = 100$, $q_i = 3/4$, $\sigma_i^2 = 1$ and $\beta_i = 1$ for all i, then skewness is negative.

3.30 (b) $q_i = 1 - e^{-\lambda_i}$.

3.31 $\mathbb{E}[\tilde{T}] = \sum_{i=1}^{n} q_i \mu_i$, the same as $\mathbb{E}[T]$;

$\mathrm{Var}[\tilde{T}] = \sum_{i=1}^{n} q_i(\sigma_i^2 + \mu_i^2)$, which is greater than or equal to $\mathrm{Var}[T]$.

Chapter 4

4.3 (a) $1/x$, (b) β, (c) $1/(2x)$;

(1) yes (decreases with wealth), no, yes (decreases with wealth);

(2) no, yes (increases with β), no.

4.4 (a) £9358; (b) (1) 1.340, (2) 1.038.

4.5 $\sigma \le £707.11$.

4.6 £12 974.

4.7 £14 366.

4.9 (a) $14/6 = 2.33$, $6/3 = 2$;

(b) gamma$(20, 9)$, mode = estimate = $19/9 = 2.11$;

(c) (1) estimate = median = 2.19, (2) (1.36, 3.30);

(d) 20/9 = 2.22.

4.10 (a) N(2, 0.29); (b) 28/53 = 0.5283, 114.96/53 = 2.17;

(c) (1.90, 2.44).

4.11 119.9, 106.7, 126.8, 109.0, 117.5.

4.12 (a) 71.98, 56.58, 85.63.

4.15 (a) 51.36, (b) 50.91.

4.16 (a) 144.9, 138.9, 162.9, 155.1.

(b) 146, 141.3, 160, 154.

(c) 147.1, 143.3, 158.2, 153.4.

4.17 (a) 4.465, 6.151, 5.322, 3.838, 6.737; (b) £352 800.

4.18 (a) (1) 0.8712; (2) 6 142 021; (3) £6453, £7533, £8195; (b) £1081, £1175.

4.19 (a) £2297, £2435, £2098; (b) £2562, £2758, £1940.

4.23 £1196, £2076.

Chapter 5

5.1 0.141.

5.2 0.757.

5.5 (b) (2) 0.0821, £1519.

5.8 (a) $Z^* \sim \text{Pa}(5, 7.2823)$;

(b) $\text{CP}(5, F_{Z^*})$, $\mathbb{E}[S_R] = £9103$, $\text{SD}[S_R] = £6648$;

(c) $\mathbb{E}[S_I] = £90\,897$.

5.9 (b) Proportional: £63 000; excess loss: £79 900.

5.10 (a) Exponential with mean 2; (b)Pa(4, 10).

(c)

	Mean	SD
Model 1	270 700	32 900
Model 2	432 000	65 700
Model 3	304 400	47 500

5.12 (b)

$$\mathbb{E}[X] = 1.4 \qquad \mathbb{E}[X^2] = 3.92$$
$$\mathbb{E}[S] = 140 \qquad \text{Var}[S] = 392$$
$$\mathbb{E}[Y] = 1.0130 \qquad \mathbb{E}[Y^2] = 1.4430$$
$$\mathbb{E}[Z] = 0.3870 \qquad \mathbb{E}[Z^2] = 1.0837$$
$$\mathbb{E}[S_I] = 101.30 \qquad \text{Var}[S_I] = 144.30$$
$$\mathbb{E}[S_R] = 38.70 \qquad \text{Var}[S_R] = 108.37$$

5.13 (a) $\widetilde{\alpha} = 3.6829$, $\widetilde{\lambda} = 3096.3$; (b) 0.160; (c) £2989.

5.14 (b) (1) 0.999, (2) 1.000.

5.15 (a) $\dfrac{\mathbb{E}[YZ]}{\sqrt{\mathbb{E}[Y^2]\mathbb{E}[Z^2]}}$; (b) $+0.471$; (c) $+0.477$.

5.16 (a) £3.5 × 10^6, £0.732 × 10^6, 0.086.
  (b) (1) £440 000;
      (2) mean = £3.1 × 10^6, standard deviation = £0.629 × 10^6,
      probability = 0.064.

5.17 $\mathbb{E}[S_I] = \dfrac{\beta\lambda\gamma}{\alpha-1}$, $\mathrm{Var}[S_I] = \dfrac{2\beta^2\lambda\gamma^2}{(\alpha-1)(\alpha-2)}$.
  Similarly for S_R, with $1-\beta$ in place of β.

5.18 (a) (1) Exponential with mean 1.
      (2) $\mathbb{E}[S_1] = 150$, var$[S_1] = 277.5$;
          $\mathbb{E}[S_2] = 105$, var$[S_2] = 198.975$;
          $\mathbb{E}[S] = 255$, var$[S] = 476.475$.
  (b) (1) 0.964,   (2) 0.932,   (3) 0.990.

5.19 (b) (1) £360, £60, £360;   (2) £325.70, £54.30, £354.30;
      (3) £294.70, £49.10, £349.10; (4) £266.70, £44.40, £344.40.
      Premium: 9.5%, 18.1%, 25.9%.
      Expected total costs: 1.6%, 3.0%, 4.3%.
  (c) 0.977, 0.971, 0.965, 0.957.

5.21 (a) 1.749, (b) 1.215, (c) 1.312.

5.22 (a) 0.714, (b) 0.789, (c) 0.556.

5.23 (c) 0.179.

5.24 (b) Expected utility: no cover −0.033, with cover −0.045, so the individual will not purchase cover.
  (c) Maximum premium the individual will be prepared to pay is £84.58.

5.27 In terms of a monetary unit of £10 million:
  (a) stop loss 0.3162, proportional 0.64;
  (b) stop loss 0.6736, proportional 0.49.
  (c) (1) 0.6762 (retention = 1.5936); (2) 0.625; (3) 0.5.

5.28 (b) (1) $\gamma/(\alpha-2)$, (2) μ; (c) 1/3.

5.32 (a) 812.5.

5.34 (a) (2) 75.9%, 56.9%, 50.6%.
  (b) (1) 89.7%, 67.2%, 59.8%;
      (2) 100%, 75%, 66.7%;
      (3) 73.7%, 55.3%, 61.4%.

Chapter 6

6.1 $R = 1/6$, Lundberg upper bound is $e^{-u/6}$.

6.2 (b) $e^{r\mu} = 1 + (1+\theta)\mu r$;
  (c) $e^{-Ru} < e^{-R_{\exp}u}$.

6.6 $R_{\exp} < R$, $e^{-R_{\exp}u} > e^{-Ru}$.

6.7 $R = 1/7$.

6.9 $R = 0.2397$, $R < R_g$, $R < R_h$.

6.11 $\varphi(u) = 1 - \frac{20}{29}e^{-u/6} - \frac{5}{203}e^{-6u/7}$.

6.12 $C = 1$, $D = -\frac{20}{29}$, $E = -\frac{5}{203}$, $\psi(u) = \frac{20}{29}e^{-u/6} + \frac{5}{203}e^{-6u/7}$.

6.14 $A = \frac{1}{2(1+\theta)}\left(1 + \frac{(3+2\theta)\sqrt{9+8\theta}}{9+8\theta}\right)$, $B = \frac{1}{2(1+\theta)}\left(1 - \frac{(3+2\theta)\sqrt{9+8\theta}}{9+8\theta}\right)$,

$R = \frac{3+4\theta-\sqrt{9+8\theta}}{2(1+\theta)}$, $\alpha = \frac{3+4\theta+\sqrt{9+8\theta}}{2(1+\theta)}$.

6.15 $\varphi(u) = 1 - Ae^{-Ru} - Be^{-\alpha u}$ with A, B, R and α as in Exercise 6.14.

6.16 $f(x) = \frac{\lambda\mu}{c}f_I(x)$, $z(u) = \frac{\lambda\mu}{c}(1 - F_I(u))$.

6.17 The Cramér–Lundberg approximation is $\frac{20}{29}e^{-u/6}$, which is always smaller than the true $\psi(u)$ for this example.

6.18 Adjustment coefficient is $R = c$, the safety loading is $\theta = \frac{1-a-b}{a+b}$, and the Cramér–Lundberg approximation is ae^{-cu}.

6.19 If $r_0 = 0.05$ then $r_n \to 0$ as $n \to \infty$.

6.20 $R = 0.1877$, but for exponential claim sizes $R_{\exp} = 0.0909$.

6.22 $\hat{R} = \begin{cases} \frac{1}{\bar{X}} - \frac{1}{c\bar{T}} & \text{if } \bar{X} < c\bar{T} \\ 0 & \text{otherwise.} \end{cases}$

$\sqrt{n}(\hat{R} - R) \to_d N(0, \sigma^2(\lambda, \mu))$, where $\sigma^2(\lambda, \mu) = \lambda^2/c^2 + 1/\mu^2$.

Confidence interval has end points $\hat{R} \pm \frac{z_{\alpha/2}}{\sqrt{n}}\sqrt{\sigma^2(\hat{\lambda}, \hat{\mu})}$, where $\hat{\lambda} = 1/\bar{T}$ and $\hat{\mu} = \bar{X}$.

References

Apostol, T. M. 1967. *Calculus, Vol. I.* 2nd edn. New York: Wiley.

Asmussen, S. 2000. *Ruin Probabilities.* Singapore: World Scientific.

Asmussen, S. 2003. *Applied Probability and Queues.* 2nd edn. New York: Springer.

Brigham, E.O. 1974. *The Fast Fourier Transform.* Englewood Cliffs, New Jersey: Prentice-Hall.

Bühlmann, H. 1967. Experience rating and credibility. *ASTIN Bulletin*, **4**, 199–207.

Bühlmann, H. and Straub, E. 1970. Glaubwürdigkeit für Schadensätze. *Mitteilungen der Vereinigung Schweizerischer Versicherungsmathematiker*, **70**, 111–133.

Casella, G. and Berger, R.L. 1990. *Statistical Inference.* Pacific Grove, California: Wadsworth & Brooks/Cole.

Christ, R. and Steinebach, J. 1995. Estimating the adjustment coefficient in an ARMA(p, q) risk model. *Insurance: Mathematics and Economics*, **17**, 149–161.

Conte, S. D. and de Boor, C. 1980. *Elementary Numerical Analysis: An Algorithmic Approach.* 3rd edn. New York: McGraw–Hill.

Cramér, H. 1994. On the mathematical theory of risk, Försäkringsaktiebolaget Skandia 1855–1930, Parts I and II, 1930, pp. 7–84. In: Martin-Löf, A. (ed.), *Harald Cramér: Collected Works*, vol. I, pp. 601–678. Berlin: Springer-Verlag.

Csörgő, M. and Steinebach, J. 1991. On the estimation of the adjustment coefficient in risk theory via intermediate order statistics. *Insurance: Mathematics and Economics*, **10**, 37–50.

Csörgő, S. and Teugels, J. L. 1990. Empirical Laplace transform and approximation of compound distributions. *Journal of Applied Probability*, **27**, 88–101.

De Pril, N. 1986. On the exact computation of the aggregate claims distribution in the individual life model. *ASTIN Bulletin*, **16**, 109–112.

De Pril, N. 1989. The aggregate positive claims distribution in the individual model with arbitrary positive claims. *ASTIN Bulletin*, **19**, 9–24.

DeGroot, M. H. and Schervish, M. J. 2002. *Probability and Statistics.* 3rd edn. Boston, Massachusetts: Addison–Wesley.

Dickson, D. C. M. 2005. *Insurance Risk and Ruin.* Cambridge: Cambridge University Press.

Dickson, D. C. M. and Waters, H. R. 1996. Reinsurance and ruin. *Insurance: Mathematics and Economics*, **19**, 61–80.

Efron, B. 1979. Bootstrap methods: another look at the jackknife. *Annals of Statistics*, **7**, 1–26.

Efron, B. and Tibshirani, R. J. 1993. *An Introduction to the Bootstrap*. New York: Chapman & Hall.

Embrechts, P. and Mikosch, T. 1991. A bootstrap procedure for estimating the adjustment coefficient. *Insurance: Mathematics and Economics*, **10**, 181–190.

Embrechts, P., Jensen, J. L., Maejima, M. and Teugels, J. L. 1985a. Approximations for compound Poisson and Pólya processes. *Advances in Applied Probability*, **17**, 623–637.

Embrechts, P., Klüppelberg, C. and Mikosch, T. 1997. *Modelling Extremal Events for Finance and Insurance*. Berlin: Springer.

Embrechts, P., Maejima, M. and Teugels, J. L. 1985b. Asymptotic behaviour of compound distributions. *ASTIN Bulletin*, **15**, 45–48.

Epperson, J. F. 2007. *An Introduction to Numerical Methods and Analysis*. Hoboken, New Jersey: Wiley-Interscience.

Feller, W. 1971. *An Introduction to Probability Theory and Its Applications, vol II*. 2nd edn. New York: Wiley.

Grandell, J. 1979. Empirical bounds for ruin probabilities. *Stochastic Processes and their Applications*, **8**, 243–255.

Grandell, J. 1991. *Aspects of Risk Theory*. New York: Springer.

Grimmett, G. R. and Stirzaker, D. R. 2001. *Probability and Random Processes*. 3rd edn. Oxford: Oxford University Press.

Grübel, R. 1989. The fast Fourier transform in applied probability theory. *Nieuw Archief voor Wiskunde*, **7**, 289–300.

Gut, A. 1988. *Stopped Random Walks: Limit Theorems and Applications*. New York: Springer.

Gut, A. 2005. *Probability: A Graduate Course*. New York: Springer.

Gut, A. 2009. *An Intermediate Course in Probability*. 2nd edn. New York: Springer.

Herkenrath, U. 1986. On the estimation of the adjustment coefficient in risk theory by means of stochastic approximation procedures. *Insurance: Mathematics and Economics*, **5**, 305–313.

Hogg, R. V. and Klugman, S. A. 1984. *Loss Distributions*. New York: Wiley.

Klugman, S. A. 1991. *Bayesian Statistics in Actuarial Science*. Boston: Kluwer.

Klugman, S. A., Panjer, H. H. and Willmot, G. E. 1998. *Loss Models: From Data to Decisions*. New York: Wiley.

Morgan, B. J. T. 2000. *Applied Stochastic Modelling*. London: Arnold.

Panjer, H. H. 1981. Recursive evaluation of a family of compound distributions. *ASTIN Bulletin*, **12**, 22–26.

Pawitan, Y. 2001. *In All Likelihood: Statistical Modelling and Inference using Likelihood*. Oxford: Clarendon Press.

Pitts, S. M. 2006. The fast Fourier transform algorithm in ruin theory for the classical risk model. *HERMIS: An International Journal of Computer Mathematics and Its Applications*, **7**, 80–94.

Pitts, S. M., Grübel, R. and Embrechts, P. 1996. Confidence sets for the adjustment coefficient. *Advances in Applied Probability*, **28**, 802–827.

Rolski, T., Schmidli, H., Schmidt, V. and Teugels, J. 1999. *Stochastic Processes for Insurance and Finance*. Chichester: Wiley.

Rudin, W. 1986. *Real and Complex Analysis*. 3rd edn. New York: McGraw–Hill.

Schmidli, H. 2008. *Stochastic Control in Insurance*. London: Springer.

van der Vaart, A. W. 1998. *Asymptotic Statistics*. Cambridge: Cambridge University Press.

Venables, W. N. and Ripley, B. D. 2002. *Modern Applied Statistics with S*. 4th edn. New York: Springer.

Verzani, J. 2005. *Using R for Introductory Statistics*. Boca Raton, Florida: Chapman & Hall/CRC.

Index

Printed in the United States
By Bookmasters